GEOGRAPHY FOR EDUCATORS
STANDARDS, THEMES, AND CONCEPTS

Susan Wiley Hardwick
Donald G. Holtgrieve
California State University, Chico

Prentice Hall, Upper Saddle River, New Jersey 07458

Library of Congress Cataloging-in-Publication Data

HARDWICK, SUSAN WILEY.
    Geography for educators: standards, themes, and concepts/Susan
Wiley Hardwick, Donald G. Holtgrieve.—[2nd ed.]
        p.   cm.
    Rev. ed. of: Patterns on our planet. c1990.
    Includes bibliographical references and index.
    ISBN 0-13-442377-1
    1. Geography.   I. Holtgrieve, Donald G.   II. Hardwick, Susan
Wiley.   Patterns on our planet.   III. Title.
G128.H35   1996   95-40940
910—dc20   CIP

Acquisitions editor: *Ray Henderson*
Assistant editor: *Wendy Rivers*
Production supervision: *Kathleen M. Lafferty/Roaring Mountain Editorial Services*
Proofreaders: *Bruce D. Colegrove, Maria McColligan*
Cover designer: *Bruce Kenselaar*
Creative director: *Paula Maylahn*
Art director: *Jayne Conte*
Manufacturing manager: *Trudy Pisciotti*

Previous edition © 1990 by Macmillan Publishing Company under the title
Patterns on Our Planet: Concepts and Themes in Geography.

 © 1996 by Prentice-Hall, Inc.
Simon & Schuster / A Viacom Company
Upper Saddle River, New Jersey 07458

Printed in the United States of America

10   9   8   7   6   5   4   3   2   1

ISBN 0-13-442377-1

Prentice-Hall International (UK) Limited, *London*
Prentice-Hall of Australia Pty. Limited, *Sydney*
Prentice-Hall Canada Inc., *Toronto*
Prentice-Hall Hispanoamericana, S.A., *Mexico*
Prentice-Hall of India Private Limited, *New Delhi*
Prentice-Hall of Japan, Inc., *Tokyo*
Simon & Schuster Asia Pte. Ltd., *Singapore*
Editoria Prentice-Hall do Brasil, *Rio de Janeiro*

With love and appreciation to
James, David, Scott, and Gordon—
geographers by assimilation

CONTENTS

5

THE THEME OF PLACE: PHYSICAL SYSTEMS I   82

6

THE THEME OF PLACE: PHYSICAL SYSTEMS II   115

7

THE THEME OF PLACE: CULTURAL SYSTEMS   150

8

THE THEME OF PLACE: ECONOMIC SYSTEMS   192

9

THE THEME OF PLACE: URBAN SYSTEMS   222

FOREWORD

In my foreword to the first edition of *Patterns on Our Planet,* I lamented the fact that most journalists and critics calling for better geography education still thought of the discipline as an inventory of state capitals. I celebrated the work that Professors Hardwick and Holtgrieve had done in this book because they gave such energy and creativity to a much broader, much more realistic definition of geography. In that edition—as they continue in this volume—they spent much time and ink demonstrating the ways in which the teacher-friendly five themes could help instructors innocent of formal geography come to grips with many additional and stimulating perspectives in geography.

In this second edition, the authors have, yet again, turned their minds to where geography might go, and might lead you, rather than look backward and chronicle where it has been—or more commonly, where it has failed to go. In this edition, a whole new player comes on board the team. There is an introduction to, and continuing explanation of, the new National Geography Standards in geography. Drawing from *Geography for Life: National Geography Standards 1994,* the authors bring a number of essential elements to the reader of this book. First, it is clear from their own familiarity with the standards that they have been deeply involved in the 27-month consensus process that brought the standards to print, and to life. Susan Hardwick was an active player in the continual iteration and reiteration that characterized the development of these 18 standards and their hundreds of applications for classroom instruction.

Second, they have worked as a team to peel away the jargon that surrounds every academic taxonomy and have introduced and explained the real geography and its real significance in these standards. Although the authors of the standards gave much attention to the creation of the clearest possible language in this innovative effort to answer the question, What should school children know and be able to do in geography?, *Geography for Educators: Standards, Themes, and Concepts* takes the standards out of the classroom and into the world as it surrounds us. This is a theme of genuine utility that was well developed in the first edition and is even more powerfully central to this edition.

In this process, geography is cast in a broad-based range of applications by the two authors. Whatever else geography is, it must be seen as a discipline that deals with the sorts of questions, problems, and confusions that face all of us every day, and in every place. Geography is a discipline that truly is for life . . . for solutions to contemporary problems, for intellectual satisfaction, and for old-fashioned curiosity.

Finally, Holtgrieve and Hardwick have given good energy to their explanations of geography's meaning and utility. Through their pens and graphics, geography is engaging and meaningful. Studying this book should cause students and professors alike to look at the view out of their car or office window or on their TV or PC monitor with new, more exploring eyes. That is a great way to see the patterns on our planet and comprehend the way that geography has caused those patterns to take on those shapes and to have developed their significance.

Christopher L. Salter
Professor and Chair
Department of Geography
University of Missouri, Columbia
Co-Chair, National Geography Standards Project,
1992–1994

PREFACE

Environmental and global education are now a part of every student's course of study in the United States and in much of the rest of the world. Surveys of what our students know, however, often demonstrate that they score low in a subject called geography. How odd this is, since geography is the study of our own planet Earth as the home of humankind. What could be more important? Our inspiration for *Geography for Educators: Standards, Themes, and Concepts* is the hope that teachers and aspiring teachers can use this work as a way of reconciling in their own minds the importance of studying our planet and its people.

Over the course of many years of teaching geographic education, conducting teacher in-service workshops, and working with geographic alliances, we saw the need for helping teachers not with "how" to teach but with "what" to teach when addressing the geography portion of their school curriculum. Many classroom teachers have expressed a feeling of being a bit overwhelmed with the rich geographic content contained in the newly released *National Geography Standards* document. In addition, many were also concerned about the upcoming National Assessment of Educational Progress (NAEP) examination soon to be given in geography classrooms.

This book is a response to this national need. Using the *National Geography Standards* and the five fundamental themes of geography, this text attempts to explain the physical, cultural, and economic systems that interrelate and operate on this dynamic planet in a straightforward manner adaptable to teachers at all levels. We hope the content contained in this book helps kindergarten through high school teachers see that not only do various components of the social studies curriculum incorporate geographical concepts but that other commonly taught subjects such as literature, mathematics, and science also have geographic dimensions. Many state social studies guidelines

in the United States now stress the interdisciplinary thematic nature of learning. This makes the subject of geography, an integrative holistic discipline, an even more important subject in today's crowded curriculum.

In support of the interdisciplinary nature of geography, this text offers a structure to the discipline based on five major themes, content woven throughout the *National Geography Standards,* and their related concepts or generalizations. The intent is to steer readers away from the traditional stereotype of geography as place memorization toward conceptual thinking about how Earth systems work. The five fundamental themes of geography have been incorporated into many state and local school curricula and publications of the major geographical associations and organizations. We hope that this book illustrates that these combined themes of *Location, Place, Human-Environment Interaction, Movement,* and *Region* offer a uniform and convenient structure for the study of any place or any topic from a geographic perspective.

Most traditional college-level geography textbooks have used an adjective-based system of subdividing the subfields of the discipline. For example, textbooks may have chapters entitled "Political Geography," "Transportation Geography," "Physical Geography," and so forth. Others have used a list of four co-traditions of geography first introduced in 1964 by William Pattison. Still other books emphasize the regional nature of geography by dividing the globe into map-based subdivisions, such as the Middle East or sub-Saharan Africa.

We recognize that most classroom teachers at the elementary and secondary school levels base their geography programs on school district guidelines or on available student textbooks. This book supplements those resources by suggesting which concepts are appropriate for the teacher's chosen regions and chapter topics. The use of Figure 1-1 (and the larger color version on the inside cover of the book), which integrates standards, themes, and concepts, should mesh this information base with other teacher resources without undue duplication or complexity.

We begin Chapter 1 with a discussion of the importance of geographic literacy in this era of increasingly important global interconnections. We suggest at the outset that part of geography's "problem" is the traditional way it has been viewed in the school curriculum. We go on to suggest that geography is dynamic, exciting, and even lucrative to those who choose to enter it on a professional basis. The remainder of the book is introduced by outlining our concept-based approach to the discipline, stressing how geography is more a way of thinking than a subject to be memorized.

Chapter 2 discusses the historic role of geography education in North American schools and offers examples of how geography is taught in other countries. The five themes and the eighteen *National Geography Standards* are then introduced as a topical structure for the following chapters.

The theme of *Location* (Standards 1 and 3 of the *National Geography Standards*) is probably the broadest of the five fundamental themes and contains some of the newer frontiers in technical applications to solving geographic problems. Chapter 3 begins with the use of the geographer's primary tool, the map, and concludes with examples of the importance of Location in history, urban affairs, economic activities, and political issues.

Chapter 4 introduces the concepts involved in the theme of *Place* (Standards 2, 4, and 6) by suggesting that places are more than simply locations. Places are often a state of mind and, as such, continually affect decision making. Chapters 5 and 6 offer a systematic way of seeing how the physical characteristics of Earth shape and create places. These physical characteristics of Place (Standard 7) form a foundation for understanding human patterns.

Chapters 7 through 9 overlay human activities onto this physical base. Chapters 7 (Standards 9, 10, and 13), 8 (Standards 11 and 16), and 9 (Standard 12) analyze the theme of Place as a result of cultural, economic, and political decision making. Since most people on Earth today live in urban places and these are the places most altered from a "natural state," a special focus is given to them in Chapter 9.

It may be seen that there are more chapters in this book that focus on the theme of *Place* than on the other themes. The reason for this is simple. Many would argue that geography is, in itself, the study of place. We agree, but add that geography is also the study of process.

Chapter 10 demonstrates the importance of relationships between human decision-making processes and the natural physical and biotic processes. Many of the results of this *Human-Environment Interaction* (Standards 8, 14, and 15) have been catastrophic, while some have provided an improved quality of life and higher standards of living on Earth. We are still learning how to balance human activity with the dynamic natural systems of our planet.

Chapter 11 also stresses process. In this case, the theme is the *Movement* of people, goods, and ideas (Standards 3 and 9). We emphasize that Earth is not a snapshot in time, but is an ongoing, ever changing complexity, driven by energy and expressed by movement and change.

Chapter 12 focuses on the last of the five fundamental themes of geography, *Region* (Standard 5). The idea of regional thinking is probably the best known geographic concept to the general public and our most frequent way of organizing and subdividing people and places on Earth. Regions are constructs for understanding global news stories, the daily weather report, and historical events. Differences in regions also provide a motive for geography's link with travel and exploration.

Finally, in Chapter 13, we present four examples of how the concepts and themes of geography can be incorporated into understanding major historical events and geographic patterns (Standards 17 and 18). Our selected case studies focus on the integration of the standards, themes, and concepts explored in prior chapters of the book. They represent examples of "geography in action".

We owe the inspiration for this book to the founding members of the Northern California Geographic Alliance, who expressed a need and encouraged us to fill it. We especially appreciate the ongoing support of Kit and Cathy Salter and our colleagues at the Department of Geography and Planning at California State University, Chico.

Thanks are also extended to our high-quality research assistants Laura Mainwaring, Jeanette Betts, Sandra Kilcollins, Kari Forbes-Boyte, and Steve Morris for cheerfully providing research and writing support for our efforts.

We are also grateful to our editors at Prentice Hall for extending encouraging hands of support for the book. It has truly been a "labor of love" for all who have supported us through the process.

Susan Wiley Hardwick
Donald G. Holtgrieve

1

THE FIELD OF GEOGRAPHY

This is the information age, a time of instant communication throughout the world. Our national policy and local decisions often depend on what we know about other places; thus we are all to some extent geographers. We are all part of a global community in which the evening news may feature floods in Bangladesh, riots in Central Africa, or warfare in Bosnia. This news video is dramatic, colorful, and aimed at a personal level. Programmers keep our eyes glued to the screen with information and drama about places from around the globe.

Over the years newspapers and other media have focused our attention upon events all over the world. In the business section of the newspaper one is certain to find information about international trade and its implications for our own lives. No one can remain insensitive to the international price of oil, to competition from Japanese electronics firms, or to the variety of foreign cars on American roadways.

Although the sports section of your newspaper might not have reported on soccer, rugby, or international wrestling 25 years ago, today's sports coverage goes even beyond these events to international games and foreign sports personalities. International athletics, once covered by the press only at Olympic events, are now a regular feature of news reports. Regional differences are evidenced in the popularity of ice hockey in the far northern latitudes and rugby in Great Britain.

Many of the largest corporations in the world are now truly multinational. Banking, resource extraction, and even retail trade today depend upon a global market. Teachers often use a trip to the local shopping center as a geography lesson for students, examining product labels to emphasize the international influence upon our everyday lives. Conversely, American products have diffused throughout the world; for example, golden-arched fast food restaurants are now in major cities on all but one of the world's continents.

The media remind us daily that we are all members of a global society, dependent upon people from other regions of our own country and from all around the world. Yet how much do we really know about other places? Do our business and political leaders know enough about the people of this planet and the surface of the planet we all share?

GEOGRAPHIC ILLITERACY

Lack of geographic knowledge about places among geography students first made national headlines and the television news in 1983 when the shocking results of a survey were announced. Professor David Helgren at the University of Miami discovered that over half of his students did not know the locations of the Arabian Sea, New Guinea, Iceland, Algeria, Kenya, Chicago, Moscow, or Capetown. An astonishing 8 percent could not even locate Miami!

This news spawned other such surveys, and it was found that the ordinary citizen of the United States had very little knowedge about places of critical importance throughout the world. Indeed, "geographic illiteracy" has become a cry of alarm about a very serious problem. According to a Gallup poll conducted in 1988:

> Americans' knowledge of world geography compares unfavorably with that of their counterparts forty years ago as well as their contemporaries in other industrialized nations. Geographic illiteracy is particularly acute among Americans 18 to 24 years old.[1]

The landmark study, conducted by the Gallup organization for the National Geographic Society, discovered that nine out of ten persons interviewed think that geographic knowledge is "absolutely necessary" (37 percent) or "important" (53 percent). Despite their opinions about the importance of geographic knowledge, however, three out of ten people interviewed could not use a map to determine direction or to calculate distance!

Combating Geographic Illiteracy

Concern over geographic illiteracy has been expressed by numerous educators, politicians, and other national leaders during the past ten years. In 1986, the California State Superintendent of Public Instruction stated that "our students are more illiterate in geography than in anything else."[2] In an address to the United States Senate in support of Geography Awareness Week, Senator Edward Kennedy stated:

> All of us in the Congress realize the vital importance of improving our educational system if we are to maintain our competitive position in the world economy. As part of that effort, we must ensure that young Americans have a clear understanding of what the world looks like and the way in which geography influences human well-being.[3]

The sponsor of the bill supporting Geography Awareness Week, Senator William Bradley, included the following in his remarks:

> We depend on a well-informed populace to maintain the democratic ideals which have made this country great. When 95 percent of some of our brightest college students cannot locate Vietnam on a world map, we must sound the alarm. When 63 percent of the Americans participating in a nationwide survey by CBS and the *Washington Post* cannot name the two nations involved in the SALT talks, we are failing to educate our citizens to compete in an increasingly interdependent world. In 1980, a Presidential commission found that companies in the United States fare poorly against foreign competitors, in part because Americans are ignorant of things beyond their borders.[4]

One of the first attempts to standardize a level of geographic literacy among students in the United States was the development of the intermediate- and secondary-level geography tests prepared by the National Council for Geographic Education in 1986. These standardized forms can be used by teachers and curriculum coordinators to assess the level of *conceptual* understanding in geography in addition to a more traditional *factual* foundation.

The strongest private-sector supporter for the passage of the Geography Awareness Week bill was the National Geographic Society. Its president, Gilbert Grosvenor, has spoken throughout the United States in support of improved geography education. In his words, "To ignore geography is irresponsible. It is just as important to business and domestic policies as it is to military and foreign policy decisions."[5] The work of the National Council for Geographic Education and the *National Geographic Society* resulted in the inclusion of geography as one of the nation's five core subjects in the National Assessment of Educational Progress (NAEP) examinations in 1994. This important national effort, along with the publication of the *National Geography Standards* in late 1994, is discussed in detail in the following chapter.

To support this effort to combat geographic illiteracy nationwide, several state departments of education, the National Science Foundation, the Department of Education, the National Geographic Society, and other educational publishers are producing innovative curriculum materials based on thematic approaches, in addition to the more traditional regionally descriptive materials. Some of these exemplary curricula are also discussed in Chapter 2.

An allied effort of the National Geographic Society has been financial sponsorship of "Geographic Alliances" across the country. These Alliances of teachers, administrators, college and university instructors, and applied geographers have carried the geography reform movement to the grassroots level in thousands of individual schools in every state in the nation. Most of the Alliances have conducted summer institutes modeled after annual National Geographic Society Summer Institutes held in Washington, DC. They also sponsor leadership training institutes and weekend workshops, publish newsletters containing teaching ideas and resources, develop teaching materials, sponsor special events (including the National Geography Bees and the Kid's Network), and produce video programs on global issues. Geographic educators hope that, through the efforts of teachers and others committed to excellence in the teaching of geography, our students will soon begin to keep pace with changes around the globe.

TRADITIONAL VIEWS OF GEOGRAPHY

Geography has been a part of the school curriculum since the first North American textbook was written on the subject in 1789. In the nineteenth and early twentieth centuries, however, geographic learning all too often was equated with place memorization. An 1860 geography text, *Mitchell's School Geography: A System of Modern Geography Comprising a Description of the Present State of the World,*

And Its Five Great Divisions, America, Europe, Asia, Africa, and Oceania, With Their Several Empires, Kingdoms, States, Territories, Etc.[6] provides examination questions that illustrate the nature of the discipline at that time:

1. What country lies between the Arabian Sea and the Bay of Bengal?
2. What country in Asia does the Arctic Circle pass through?
3. What islands belong to Great Britain?
4. What nation lies east of the Kalahari Desert?
5. What river forms the boundary between New York and Canada West?

This foundation of geography as location-and-place memorization is a part of the geographic tradition that extends back to its Greek origins. The root word *geo* meaning "earth" was attached to *graphos* meaning "to write about." Naturally there was a need for exploration and cataloging of new knowledge about location from the period of the Greeks through the Age of Exploration into the modern period. This ultimately became the mission of geography.

The tradition of geography as place location and mapping formed a logical beginning for the generalist scholar before the Scientific Revolution. These generalists contributed encyclopedic works, such as Alexander von Humboldt's *Kosmos,* published in four volumes between 1845 and 1862. This collection of maps and narrative descriptions would later be used to promote a more systematic understanding of differences among regions and peoples. An example of the synthesizing of geographic information is Friedrich Ratzel's *Anthropogeographie,* published from 1882 to 1891.

The tradition of geography as travel and exploration is still evident in the numerous travel videos, television programs, and articles in popular magazines, and in the immense current interest in tourism. The *National Geographic* magazine, an institution in American journalism, has a worldwide circulation of over 10.5 million! We all share a fascination with places beyond the horizon.

As information from places around the world was compiled in the late nineteenth and early twentieth centuries, the scope of geography was broadened by making generalizations and comparisons. Many of these generalizations stressed the influence of the physical environment (namely climate) on human activity.

The earliest scholars were generalists who accumulated information without regard for specific disciplines; they did not label themselves "historians" or "geographers."

Thus interpreters of civilization such as Herodotus—and later Toynbee—included a discussion of environmental influences on human activity as a part of their interpretation. This theme goes back to the time of Aristotle, but the philosophy of **environmental determinism** gained the most acceptance during the early part of the twentieth century in the writings of Ratzel, Ellen Churchill Semple, and Ellsworth Huntington. (Note: Words in **bold** type appear in the glossary.) Semple wrote:

> Climate, undoubtedly, modifies many physiological processes in individuals and peoples … influences their temperament, their energy, their capacity for sustained or for merely intermittent effort, and therefore helps determine their efficiency as economic and political agents.[7]

Although some of the generalizations of the environmental determinists were based on sound observations and good logic, their conclusions were fundamentally inaccurate and ultimately racist. For example, Ellsworth Huntington made the case that the world's most influential civilizations arose in the mid-latitudes and that the mid-latitudes experienced the most seasonal variation in climate. Huntington then incorrectly concluded that the variation in seasons inspired human creativity and civilization, despite an early lack of material culture in such mid-latitude locations as Tasmania, Tierra del Fuego, and California.

Today it is recognized that there were as many creative innovations in tropical and subtropical climates as in the mid-latitudes. This environmental deterministic viewpoint relied on overgeneralizations and selected case studies but nevertheless appeared in school books and in popular literature under the heading "Geography" for many decades.

From the 1920s to the 1950s, geography in the United States evolved into a problem-solving discipline as well as an academic research field, as was the focus of the discipline in Europe. Land use surveys and resource inventories were conducted in the United States, and global issues were addressed in attempts to construct world peace. Following World War I, for example, geographer Isaiah Bowman advised world governments on boundary issues and contributed to the organization of the League of Nations. Many government employees were involved in intelligence gathering and mapmaking. The global effort was greatly aided by geographers.

During and immediately after World War II, geographers continued to advise international policymakers on matters dealing with national security and geopolitics. Today many geographers are employed by the federal government in agencies such as the USDA Forest Service, the U.S. Geological Survey, the Department of Commerce, and the Department of Defense.

The popular conception of geography in the 1960s and 1970s had two foci: issue-based concerns and attention to quantitative techniques. Some geographers such as William Bunge and David Harvey worked in both camps. Issues dealing with race relations, housing, women's rights, and the environment were seen to be action-agenda items for geographers and not just topics in textbooks. The contributions of geography to each of these issues is discussed in later chapters.

For the most part, however, these contributions have not been clearly attributed to geography alone. Working at grass-roots levels in the areas of urban planning, ethnic consciousness, patterns of crime, feminist issues, and historic preservation, geographers often have been members of interdisciplinary teams. With this multifaceted involvement, a generally accepted definition of geography ceased to exist. It still was presented in textbooks as regional description, but its practitioners were quite diverse in their own understandings of the field. Some studied regions of Earth in a descriptive style, while others insisted on quantifying multitudes of variables, usually in an urban context.

In the 1990s, the scope of geography is finally becoming clearer and more visible to the public. Almost all geographers—researchers, teachers, and practitioners—now identify themselves as part of a unified team. Today it is difficult to find public agencies or large companies that are involved with land or resources that do not have geographers in their employ.

GEOGRAPHERS IN ACTION

Although U.S. geographers are modest in number (membership in the Association of American Geographers totals only about 8000), their impact can be felt at most levels of business, government, and education. The boxes in this chapter, and others throughout the book, illustrate the variety of geographers' work experiences.

A CONCEPTUAL VIEW OF OUR PLANET

You have been introduced to people who have an interest in space and place on the surface of Earth. This is modern geography. In the past geography was necessarily descrip-

MEET THE GEOGRAPHER
The Excitement of Geography in the Classroom

"Everyone can relate to geography because everyone lives on this planet and can identify with some part of Earth."
Steven R. Herman, geography teacher.

Steve Herman became a geographer at age two when he first climbed out of his high chair and headed north. Although restrained at that time, a childhood interest in travel and maps navigated his career choice toward geography. Family trips and a careful background reading about the places visited created a dedicated geographer by junior high school age.

Steve remarked recently that he always wanted to teach, to share his enthusiasm for places. He augmented his formal education in geography at the university level with courses in foreign languages and history. In fulfilling his master's degree requirements, he focused upon integrating geography with other subjects, an approach that has contributed in large measure to his success in the classroom.

On the first day of school, Steve's students receive a base map of the world, which they use every day to relate lesson topics to current world affairs. "I try to integrate art, music, and literature with geography. For example, singing "It's a Long Way to Tipperary" presents an opportunity to study the cultural and physical landscape of Ireland. Teachers must relate geography to the student's own experience. This can be done through music, sports, and current events."

Steve Herman is concerned about the lack of required geography in his state's curriculum. In an effort toward curriculum reform he has encouraged his regional Geographic Alliance to lobby at the state capital.

tive, but today it is more analytical and somewhat predictive.

In most cases the geographic viewpoint is through a wide-angle lens. This holistic approach is neither unique to geography nor new to scientific analysis. As mentioned at the beginning of this chapter, early scientists and travelers accumulated as much information as possible without regard to subject disciplines such as astronomy or biology. For example, Herodotus's *Histories* contain vivid and accurate descriptions of the stage upon which historical events transpired; he makes political history come alive within the context of geography. Von Humboldt often is regarded as the first of a great tradition of geographers who combined field work with analysis to gain an understanding of regions.

In the late nineteenth and twentieth centuries the discipline of geography became established at universities along with other commonly taught subjects, such as history, literature, and the natural sciences. The establishment of university departments in Europe and later in North America had the effect of isolating these disciplines to a degree. The generalist was at a disadvantage under this system, in that depth of knowledge in a particular field was emphasized rather than expanding interests across departmental barriers.

Early twentieth-century geography courses were usually offered within departments of geology or earth sciences. William Morris Davis, often considered the founder of American geographic thought, taught geology at Harvard University from 1876 to 1912. Unlike his peers, his approach to geology was based on *processes that shaped Earth's surface,* a new and radical method at the time. Eventually, departments of geography that were structured to emphasize physical-geography processes were created in the United States and Europe.

By the mid-twentieth century, geographers had become specialists in many subfields within the discipline. These subfields often overlapped with other disciplines. An inventory of the status of American geography in 1954 categorized some subfields of geography as economic, historical, and physical. In the 1950s, most geographers agreed on the importance of a regional approach, which at that time meant compilation of a descriptive inventory of all physical and human characteristics of a selected area.

Since that time, however, most geographers have developed kinships with one or more related disciplines. This

MEET THE GEOGRAPHER
Environmental Planning: The Geographic Perspective

*"The geographer is a generalist, one who can see the big picture in a land-use planning project. I enjoy the excitement of frontline work, and I like to see the effects of sound decision making on the landscape."
Carolyn Cole, geographer/environmental planner.*

Carolyn Cole grew up in west Texas on flat land with few trees and little surface water. Perhaps because of the severe plains environment, she developed a strong sense of place at home and in her imagination. "I remember wondering what large bodies of water might be like when I looked at maps. To someday live in a green environment became an essential for me."

In college Carolyn explored art, architecture, the natural sciences, and urban studies. It was geography that brought them all together for her in a most meaningful way. Her studies in geography included urban development, an area she found particularly interesting. About the time Carolyn graduated, the National Environmental Policy Act (NEPA) had just been passed, and the need for environmental impact statements related to new development projects became apparent.

With guidance from her professors, Carolyn became a team member and later president of an environmental impact consulting firm. She subsequently joined an even larger company as project director, working with other planners as well as biologists, hydrologists, geologists, landscape architects, and cartographers.

A typical week for Carolyn includes field work or team meetings where field data are mapped and analyzed. Much of her time is spent synthesizing information into comprehensive documents used by policymakers and developers. Many of her projects are quite controversial, involving sensitive habitats and endangered species. "I feel on the cutting edge of public decision making and sense an urgency in my work," Carolyn says.

This geographer feels that her geographic training has enabled her to work well in an interdisciplinary setting where she can serve as a guide for team specialists. Her wide range of interests, both indoor and outdoor, have proven to be an asset in seeing projects translated into sound environmental practices.

is evident in the remarkable number of geography specialty groups now a part of the organization of the Association of American Geographers (AAG) and other professional geographical organizations (Table 1-1).

Another way to measure the diversity of geography is to assess the number of topical proficiencies professed by the members of the Association of American Geographers. Members may list up to three proficiencies on their membership application (selected from a list of 51). AAG members in 1994 declared various topical specialties, with the most prevalent being geographic information systems (1165 members), urban geography (758 members), cartography (557 members), and geographic education (437 members). It is significant to note that the geographic education specialty group has grown enormously in recent years due to the national reform ef-

forts in geography education discussed in this and the following chapter.

The common thread that binds geography and geographers in the latter part of the twentieth century is a focus upon land, space, and the interaction of people and environments on Earth. This essence of geography remains true to its roots, that is, a "description and analysis of the surface of Earth." An informal definition often heard is, "if you can map it, it must be geography."

Geography combines a description of places with the formulation of principles and concepts. It promotes an understanding of patterns, processes, and the resultant landscapes on the planet. The relative significance of our study of Earth's features is determined by human needs. Change on the surface of the land often is affected by human activity that is based on these perceived needs. This makes under-

TABLE 1-1
Formalized specialty groups in geography, 1994

Association of American Geographers	*Institute of British Geographers*	*Canadian Association of Geographers*	*International Geographical Union*	
Specialty Groups: Africa Aging American Ethnic American Indians Applied Asian Bible Biogeography Canadian Studies Cartography China Climate Coastal and Marine Contemporary Agriculture Cultural Ecology Cultural Geography Energy and Environment Environmental Perception European Geographic Information Systems Geographic Perspectives on Women Geography Education Geography/Religions and Belief Systems Geomorphology Hazards Historical Human Rights Industrial Latin America Mathematical Models Medical Microcomputers Political Population Recreation, Tourism Regional Development and Planning Remote Sensing Rural Development Russia, Central Europe, and Eastern Europe Socialist Trasportation Urban Water Resources	**Study Groups:** Biogeography Developing Areas Geography and Planning Geomorphology Higher Education Historical Geography Industrial Activity and Area Development Medical Geography Political Geography Population Geography Quantitative Methods Rural Geography Social Geography Transport Geography Urban Geography Women and Geography History and Philosophy of Geography	**Special Interest Groups:** Environmental Impact Assessment Geography of Parks, Recreation & Tourism Industrial Geography Marine and Coastal Zone Management Medical Geography Rural and Urban Fringe Canadian Women in Geography	**Commissions:** Geographical Education Geographical Data Sensing and Processing Geographical Monitoring and Forecasting Measurements, Theory, and Application in Geomorphology Mountain Geoecology Population Geography Urban Systems in Transition **Working Groups:** Resource Management in the Drylands Cartography of the Dynamic Environment Environmental Atlases Mathematical Models Tropical Climatology and Human Settlements Landscape Synthesis The Great World Metropolitan Cities **Study Groups:** Climatic Change Topoclimatological Investigation and Mapping World Political Map Development in Highlands and High Latitude Zones Geography and Public Administration Famine and Food Crisis Management	Changing Rural Systems Coastal Environment Industrial Change International Division of Labour and Regional Development The Significance of Periglacial Phenomena Geography of Tourism and Leisure Comparative Research in Food Systems of the World History of Geographical Thought Geomorphological Survey and Mapping International Hydrological Programme Geography of Transport Dynamics of Land Use Systems Urbanisation in Developing Countries Energy Resources and Development Man's Impact on Karst Areas Geography of Commercial Actvities Geography of Telecommunication and Communication Map Use Environmental Perception in Resource Management Marine Geography

Source: Association of American Geographers, 1994. Used with permission.

Note: The status and activities of most of the specialty groups in the Association of American Geographers (listed in the first column) are examined in detail in the definitive compilation by Gary L. Gaile and Court J. Willmott, Geography in America *(Columbus, OH, Merrill, 1989).*

MEET THE GEOGRAPHER
Location Analyst at Work

"There are many people in business who do geography everyday but they don't call it that." Virginia Oliver-Hornbeck, geographer/site location analyst.

Virginia Oliver-Hornbeck discovered geography in a required college course, then explored it in greater depth in elective courses. She finally ended up majoring in the subject, with an emphasis in historical geography. She is now vice-president and partner in Area Location Systems, a location and market-research consulting firm in the Los Angeles area.

Virginia says that there are three prerequisites to success in applied geography: common sense, knowledge of economic geography and location theory, and the ability to apply geographic principles to solve a client's problems.

Virginia's clients most often are financial institutions and retail operators. Her teams of analysts evaluate various locations for potential retail outlets or branch offices. This evaluation is done from the perspective of regional, county, city, and individual site levels. The company has compiled computer databases to assist in analyzing local population and economic factors throughout California. They have a database for use by financial institutions that details site characteristics of the approximately 7000 bank and savings-and-loan branches in California.

Maps, which are always necessary in reports prepared for clients, are prepared using a complex computer automated design (CAD) program. Clients are becoming aware of the usefulness of maps in analysis and presentation, and their demand for maps constantly increases.

According to Virginia, success in her field requires the ability to think through problems in a geographic or spatial sense. She says that geographic analysis can be beneficial to most companies and is sure to become an integral part of business operations when the value of such thinking is realized.

standing Earth in relation to human societies a vitally important part of general education. An "Earthmanship" approach can be taken—and should be taken—within many subjects taught in schools.

Geography is a "linking" subject. It unites the natural sciences, humanities, and social sciences. It is a synthesizing discipline, blending scientific and humanistic perspectives. A geography convention may include meetings of specialists as well as assemblies on topics likely to be of common interest. Very often there are animated discussions about the nature of geography as a discipline. Such conferences usually end in a spirit of renewal and a reaffirmation that elements of the natural and human environments of Earth are related in space and must be studied as an interrelated system. *The big picture must remain as the essence of geographic thought.* As we enter the next millenium, we are also entering a global economy, a period of the highest international migration in history, global environmental concerns, and almost universal communications capabilities. In such circumstances, the analyst as generalist is very necessary.

STANDARDS, THEMES, AND CONCEPTS IN GEOGRAPHY

Thus far we have presented geography as a description of the land, an analysis of Earth features, and an understanding of the relationships of natural and human environments. This complex web of facts and interrelationships can be made more meaningful by identifying universal concepts that apply to the phenomena studied by the geographer. For example, if we are studying the economic geography of South Africa, we will try to identify the physical processes occurring there and the human decisions that create local economies in the region. One of many such processes is the mining of diamonds. This process may be conceptualized as **resource extraction**. The concept of resource extraction also may be applied to forestry, fishing, or other such activities for comparison.

Other concepts are developed around the marketing of

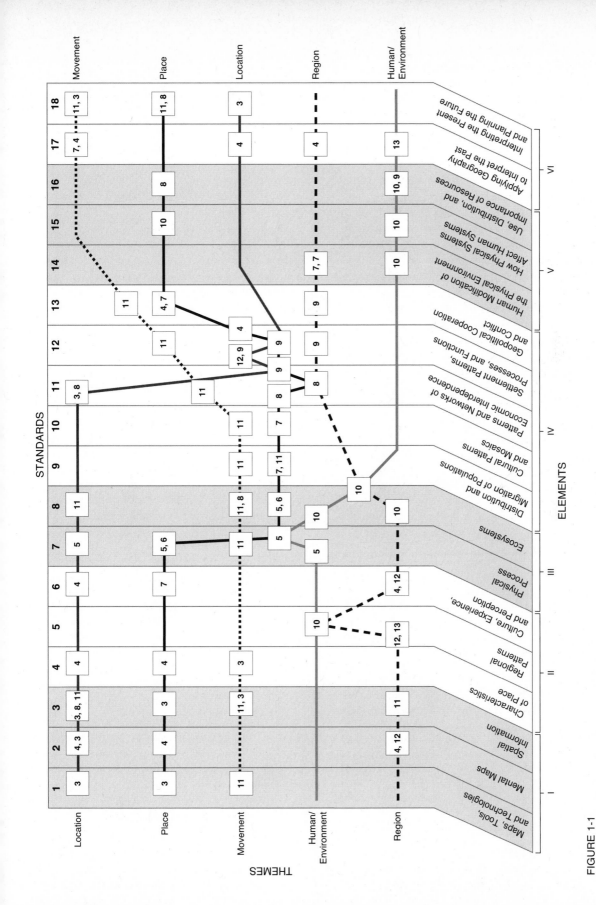

FIGURE 1-1
Standards, themes, and concepts in geography. Numbers refer to chapters in this book.

9

these products, including their transport and fabrication into other materials. Concepts are here defined as "generalizations that help us understand how Earth's natural systems work and how human activity takes place on its surface."

Perhaps we may decide to study the cultural geography of a certain region. If we begin by identifying the various ethnic groups in the area, this may lead to the concept of **cultural pluralism**. Concept building is more than mere labeling, however. It is a means by which we can generalize the specific into meaningful and somewhat predictable patterns. Concept building and association become geographic analysis.

In the following chapters you will be presented with many geographic facts, but within each chapter the concepts that give meaning to the facts are identified. The emphasis throughout the text is upon the *conceptual nature of geography* rather than upon geography as description. (This is not to minimize the necessity for students to know basic information. One must have an assemblage of facts at hand before concepts can be built from them!)

To summarize the relationships between the five fundamental themes of geography, the *National Geography Standards* categories, and the concepts stressed in this textbook, Figure 1-1 is offered here as a model linking various ideas into a total understanding of geography. This chart is not meant to be definitive, but it does represent how the various components of geography may be interconnected to form the big picture. In this model, the numbers of the chapters containing basic concepts are placed on one or more of five tracks, with each track representing a larger theme. The themes are keyed to the table of contents of this book. In the background are shaded indicators of where the themes and concepts fall into the purview of each of the eighteen geography standards.

Within the following chapters, these themes, standards, and concepts are defined and further developed. In Chapter 2, the concepts are grouped into the previously mentioned five fundamental themes of geography, and the themes then become the structure for the chapters that follow. Any place on Earth may be studied by following the five themes and identifying examples of appropriate concepts. Use of these five themes has become the prevalent method of geographic study in grades from kindergarten through high school in the last several years, especially in North American schools. The *National Geography Standards* list specific examples of knowledge and skills to be gained by fourth-, eighth-, and twelfth-grade students. The appropriate standards are stated at the beginning of each chapter to help clarify their relationship with the themes and concepts discussed in that particular section of the book. We hope that this interrelating of concepts, themes, and standards makes teaching and learning geography enjoyable, easy to follow, and interesting.

FURTHER READING

Among the first professional geographers to make known the state of geographic ignorance among university freshmen were James Curtis and David Helgren at the University of Miami. Their findings appeared on national television news and newspaper wire services. Helgren's summary of how geographic ignorance came to be a national concern appeared in "Place Name Ignorance is National News," *Journal of Geography* 82 (1983): 176–178. Helgren has since worked to remedy the situation with Jack McClintock, by publishing *Everything is Somewhere: The Geography Quiz Book* (New York: William Morrow, 1986).

A special committee on geographic education with representatives from the National Council for Geographic Education (NCGE), the American Geographical Society, the National Geographic Society, and the Association of American Geographers was created to address the need for more geographic education in American schools. This committee is known as GENIP (Geographic Education National Implementation Project). Their publications on geographic education may be ordered through the NCGE office at the Indiana University of Pennsylvania in Indiana, Pennsylvania.

A strong commitment to solving the problem of geographic illiteracy has been assumed by the National Geographic Society. Its various programs and policies were outlined in various editorials from 1986 to the present. The Association of American Geographers, the National Council for Geographic Education, and the American Geographical Society have also participated in several projects leading to new teaching materials, improving geography teaching techniques, and getting more geography into the public school curricula.

Two histories of the National Geographic Society were made available in its centennial year: Howard Abramson, *National Geographic: Behind America's Lens on the World* (New York: Crown, 1987), and C. D. B. Bryan, *The National Geographic Society: One Hundred Years of Adventure and Discovery* (New York: Harry N. Abrams, 1987).

A survey of geographic ideas from the earliest recorded civilizations to the present is found in Geoffrey J. Martin and Preston E. James, *All Possible Worlds: A History of Geographical Ideas*, 3rd ed. (New York: John Wiley, 1993). An in-depth analysis of how professional geographers think and act today is *Geography's Inner Worlds: Pervasive Themes in Contemporary American Geography*, edited by Ronald F. Abler, Melvin G. Marcus, and Judy M.

Olson (New Brunswick, NJ: Rutgers University Press, 1992). George J. Demko with Jerome Agel and Eugene Boe have assembled an enjoyable collection of geography in action stories dealing with the value of geographic knowledge in their recent book *Why in the World? Adventures in Geography* (New York: Anchor Books, 1992).

A handsome, full-color, "coffee-table book" filled with accurate and fascinating geography of the world is *The Real World: Understanding the Modern World Through the New Geography*, edited by Bruce Marshall (Boston: Houghton Mifflin, 1991). Another book written for the popular audience is *Don't Know Much About Geography: Everything You Need to Know About the World But Never Learned*, by Kenneth C. Davis (New York: Avon Books, 1992).

A survey of the tradition of geography from its beginnings to the present would fill libraries. However, a selection that represents the richness of geographic thought may be sampled with the following works:

- Isaiah Bowman, *The New World: Problems in Political Geography* (New York: World Book, 1921), demonstrated how geographic thinking helped society rebuild in the aftermath of World War I.
- Friedrich Ratzel, *Anthropogeographie*, vols. I and II (Stuttgart: J. Engelhorn, 1882–1891), related the then-known principles of anthropology to human activity on Earth.
- Ellen Churchill Semple, *Influences of Geographic Environment: On the Basis of Ratzel's System of Anthropogeographie* (New York: Henry Holt, 1911), explained differences among traditional societies in terms of location, area, boundaries, and environmental factors. Examples of her selected environments and how groups of people adapted to them include coast peoples, island peoples, mountain peoples, and those on steppes and deserts.
- Vidal de la Blache, *Principles of Human Geography* (New York: Henry Holt, 1926), produced a very popular work wherein the basis of population, transportation, and what he called the "elements of civilization" were explained.

- Ellsworth Huntington, *Mainsprings of Civilization* (New York: John Wiley, 1945), wrote one of the most widely read books about geography in mid-century. In it he contended that the inherited characteristics of peoples were influenced over time by the physical environment and that current culture and civilization is a product of the environment, particularly climate.
- The range and depth of cultural geography at mid-century is collected in *Land and Life: A Selection of Writings of Carl Ortwin Sauer*, edited by John Leighly (Berkeley: University of California Press, 1963).

Geography as a social scientific discipline was inventoried and explained in two mid-century works. Preston James edited *American Geography: Inventory and Prospect*, published by the Association of American Geographers in 1954. Also published by AAG was Richard Hartshorne's monograph *Perspectives on the Nature of Geography* (1959), a supplement to his earlier landmark monograph, *The Nature of Geography* (Washington, DC: AAG, 1939). A related endeavor in 1989, edited by Gary L. Gaile and Cort J. Willmott, was *Geography in America* (Columbus, OH: Merrill).

A summary of how contemporary geographers think and work, written for nongeographers, is Peter R. Gould's *The Geographer at Work* (London: Routledge and Kegan Paul, 1985). A survey of the contemporary field written particularly for senior- and graduate-level students is *The Student's Companion to Geography* by Alisdar Rogers, Heather Viles, and Andrew Goudie (Cambridge, MA: Blackwell, 1992). Prospects for the future of the discipline are addressed in Ronald J. Johnston's *A Question of Place: Exploring the Practice of Human Geography* (Oxford: Blackwell, 1991). A second book on the future of the discipline, edited by the same author, is *The Future of Geography* (New York: Methuen, 1985).

Geography as an employment field is described in publications of the Association of American Geographers (Washington, DC). These include "Geography as a Discipline" (1973), "Geography: Tomorrow's Career" (1981), and "Careers in Geography" (1995).

ENDNOTES

[1] George Gallup, Jr., and Alec Gallup, "Geographic Illiteracy on the Rise in the U.S.," *San Francisco Chronicle* (September 21, 1988): A-18.

[2] Bill Honig, "California's Educational Reform," speech at PLACE Conference, University of California at Los Angeles, 1986.

[3] U.S. Senator Edward Kennedy, speaking in support of Geography Awareness Week, *Congressional Record* (June 9, 1987): S-7780–7781.

[4] U.S. Senator William Bradley, speaking in support of Geography Awareness Week, *Congressional Record* (June 9, 1987): S-7780.

[5] Gilbert Grosvenor, "Teachers Gather to Create Comeback of Geography," National Geographic Society press release, July 6, 1987.

[6] S. Augustus Mitchell, *Mitchell's School Geography: A System of Modern Geography Comprising a Description of the Present State of the World, and its Five Great Divisions, America, Europe, Asia, Africa, and Oceania, with Their Several Empires, Kingdoms, States, Territories, etc.* (Philadelphia: Butler, 1860). These are sample questions (from pp. 7–74) typical of the time.

[7] Ellen Churchill Semple, *Influences of Geographic Environment* (New York: Henry Holt, 1911): 609.

2 GEOGRAPHY IN THE SCHOOLS

Finnish-American school children in Rocklin, California, 1910. (Photo courtesy of Roy Ruhkala.)

Geography is the study of people, place, and environments and the relationship among them. Geographically informed persons understand and appreciate the mosaic of the interdependent worlds in which they live. While a knowledge of geography is enjoyable in itself, it has practical value of spatial and environmental perspectives to life situations from local to global scale.

National Geography Standards 1994, p. 1

It is second period for fifth graders at Anderson School in central Kentucky and time for geography. With textbooks ready, maps hanging on the wall, and a globe in the corner of the classroom, the teacher begins a lesson on major cities of the United States, emphasizing their importance as pop-

ulation centers, political centers, and cultural magnets. The lesson ends with students locating the twenty largest cities on a wall map.

Meanwhile, a first-year lower secondary school class in Miazaki Prefecture, Japan, is analyzing world oil production and the advent of supertankers in global oil distribution. The discussion is lively but orderly. The children refer to their individual copies of a small and colorful world atlas as the teacher guides the class through thematic maps illustrating the concepts discussed.

North of Sydney, Australia, in a physical geography and environmental camp at Yarrahappini, a teacher is hiking through the eucalyptus and scrub forest with a squad of upper primary school students trailing behind. The instructor is pointing out the effects of erosion and sedimentation along a streambank. One student notes that undercutting on one side of the stream is matched by deposits of sand on the other side. The processes of valley deepening and widening are occurring before their eyes.

Each of the three class activities is geography in a traditional setting. One is descriptive, one is analytical, and one is inductive. Each of the three is a valid but somewhat incomplete approach to geographic education. Teachers and curriculum specialists have seldom agreed upon what constitutes a proper fundamental education in geography. Most would agree that skills are important, particularly map reading. Acquisition of a basic database of locations and terminology is also necessary. Most teachers would like to see students develop a positive attitude about other places and peoples. Beyond these points, however, there is little agreement as to what the curriculum should include.

In this chapter we offer reasons why geography is varied in content throughout the world and why it has changed significantly in the past few years in North America. We conclude with a status report on geography education along with predictions for the remainder of the decade. We introduce the five fundamental themes of geography, anticipating that these themes will constitute the structure of geography to be taught in the next generation's classrooms. These geographic themes and associated concepts are then organized around the structure provided by the 1994 *National Geography Standards* document, which is a result of work accomplished for the national GOALS 2000 Project. In total, the structure of this chapter reflects an approach to organizing geography education that most schools will eventually use in the coming decades.

GEOGRAPHY EDUCATION IN THE WORLD TODAY

As we noted in Chapter 1, the state of geography education in schools in North America has been severely criticized in recent years. A review of the literature, however, reminds us that criticism of this type is nothing new. The need for more geographic knowledge among our population was pointed out as early as 1903. In conducting a historical survey of the state of geographic education, the most important thing we learned was that conditions have not substantially changed until very recently. Our population is still woefully ignorant about other places in the world.

We wondered if teachers in other countries have similar concerns about geography as taught in their schools. When we briefly surveyed geography curricula in some foreign schools, we found similarities and differences in scope and depth of subject matter taught, both among the countries themselves and as compared to the curriculum in the United States. The following brief discussions of the status of geography education in Russia, Western Europe, Canada, Australia, Africa, the Middle East, and Japan are presented here as case studies for comparison with the section on geography education in the United States that follows.

Geography in Russia

Considering the size and diversity of Russia, it is not surprising that students in Russian schools receive several years of intensive geography training.[1] From grades six to ten, geographic concepts are presented as a part of the social studies curriculum. Grade six emphasizes physical geography, grade seven presents a world regional course entitled "Continents and Oceans," grades eight and nine focus upon the geography of the former USSR, and grade ten presents the economic and social geography of the world. This strong, broad-based curriculum of 357 class hours of geography between grades six and ten has been structured by government at the national level and is required in all schools across the country.

Some major topics covered in the sixth-grade introductory physical course include map reading and interpretation, elements of site surveying, distance and direction of coordinates, landform features, physical processes, the hydrosphere, the hydrologic cycle, weather and climate, the biosphere, interactions among the components of the natural world, population geography, and national boundaries on Earth.

The seventh-grade class on the geography of the continents and oceans includes the following topics: nature of the lithosphere and terrestrial relief; global oceans; Earth's atmosphere and climate; natural zoning (biomes); human mastery of Earth; regional studies of Africa, Australia, the Antarctic, South America, North America, and Eurasia, and the Atlantic, Pacific, Indian, and Arctic oceans; and the meaning of the "geographical shell."

"The Geography of the Former Soviet Union" taught in grades eight and nine is a descriptive regional course including physical, economic, and social geography. The final required class in tenth grade is titled "The Economic and Social Geography of the World." Major topic headings for the course are natural resources and ecological problems, world population issues, world economies and the division of labor, and global problems of mankind. Countries that were formerly socialist are emphasized and compared with capitalist economic systems. The developing countries of the world are given special attention.

Western travelers' accounts upon returning from Russia lead us to believe that the average Russian student's working knowledge of geography is probably no greater than his or her counterpart in the United States, perhaps owing to a lack of travel opportunities. Maps are difficult and expensive to purchase in most parts of the country (and often contain inaccurate information gathered and mapped during the Soviet period). An American visitor to Russia related that his Tartar traveling companion on the trans-Siberian Railroad east of the Ural Mountains enjoyed the visitor's National Geographic Society map of the region. This resident of Russia proclaimed to our friend: "This is the best map I've seen of the area ... it fills in all the empty spaces." [2]

Recent field work by one of your authors in Russia revealed that mapmakers during the Soviet period intentionally put incorrect information on their maps. Apparently Russians may be almost as geographically illiterate as people from the United States! According to Professor John Adams, who taught geography at Moscow State University in 1989, Russian education experts are just as concerned about geographic illiteracy as are their North American counterparts.

Further reinforcing this impression of Russian geographical illiteracy is the 1989 study conducted jointly by the Gallup Organization, Inc. and the Russian Academy of Sciences. [3] This study compared adults in ten countries to determine their geographic knowledge. Among other things, the results showed that both Russian and American citizens know embarrassingly little about world geography and place locations.

For example, Russians correctly identified an average of only about 7.4 countries and major water bodies out of 16 on a world map, placing their geographic knowledge at the bottom, along with Mexicans. American adults ranked only slightly higher, with 8.6 correct on average. Swedes topped the comparison with an average of 11.6 correct.

Among the younger adults surveyed (18 to 24), the Russian score of 9.3 put them in fourth place, tied with Italians and Canadians. American young adults came in last, averaging only 6.9 correct. Pollsters said the Russian results were dragged down by those over 55 who grew up

before World War II, when many Soviet maps were sketchy and deliberately distorted.

Geography in Western Europe

Schools in the United States frequently have been judged inferior to those in the British system. In the British curriculum, geography is taught as a special subject for five years to children of ages 11 to 15; thereafter, it becomes optional. Some students bring seven years of geography study to the university from the high school. Success is measured by terminal examinations. Concepts to be mastered include the basics of physical geography, map reading, and the rudiments of location theory. Field work is usually taught. Humanistic geography and behavioral geography are newer perspectives that are currently emphasized. Employment opportunities for geographers exist, so it is a popular major in university, and the profession is more generally known and respected by the general public than in the United States.

We conclude that the average citizen of the British Isles and most of Western Europe is much more geographically literate than the average citizen of most other parts of the world. A survey of European geography textbooks at the secondary level revealed concepts and principles that are more often taught in first-year university courses in the United States.

In our recent travels to Western Europe, we saw crowds of workers on lunch break crammed into bookstores in major cities discussing maps and current events. Displays of atlases and travel books decorated bookstore windows. Conversations assumed from their tone and content a solid geographical education. In spite of the relative success of their geography programs, however, educational and government leaders in Great Britain and on the Continent are still concerned about the continuing strength of the geography programs in their schools. This is largely in response to recent efforts at integrating geography into a more general social studies framework. Elements of each curriculum are constantly being restructured to keep pace with a changing world.

Geography in Canada

In Canada, geography has been adapted from earlier programs in Britain and France and has been established in the curriculum through this century. Although province-by-province variations abound, world regional geography is the most common course taught in the Canadian social studies curriculum. The "Geography of Canada" is taught in almost every province, but at different grade levels.

Canadian evaluators feel that one of the weakest links in the system is the unqualified geography teacher. The use of specific teaching themes and more field work has been recommended along with increased emphasis on urban geography and decreased emphasis on agricultural geography. High school geography is primarily taken on an elective basis in Canada and is not considered an important part of the core curriculum.

At a recent meeting of the National Council for Geographic Education, we discussed the status of geographic education with numerous Canadian geography teachers. They too expressed concern about their students' lack of geographic knowledge.

Geography in Australia

In Australia, the structure of geography within the social studies curriculum is quite similar to that found in Canada. Each school curriculum is determined by local school boards within each of the Australian states. Although the curricula are not uniform, major common themes tend toward geographic foundations of history, analysis of Australian society, surveys of the world's people, and the place of local communities within a larger national and global context.

A survey of how geography is taught in Australia has not been undertaken, but our sense of the "state of the art" is that the subject is descriptive and social studies–oriented. In Figure 2-1, the class is studying global location patterns.

Geography in Africa, the Middle East, and Latin America

Most geography taught in Africa south of the Sahara is based on the original colonial curriculum. For example, in Nigeria geography is considered a fundamental subject of study but it is included in "General Studies," which also incorporates such other subjects as nature study and physical science. Geography in North Africa is generally taught in every grade at the primary level, but is limited to a one-year course at the secondary level.

In some countries, such as Saudi Arabia, the requirement for geography is based on which "stream" (or "track") students may be on from year to year. For example, only liberal arts–track students are required to study geography for more than two years. Geography is studied for five years in Pakistan and three years in Iran at the middle school level.

Although a very small percentage of students in India proceed to the secondary level, those who do receive a curriculum based on the British model. Geography is a popular subject at the university level, although only 4 percent of the university-age population in India receives a higher education.

Most Latin American countries require classes in geography at least half of the year in middle school social studies and one year in high school. In Latin America, physical and economic concepts of geography are emphasized.

Geography in Japan

On a visit to Japanese schools, we saw geography taught from age six upward in a social studies framework. Figure 2-2 shows a page explaining the concepts of world trade, taken from a fourth- to sixth-grade school atlas.

Geography content was weighted equally with other

FIGURE 2-1
Geography class in Melbourne, Australia, prepares to discuss concepts learned on a recent field trip to the "Outback." (Photo courtesy of Edward L. Myles.)

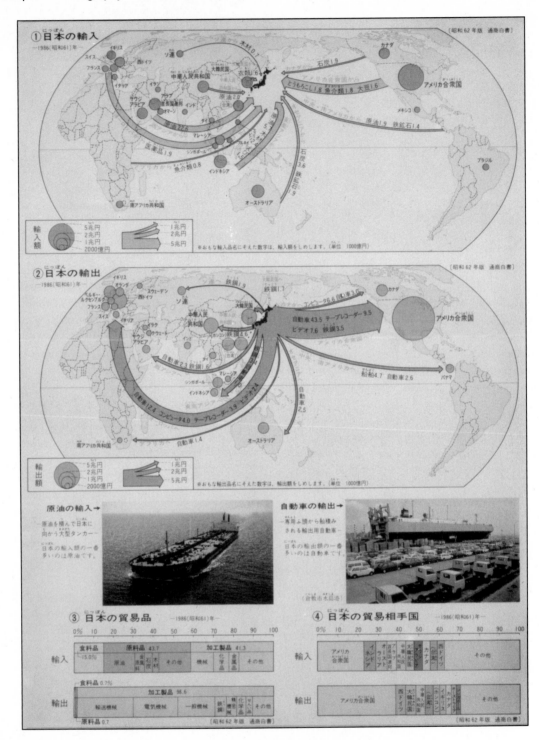

FIGURE 2-2
*Page from a school atlas for fourth- through sixth-grade geography students in Japan illustrates
their level of understanding of world trade patterns. (© Teikoku-Shoin Co., Ltd. Reprinted with permission of the publisher.)*

FIGURE 2-3
Location maps such as this one posted in a park are found throughout Japan since many streets are not named and building numbers are not always consistent. These maps are found in most neighborhood and city districts. (Photo by author.)

social studies subjects such as civics and history. Physical geography was included in natural science courses. In lower secondary schools in Japan, geography is universally taught as a self-contained subject. The emphasis is on world regions, Earth as an arena for life, the geography of Japan, population, resources, and trade. Upper secondary schools in Japan include geography within the social studies curriculum along with world history, Japanese history, and politics. Many of the same subtopics of geography are presented, with more detail and complexity added in the upper grades.

According to the official course of study for upper secondary schools in Japan, geography should "make students understand regional features of people's lifestyles in the world, and their trends in connection with the natural and social environments of each area, cultivate the geographical recognition toward the present-day world, and think of the position and the role of Japan in the international society." [4]

In sharp contrast to Russia, maps in Japan are commonplace (Figure 2-3). They appear as billboards in almost every urban neighborhood and, because of a lack of a grid system in urban areas, are absolutely essential for locating an address anywhere in the country. Personal map drawing is a skill for which training begins in the first grade. All Japanese students have an atlas at hand throughout their school career.

An Observation on the Status of Geography Education

Based on our comparative observations of geographic education around the world, we suggest that the United States needs more geography taught at all levels. Understanding basic concepts and principles of Earth systems—physical, cultural, economic, and political—is necessary for communication among nations about Earth and its people.

GEOGRAPHY EDUCATION IN THE UNITED STATES

Geography has been taught in public schools in the United States ever since Jedediah Morris first codified the subject in 1784. Various editions of his textbooks were the mainstay of geography courses for most of the nineteenth century. His books, *The American Geography* and *American Universal Geography* were primarily descriptions of the known world at the time.[5] Although Morris's *Geography Made Easy* proves inadequate by today's standards, it made geography popular in the primary school curriculum in eighteenth-century America (Figure 2-4).[6] His books discussed physical, cultural, and economic characteristics of places—with heavy moral and religious overtones.

It was probably the immense popularity of Samuel Augustus Mitchell's textbooks, however, that defined popular geography as the study of place locations and lists of these to be memorized. The discipline of geography has been trying to recover from this limited approach ever since. An all-too-familiar criticism of geography as it was taught in 1864 came from a Maine school teacher as follows:

> Probably no branch of study in our common schools is so poorly taught or so indefinitely understood as geography. Scholars often commit to memory many detached facts, unimportant descriptions, and long lists of names of towns, capes, gulfs, rivers, etc., but gain no conception of the principles and laws that underlie this important science.[7]

Several nationwide efforts to integrate geography with other subjects within the curriculum and to make it more relevant to learners were largely unheeded by local school districts. One such study, the National Education Association's "Committee of Ten Report" in 1894, says that geography, "although in name not a new study, it is in reality new as it is presented by the conference upon geography." [8]

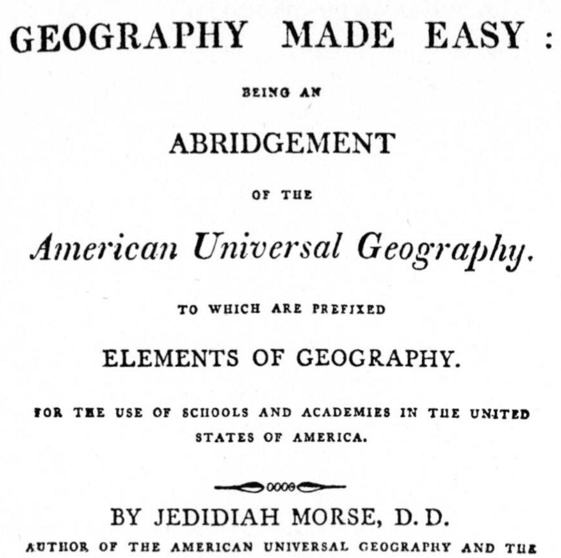

GEOGRAPHY MADE EASY :

BEING AN

ABRIDGEMENT

OF THE

American Universal Geography.

TO WHICH ARE PREFIXED

ELEMENTS OF GEOGRAPHY.

FOR THE USE OF SCHOOLS AND ACADEMIES IN THE UNITED
STATES OF AMERICA.

BY JEDIDIAH MORSE, D. D.

AUTHOR OF THE AMERICAN UNIVERSAL GEOGRAPHY AND THE
AMERICAN GAZETTEER.

There is not a son or daughter of Adam, but has some concern both
in Geography and Astronomy.   DR. WATTS.

ILLUSTRATED WITH A MAP OF THE WORLD, AND A
MAP OF NORTH-AMERICA.

Seventeenth Edition,

AND FIFTH OF THIS NEW ABRIDGEMENT.

BOSTON :

PUBLISHED BY THOMAS & ANDREWS.

SOLD AT THEIR BOOKSTORE, NO. 45 NEWBURY-STREET; BY EAST-
BURN KIRK, AND CO. NEW-YORK, M. CAREY, PHILADELPHIA; AND
BY THE PRINCIPAL BOOKSELLERS IN THE UNITED STATES.

J. T. BUCKINGHAM, PRI
Nov. 1814.

FIGURE 2-4
Title page from Geography Made Easy *by Jedidiah Morse,
1814. (Photo by author.)*

The report declared intellectual discipline to be the chief goal of secondary education and that this view should be developed by drawing upon the written advice of nine subsidiary conferences, one of which was on geography. The conference on geography of 1893, for its part, spoke out against any further pursuit of a survey-type of world geography then in fashion and recommended in its stead one or more courses in earth science, embracing physiography, geology, and meteorology. By 1900, earth science—weight-ed heavily in favor of physiography—had become the leading science course in the American high school.[9]

In the 1850s and 1860s, the teachings of Arnold Guyot influenced college and secondary school geography curricula with a stronger emphasis on physical concepts. According to a recent article, "Voices for Reform in Early Geographic Education":

> Guyot was convinced the earth was a theater created for the enactment of human drama. History was shaped by and played out against the size, shape, and physical geography of continents.[10]

The emphasis of geography, first as a study of Earth's physical features and processes, and second as human influences, became the primary direction for North American geography at the turn of the century, led by William Morris Davis at Harvard University. The 1890s have, in fact, been called the "Physiographic Era." Davis's own writings and his advice to students and educators were based on known principles of geology and physiography with suggestions that human activity on Earth was strongly influenced by physical geography processes. This theme was criticized by many as Social Darwinism but was nevertheless adapted and transmitted by many educators.

Early geography appeared in the public school curriculum primarily as an earth science component of general science courses. However, by 1916, the newly created framework called "Social Studies" began to lay claim to geography. Social Studies at this time included economics, civics, and history, as well as geography. Geographer Preston James, noted recorder of the history of geography, stated that a typical secondary school social studies program in 1916 included geography and history at the seventh-grade level, American history in the eighth grade, commercial and vocational education in ninth grade, world history in tenth grade, American history again in eleventh grade, and problems of democracy in the final senior year of high school. At the same time, many universities and colleges in the United States were in the process of dropping geographic education–oriented courses from their catalogs. In addition, numerous journal articles and other suggestions for teaching from this era centered on "geographic influences"; that is, physical geographic influences on events and people in various places in the world.

We reviewed the contents of the major geography education publication in the United States, the *Journal of Geography*, for each decade of the twentieth century on the assumption that this journal would be an indicator of trends in geographic education and would supply clues to major concepts taught in each ten-year period. According to the articles in the journal, the emphasis in teaching during the 1920s focused upon gathering new teaching materials and suggestions for lessons on field work and mapping. The major content focus during this decade and the succeeding one was on economic and commercial geography.

The 1930s were a time of increased study of individual nations in the world, with much more of a global perspective. Mapping skills again were emphasized, and commercial geography continued to be stressed. The decade of the 1940s was bisected by World War II, and what was learned in that decade would not appear in school curricula until the early 1950s. Post–World War II geography emphasized geopolitics, global understanding, and world regional concepts. Preston James exhorted teachers of the postwar era that geography must be integrated gradually into the student's experience, the use of maps must be taught more adequately, the historical element of geography must be reorganized, and global geography and conservation must be emphasized. The technology of air photography and improved mapping techniques made possible the integration of more geographic techniques in the classroom.

As a result of the launching of the Soviet satellite *Sputnik* in 1957, considerable concern was expressed during the late 1950s about the state of education in the United States as a whole compared with other countries, particularly the Soviet Union. Deficiencies were noted, especially in science and math programs. Massive infusions of funding and research effort in the late 1950s and early 1960s favored math and science instruction at the expense of the social sciences. Numerous opinions suggested "new" frameworks for classroom instruction. Some were learning-based, others were content-based, but all diluted specific subject matter areas such as geography and anthropology into a broad milieu known as "social studies."

The importance of the individual disciplines was replaced by a concern for community studies and societal issues. Many teachers were never exposed to geographic concepts and assumed that teaching geography simply meant selecting facts about places and requiring students to commit them to memory. This is the geography that we suspect many of our readers grew up with.

The education profession was blessed with a number of significant contributions by geographers in the late 1960s. One of the most important directions given to the discipline for teaching at the precollegiate level was provided by William B. Pattison in 1964. He suggested that, as an alternative to the competing monastic definitions that had been geography's lot, geography could be structured around four distinct but affiliated traditions, "operant as binders in the minds of members of the profession":

- The spatial tradition
- The area studies tradition
- The man-land tradition
- The earth science tradition[11]

The spatial tradition traces geography's interest in location and movement from earliest times to the present. It is focused upon map study and quantified aspects of geographic study, such as location analysis and statistical relationships among spatial variables.

The area studies tradition is also represented in antiquity and centers its focus on the comprehensive knowledge of places large and small. As espoused in Association of American Geographer's presidential addresses since the 1950s, these area studies are sometimes equated with the concept of **region**. Preston James's comments on a further understanding of the regional concept (1952) and John Fraser Hart's address on regional geography as the "highest form of the geographer's art" (1982) underscore the importance of Pattison's second tradition.[12]

As noted earlier in Chapter 1, the man-land tradition (more recently labeled "human-environment interaction"), also reaches back into antiquity as a focus of geographic study. Questions of environmental influences and environmental impacts have been addressed under this theme. This tradition of attempting to correlate human activity to environmental circumstances is probably the most understood and recognized of the traditions by nongeographers. The human-environment tradition also includes human ecology, resource perception and use, and humans as modifiers of Earth.

The earth science tradition, as its name implies, embraces the study of Earth and its physical attributes, including its relationship with the sun. "Introduction to Physical Geography" is still the most frequently offered course in most colleges and universities in the United States.

Until recently, the four traditions of geography often were used as a framework for geography curricula at the school and college levels and provided structure for several major textbooks surveying the subject. Although they offer one means of organizing a diverse field, they do not address geography's potential for exploration and discovery. Some have also noted the omission of the time element—historical geography—and the importance of cartography.

Government-sponsored teaching institutes, revised textbooks, and amended course outlines began to expand the scope of what geography is and how it is taught. Increased awareness and attention to the problem of cities and the environment have created issue-based programs within the schools. In the 1960s and early 1970s, the High School Geography Project became a most significant and large-scale alliance of professional geographers with classroom teachers.

The High School Geography Project

The High School Geography Project (HSGP) was first proposed by members of the Association of American Geographers in 1961. A steering committee was formed and classroom teachers became involved in the massive effort. The HSGP was defined by its creators as:

> A course content improvement program in geography sponsored by the Association of American Geographers and supported by the National Science Foundation. The project's goal is the development of new geography teaching materials at the tenth-grade level. Current work is concentrated on the development of materials following a course outline on the settlement theme.[13]

The HSGP officially began in 1961 with funding from the Ford Foundation. Initial tasks were to convene and receive guidance from an advisory committee of professional geographers. For the next few years, advisory papers, preliminary content outlines, and potential teaching activities were produced and reviewed by the project staff, consultants, and committee members. In 1964 new funding and project direction were obtained from the National Science Foundation. After three sets of field trials and after years of additions, deletions, and revisions, the project materials (including six sets of teaching aids) were published as *Geography in an Urban Age* in 1970. The units were entitled "Geography of Cities," "Manufacturing and Agriculture," "Cultural Geography," "Political Geography," "Habitat and Resources," and "Japan."

After public distribution and use in the early 1970s, the High School Geography Project was generally felt to contain both successes and limitations. Major strengths of the project were that:

- It promoted individual thinking by use of the inquiry method.
- It stressed focused learning over broad course "coverage."
- It was interesting to both students and teachers.
- It extended beyond past preconceived notions of geography into more realistic definitions and concepts.
- It showed how geographic principles can be used in problem solving.

- It created an interest in geographic education within the geographic and educational communities.

The HSGP's major difficulties were that:

- It did not fit the adopted curricula of most states.
- Its most effective use by social studies teachers required training in geography, which was not generally available to them at that time.
- Only the unit on Japan was focused outside of North America, whereas most secondary geography had a global orientation.
- Communication between university geographers and classroom teachers was limited and often contradictory. Even the basic definition of the term "geography" was often debated among these various groups of educators.

Unfortunately, we contend that the High School Geography Project also did not contain sufficient material relevant to pressing social and political issues of the time. Themes in ethnic relations, gender, poverty, crime, and other social concerns, for example, were almost entirely absent from the course materials. Despite reform efforts, geography was not attuned into current events and issues in time to make an impact on student learning.

Issue-based learning became important to geographers when eco-awareness entered the American classroom in the 1970s. At last geography had a rallying point. Curricular emphasis in conservation, pollution, and resource management began to elevate the importance of geography in the schools.

Geographers have been criticized for not rallying to the cause of environmental education in time to make it a focal point of the discipline in the 1970s and 1980s, but, at that time, professional geographers were fragmented in their research interests and philosophical goals. Some continued to work with statistical analysis and quantitative techniques, making predictions about research outcomes; others stressed social and cultural topics in a more qualitative mode of analysis. Meanwhile, some geographers did begin to turn their attention toward environmental concerns, but most failed to communicate with educators in the schools.

Since geographers could not agree about the goals and concerns in their field among themselves, it is no wonder educators and curriculum specialists had given up trying to incorporate geographic concepts into the social studies framework.

The beginning of the 1980s excluded most subjects not considered fundamental to a new "back-to-basics" philosophy in education in the United States. Reading and mathematics dominated the curriculum at the expense of the arts, humanities, and social sciences. Perhaps it was true that "Johnny" (and "Sara") did learn to read, but they certainly did not learn to read maps! In the 1990s priorities changed again.

CURRENT STATUS
OF GEOGRAPHY EDUCATION
IN THE UNITED STATES

Teachers are busy people and geography teachers are no exception. Those who teach geography may or may not be formally trained in the subject, and few geography teachers enjoy the luxury of teaching only that subject. Professional stimulation is generally limited to yearly contacts with the state geographical society, regional social studies teacher's meetings, or subscriptions to journals such as the *Journal of Geography, Social Education,* or *Social Studies.* These inputs compete with other professional associations and other educational publications.

The gap between professional geographers—who are usually university professors or persons in the public sector or in private business—and classroom teachers often has seemed a deep and unbridgeable chasm. Fortunately, recent successes in getting geography in the public eye and raising interest for the subject in schools are closing this gap.

The beginning of the current surge of energy and increased communication within the geography community began in 1983 in California with the creation of an informal alliance of teachers, administrators, and university geographers with the express purpose of making their concerns felt at the state level where social studies curriculums were being revised. One of the founders of this Geographic Alliance movement, Christopher Salter, recalls:

> In trying to find a voice to speak for geography in the state [California], it became apparent that we did not have an organization ready to do the presentations, the lobbying, and the advocacy essential for state educational decision makers to give some attention to geography as a precollegiate course. In an effort to rectify that, a decision was made at the University of California, Los Angeles, to convene a population of educators linked not by a common teaching level—as is traditional—but rather by a common affection for geography.[14]

Alliance members were successful in providing needed input into curriculum mandates and to sponsor in-service activities such as workshops, conferences, and presentations for teachers (Figure 2-5). A large number of social studies educators not previously familiar with geography were captivated.

FIGURE 2-5
Social studies teachers share ideas and teaching strategies at a Geographic Alliance workshop. (Photo courtesy of Muncel Chang.)

The Alliance concept caught the attention of the National Geographic Society, which at the same time was publishing editorials about the plight of geographic illiteracy in the country. Similarly, television news and talk shows were presenting shocking statistics that revealed a real lack of geographic knowledge among the American public and university students in particular. As with most news items, syndicated versions of such stories swept the nation. Geographic literacy became a national issue—at least for a few weeks. This was enough to attract the attention of politicians as well as policymakers at the National Geographic Society, the largest nonprofit and nongovernmental geographical organization in the world. In 1985, President Gilbert M. Grosvenor launched the Geography Education Program with a mission to revitalize the teaching and learning of geography in the nation's kindergarten through twelfth-grade classrooms.[15] Due to the support of this enormous institution in Washington, DC, by 1985, there were 14 Geographic Alliances and by 1988 there were 27 of these organizations in the United States. By 1993, every state had its own Geographic Alliance, with California having two—one in the north and one in the south.

The basic role of each individual Alliance has been to coordinate the grass-roots energies of classroom teachers, academic geographers, administrators, and students toward the increase and improvement of geographic education. The Alliances provide an effective forum in which discussions can be held on curriculum guidelines, the availability and use of classroom materials, new and effective presentation methods, and professional outreach activities. Most of these Alliances are taking advantage of the multiplier effect and offering their own summer institutes, staffed with local university professors and augmented with Teacher-Consultant graduates of National Geographic Society summer institutes (SGI), which began in 1986.

Another major component of the involvement of the National Geographic Society in improving geographic education nationwide has been their multiyear Teaching Geography Project, first funded by the Department of Education's National Diffusion Network in 1987. This project's three objectives include:

- Introducing educators to fundamental geographic themes
- Sharing hands-on teaching strategies
- Informing teachers of available follow-up resources and opportunities in teaching geography

The Teaching Geography Project significantly enhanced the efforts of the Alliance network and Teacher-Consultants as well as encouraged the continued and vital collaboration between university-level geographers and classroom teachers. It is only one part of the effort of the National Geographic Society's Education Program, however. Other successful projects designed to enhance the status of geography education and improve its teaching in the nation's schools include summer institutes to develop leadership skills among teachers (SLIs) and on-site workshops on water (in California in 1993) and on wilderness (in the Southwest in 1994). Most recently, their urban outreach program offers geography institutes in urban areas to encourage local teachers to become more involved in their state's Geographic Alliance network.

As mentioned in Chapter 1, many new teaching materials have been developed to support these in-service efforts to support the reinvigorated teaching of geography that has been taking place in thousands of the country's schools since 1985. In-service teacher training classes, teaching materials, newsletters, lesson plans, teacher-produced videos, and special presentations at professional meetings are typical activities undertaken by today's geography teachers. The rejuvenation of geography and the revolution in the geographic content of courses and the methods of teaching in the schools continue to gain momentum in the 1990s.

Educational publishers have responded by identifying the need for teacher materials and improved textbooks. Social studies teachers can now choose among several vendors and a variety of materials at regional meetings. Computerized instruction, several new multimedia projects, and improved maps and atlases have augmented the several new textbooks that are now available. Public awareness about the importance of geography education remains high and support continues to be strong at national, state, and local levels of educational decision making.

This heightened awareness has stimulated educators and geographers to reexamine the basic content of the discipline that a "geographically literate" person should possess. Traditional forms of presenting geography have been regional or topical. Often, regional courses at the global level or even community level have attempted to include everything that could be mapped or described about the area, such as soils, landforms, climate, vegetation, land use, transportation, population characteristics, and so forth. These encyclopedic presentations were often overloaded with information. This approach made difficult the selection and assembly of important pieces of information into meaningful interpretations beyond the descriptive level. This malady is easily corrected with instructor attention to issues and themes, and with the identification of overriding concepts that tie the understanding of the selected themes and concepts into a cohesive whole.

The topical approach, in which thematic depth is preferred over breadth of subject matter, also has its shortcomings. Much geography taught at the university level is based on research fields of interest and categorizations that do not always lend themselves to elementary or secondary classrooms. To really understand the geography of agriculture, for example, a student must understand concepts in the larger domains of cultural, physical, and economic geography. Isolated research topics often hold little interest for students unless it can be demonstrated that they are part of a meaningful whole, with application toward issues and current concerns within a particular region.

In addition to the regional and the topical approaches, a third approach is possible. Presentations on geographic thinking from a conceptual basis have also been attempted over the years. Jan Broek's *Geography: Its Scope and Spirit* has undergone several revisions and remains a classic introduction to the field for teachers as well as practitioners.[16] It stresses the development of geographic thought, viewpoints, and methods. A similar successful publication by Rhoads Murphey, entitled *The Scope of Geography*, explains basic concepts in spatial interaction, environmental systems, and parallel concepts in the social sciences.[17]

The most recent conceptual approach to introducing geography to nongeographers was carefully constructed and agreed upon by a committee of academic geographers and classroom teachers. Representatives from the Association of American Geographers and the National Council for Geographic Education distilled the unique essence of geographic thinking into five fundamental themes. Use of these themes has met with unprecedented success in catching the attention of the public and geography teachers across the nation.

Implementing geography curriculum based on these themes has been the goal of the National Council for Geographic Education, the Association of American Geographers, the American Geographical Society, the Geographic Education National Implementation Project (GENIP), and the Geographic Alliance movement (see Figure 2-6).

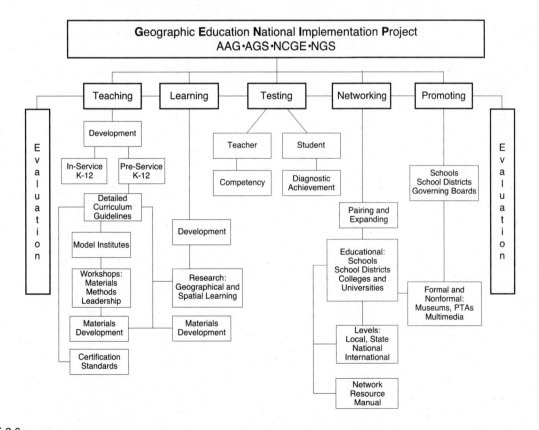

FIGURE 2-6

GENIP'S Interrelated Responsibilities and Activities. (From Salvatore J. Natoli, "Guidelines for Geographic Education and the Fundamental Themes in Geography, Journal of Geography 93 *(Jan.–Feb. 1994): 5. Used with permission.)*

THE NATIONAL STANDARDS AND FIVE FUNDAMENTAL THEMES IN GEOGRAPHY

The geographic themes lend themselves to the study of almost any place. Taken together they utilize the advantages of both the topical and regional approaches to geographic thinking, while minimizing their limitations. The definitions of each of the five themes listed here are taken directly from the Association of American Geographers and the National Council for Geographic Education's *Guidelines for Geographic Education:*[18]

1. Location: Position on Earth's Surface

Absolute and relative location are two ways of describing the positions of people and places on the Earth's surface.

2. Place: Physical and Human Characteristics

All places on Earth have distinctive tangible and intangible characteristics that give them meaning and character and distinguish them from other places. Geographers generally describe places by their physical or human characteristics.

3. Relationships within Places: Humans and Environments

All places on Earth have advantages and disadvantages for human settlement. For example, high population densities have developed on flood plains, where people could take advantage of fertile soils, water, resources, and opportunities for river transportation. By comparison, population densities are usually low in deserts. Yet flood plains are periodically subjected to severe damage, and some desert areas, such as Israel, have been modified to support large population concentrations.

4. Movement: Humans Interacting on Earth

Human beings occupy places unevenly across the face of Earth. Some live on farms or in the country; others live in towns, villages, or cities. Yet these people interact with each other; that is, they travel from one place to another, they communicate with each other, or they rely upon products, information, or ideas that come from beyond their immediate environment.

The most visible evidences of global interdependence and the interaction of places are the transportation and communication lines that link every part of the world. These demonstrate that most people interact with other places almost every day of their lives. This may involve nothing more than a Georgian eating apples grown in the state of Washington and shipped to Atlanta by rail or truck. On a larger scale, international trade demonstrates that no country is self-sufficient.

5. Regions: How They Form and Change

The basic unit of geographic study is the region, an area that displays unity in terms of selected criteria. We are all-familiar with regions showing the extent of political power such as nations, provinces, countries, or cities, yet there are almost countless ways to define meaningful regions depending on the problems being considered. Some regions are defined by one characteristic such as a governmental unit, a language group, or a landform type, and others by the interplay of many complex features.

For example, Indiana as a state is a governmental region, Latin America is a region where Spanish and Portuguese as major languages can be a linguistic region, and the Rocky Mountains as a mountain range is a landform region. A geographer may delineate a neighborhood in Minneapolis by correlating income and educational levels of residents with the assessed valuation or property or tax rate, or distinguish others by prominent boundaries, such as a freeway, park, or business district. On another scale, we may identify the complex of ethnic, religious, linguistic, and environmental features that delineate the Arab World from the Middle East or North Africa.

These five themes were taken one step farther when the nation's governors and the Bush Administration devised the National Education Goals at an education summit in Virginia in 1989. These goals identified geography as one of the five core subjects for schools in the United States. Out of this grew, first, the development of a document for assessing the knowledge of every fourth-, eighth-, and twelfth-grade student in the United States in each of these core subjects. The document guiding this national assessment was published in 1993 and is entitled *Geography Framework for the 1994 Assessment of Educational Progress.* It was developed by the NAEP Geography Consensus Project, a committee of twenty-six academic geographers and classroom teachers from across the country. The NAEP document links the five themes of geography into a three-tiered structure, as shown in Figure 2-7. This diagram summarizes the interconnections between the three NAEP content outcomes: Space and Place, Environment and Society, and Spatial Dynamics and Connections and the five fundamental themes of geography.

INSTRUCTION ──────▶ ASSESSMENT

**1994 Assessment
Framework**

Organizing Ideas

Five Themes of Geography*	**Three Content Outcomes**
LOCATION Position on Earth's Surface	**SPACE AND PLACE** Knowledge of geography related to particular places on Earth, to spatial patterns on Earth's surface, and to physical and human processes that shape such patterns.
PLACE Physical and Human Characteristics	
HUMAN/ENVIRONMENT INTERACTION Relationships within Places	**ENVIRONMENT AND SOCIETY** Knowledge of geography related to the interactions between environment and society.
MOVEMENT Humans Interacting on Earth	**SPATIAL DYNAMICS AND CONNECTIONS** Knowledge of geography related to spatial variations and connections among people and places.
REGIONS How They Form and Change	

INSTRUCTION ◀────── ASSESSMENT

*Guidelines for Geographic Education–Elementary and Secondary Schools.
Joint Committee on Geographic Education of the National Council for Geographic
Education and the Association of American Geographers, 1984.

FIGURE 2-7
*The Five Themes of Geography and NAEP Content Outcomes. Geography instructional themes
and content learning outcomes for the 1994 NAEP Assessment Framework. (From* Geography
Framework for the 1994 National Assessment of Educational Programs *(NAEP, 1993), p. 5.)*

Following the work of the Geography Consensus Committee to develop this national assessment framework, *National Geography Standards* in geography were developed as an integral part of President Clinton's GOALS 2000 Project. They were designed to equip U.S. students with the knowledge and skills necessary to compete in an informed labor force and to serve as educated citizens. Geography's "day in the sun" as a central part of this national educational reform effort had finally arrived!

Organizing the development and dissemination of these voluntary *National Geography Standards* proved to be an enormous undertaking. According to the *Standards* document, "These Standards reflect the belief that geography must be as rigorously taught in the United States as in all other countries." [19] The numerous committees of academic geographers and classroom teachers who developed the Standards were overseen by representatives of the public policy, business, and education communities.

Geography for Life: The National Geography Standards was published in late 1994. This important document provides specific guidelines about what students should know in geography after completing grades four, eight, and twelve. The criteria used in developing the Standards was based on those identified in the National Education Goals, which stated that the Standards should be:

- World class (of high quality, competitive, and challenging)
- Focused on the most important knowledge and skills
- Useful in students' lives
- Clear and usable
- Accessible
- Adaptable and flexible
- Appropriate for different levels of students' development
- Reflect a broad consensus within geography
- Reflect the current state of scholarship within geography

The eighteen Standards in *Geography for Life* incorporate both the five fundamental themes of geography and the content outcomes of the NAEP assessment document. These *National Geography Standards* are used to contextualize and conceptualize the geography content provided in this book. They are grouped into five broad categories including Seeing the World in Spatial Terms, Places and Regions, Physical Systems, Human Systems, Environment and Society, and Applying Geography. The diagram shown in Figure 1-1 in the previous chapter integrates the five fundamental themes with these eighteen Standards and the concepts discussed throughout this book. It will be helpful to refer to it often.

The publication of the *National Geography Standards* document provides only one example of the flurry of activity in geographic education in the 1990s. Other federally funded curricular projects, such as the Association of American Geographers' ARGUS project *(Activities and Readings in the Geography of the United States)* and GIGI *(Geographic Inquiry into Global Issues),* were developed and disseminated during this period of growth in the field as well. The National Council for Geographic Education has also made major gains in its membership and national impact in the 1990s. This organization has focused its attention on leadership of the *National Geography Standards* project and the continuation of curricular development and dissemination. In the late 1980s and 1990s, NCGE published a series of special publications on various critical issues in geographic education, including Slater's *Learning Through Geography* and others. This organization has also taken the lead in addressing the needs of underrepresented students in today's multicultural geography classrooms through support of the four-year *Finding a Way: Encouraging Young Women in Geography* project, which has developed and disseminated learning activities and teaching strategies to encourage young women and minority students in geography.

As is shown in a 1994 publication by James F. Marran, a classroom teacher who worked on the High School Geography Project in the 1960s, the NAEP assessment committee in the 1980s, and the *National Geography Standards* project in the 1990s, the "old geography" and the "new geography" are vastly different. Table 2-1 summarizes these changes in the field and offers hope for the future of geographic education all across the nation.

The remaining chapters of this book are structured around each of the fundamental themes of geography and the *National Geography Standards*. In this text, the themes and related concepts and standards appear as follows:

The theme of Location is the focus of Chapter 3, and it is referenced in subsequent chapters as a foundation for geographic thinking. This chapter emphasizes concepts contained in the following *National Geography Standards:*

Standard 1: *The geographically informed person knows and understands how to use maps and other geographic representations, tools, and technologies to acquire, process, and report information from a spatial perspective.*

Standard 3: *The geographically informed person knows and understands how to analyze the spatial organization of people, places, and environments on Earth's surface.*

The theme of Place is the structure for Chapters 4 through 9, with the appropriate physical, cultural, eco-

TABLE 2-1
The "old" and the "new" geography

Table 1 *Components of the OLD Geography*	*Table 2* *Components of the NEW Geography*
Oriented on Specific Place/Location	Emphasis on Spatial Relationships
Structured on the Recall of Information	Encourages Problem Solving
Fact-Based Objective Testing	Connected to Critical Thinking Skills
Limited Skill Development	Depth Replaces Breadth
Teacher Directed/Teacher Shaped	Collaborative Learning Strategies
Textbook Driven	Research Based
Student as Segregated Learner	Adaptable to the NEW Technology
Minimal Problem Solving	Observation Through Field Work
Hooray! It's Field Trip Day	Human/Environmental Interaction Emphasis
Regional Emphasis	Framework/Standards Driven
Ethnocentric/Nationalistic Bias	

Source: Marran, Journal of Geography *93 (Jan.–Feb. 1994), pp. 8–9.*

nomic, and urban characteristics of place offered in each chapter. The following *National Geography Standards* are emphasized in each of these chapters:

Chapter 4:

Standard 2: *The geographically informed person knows and understands how to use mental maps to organize information about people, places, and environments in a spatial context.*

Standard 4: *The geographically informed person knows and understands the physical and human characteristics of places.*

Standard 6: *The geographically informed person knows and understands how culture and experience influence people's perception of places and regions*

Chapters 5 and 6:

Standard 7: *The geographically informed person knows and understands the physical processes that shape the features of Earth's surface.*

Chapter 7:

Standard 9: *The geographically informed person knows and understands the characteristics, distribution, and migration of human populations on Earth's surface.*

Standard 10: *The geographically informed person knows and understands the characteristics, spatial organization, and complexity of Earth's cultural mosaics.*

Standard 13: *The geographically informed person knows and understands how forces of cooperation and conflict among people shape human control of Earth's surface.*

Chapter 8:

Standard 11: *The geographically informed person knows*

and understands the patterns and networks of economic interdependence on Earth's surface.*

Standard 16: *The geographically informed person knows and understands the idea of "resource" and the changes that occur in the use, distribution, and importance of resources.*

Chapter 9:

Standard 12: *The geographically informed person knows and understands the processes, patterns, and functions of human settlement.*

The theme of Human-Environment Interaction is the focus of Chapter 10, which stresses an environmental emphasis in human decision making. The following *National Geography Standards* structure the concepts presented as a part of this theme:

Standard 8: *The geographically informed person knows and understands the characteristics and spatial distribution of ecosystems on Earth's surface.*

Standard 14: *The geographically informed person knows and understands how human actions modify the physical environment.*

Standard 15: *The geographically informed person knows and understands how physical systems affect human systems.*

The theme of Movement is introduced in Chapter 11. This theme focuses on each of the following *National Geography Standards:*

Standard 3: *The geographically informed person knows and understands how to analyze the spatial organization of people, places, and environments on Earth's surface.*

Standard 9: *The geographically informed person knows and understands the characteristics, distribution, and migration of human populations on Earth's surface.*

The theme of Regions and the importance of regional thinking are stressed in Chapter 12. The following *National Geography Standard* is important in structuring the concepts presented in this chapter:

Standard 5: *The geographically informed person knows and understands that people create regions to interpret Earth's complexity.*

Chapter 13 integrates all the themes and concepts back into the whole of geography by offering applied case studies that we call "Geography in Action." The final two

National Geography Standards pertain to this final chapter and are stated as follows:

Standard 17: *The geographically informed person knows and understands how to apply geography to interpret the past.*

Standard 18: *The geographically informed person knows and understands how to apply geography to interpret the present and plan for the future.*

With the *National Geography Standards* and the five fundamental themes of geography in mind, and with your concept diagram in hand, you are invited into the experience of geography. We will begin with the "basics" in the following chapter.

FURTHER READING

Introductory university textbooks are often excellent reviews of the content of the geography courses taught there. One of the most widely used survey texts in university settings is Harm J. de Blij's *Geography: Regions and Concepts* (New York: John Wiley, 1992). Also useful as background in conceptual thinking about geography is Christopher L. Salter and C. F. Kovacik's *Essentials of Geography* (New York: Random House, 1988).

More academic surveys of the nature of the discipline and the variety of subjects studied by geographers are found in Jan O. M. Broek's *Geography: Its Scope and Spirit* (Columbus, OH: Merrill, 1965), Roger Minshull's *The Changing Nature of Geography* (London: Hutchinson, 1972), and Rhoads Murphey's *The Scope of Geography* (New York and London: Methuen, 1982).

Two important works in nineteenth- and early twentieth-century geography are Ellen Churchill Semple's *American History and Its Geographic Conditions* (Boston: Houghton Mifflin, 1903), Arnold Guyot's *The Earth and Man: Lectures on Comparative Physical Geography in its Relation to the History of Mankind* (Boston: Gould, Kendall, and Lincoln, 1849).

William Pattison's "The Four Traditions of Geography" was published by the National Council for Geographic Education in its professional paper Number 25 in May 1964. It is reprinted in the *Journal of Geography* 63 (May 1964): 211–216. Two early compilations of readings for geography teachers are Phillip Bacon, ed., *Focus on Geography: Key Concepts and Teaching Strategies* (Washington: National Council for the Social Studies, 1970), and John M. Ball, John E. Steinbrink, and Joseph P. Stoltman, eds., *The Social Sciences and Geographic Education: A Reader* (New York: John Wiley, 1971).

Other worthwhile references for teachers of geography are *Directions in Geography: A Guide for Teachers* by Gail S. Ludwig et al. (Washington, DC: National Geographic Society, 1991); Bruce Marshall's *The Real World: Understanding the Modern World Through the New Geography* (London: Houghton Mifflin, 1991); *Learning Through Geography* by Francis F. Slater (recently

reissued by the National Council for Geographic Education, 1993); *Geographic Themes and Challenges* by Tony Burley and Jim Latimer (Edmonton, Alberta: Arnold, 1990); and *Missing the Magic Carpet: The Real Significance of Geographic Ignorance,* by Christopher L. Salter (Princeton, NJ: Educational Testing Service, 1990).

A provocative alternative approach to geography teaching may be found in David Boardman's *Graphicacy in Geography Teaching* (London: Croom Helm, 1983). Boardman's *New Directions in Geographical Education* (London: Falmer Press, 1985) summarizes trends in geographic education, offers sample projects for students, and examines values in teaching geography.

The High School Geography Project is a collection of textbooks, teaching materials, and teacher's guides collectively entitled *Geography in an Urban Age* (New York: MacMillan, 1975). Although this collection is currently out of print, it is still valuable for providing teaching ideas and is usually available in university libraries and in most American school district offices.

A collection of excellent articles on the current status of geography education in the United States, presented at the Summit in Geographic Education at San Marcos, Texas in 1993, are included in a special issue of the *Journal of Geography* 93 (Jan.–Feb. 1994).

The impact of the Geographic Alliance movement in the United States is summarized by Christopher L. Salter in "The Nature and the Potential for a Geographic Alliance," *Journal of Geography* (Sept.–Oct. 1987): 211–215. The larger context of geography within the process of educational reform is expanded in Salter's "Geography and California's Educational Reform: One Approach to a Common Cause" in the *Annals of the Association of American Geographers* 76 (1986): 5–17. Professor Salter's impact on geography in the United States is portrayed in Ellen K. Coughlin's "Reformer Maps Plan for Geography Professors to Aid School Teachers" in the *Chronicle of Higher Education* (Nov. 18, 1987): A-3. The same issue of this journal included an article by

Ronald F. Abler entitled "It's Time to Make Geography a Vital Science Again" (p. A-52).

The Joint Committee on Geographic Education, a collaboration of professional organizations in geography, published its major policy statement entitled *Guidelines for Geographic Education: Elementary and Secondary Schools* (Washington, DC: AAG 1984). A special report on kindergarten and sixth-grade geography featuring themes, key ideas, and learning opportunities, was subsequently published by this same group in 1987. Two years later, GENIP published a similar effort for secondary schools entitled *7–12 Geography: Themes, Key Ideas, and Learning Opportunities* (Skokie, IL: Rand McNally, 1989). Further information about the status of geographic education in the United States may be obtained by contacting the Geographic Education National Implementation Project (GENIP).

Other useful sources on the topic of geographic education include:

- J. B. Binko, "Spreading the Word about Geography: A Guide for Teacher-Consultants in Geography Education" (Washington, DC: National Geographic Society, 1989).
- Mark H. Bockenhauer, "The National Geographic Society's Teaching Geography Project," *Journal of Geography* 92 (May–June 1993): 121–124.
- David A. Hill, *Rediscovering Geography: Its Five Fundamental Themes*, NASSP Bulletin 73, No. 521 (1989): 1–7.
- Michal Levasseur, *Finding a Way: Annotated Bibliography on Underrepresented Groups in Geography* (Indiana, PA: National Council for Geographic Education, 1993).
- James F. Marran, "Discovering Innovative Curricular Models for School Geography," *Journal of Geography* 93 (Jan.–Feb. 1994): 7–10.
- NAEP Geography Consensus Project, *Geography Framework for the 1994 Assessment of Educational Programs* (Washington, DC: National Assessment Governing Board, 1993).
- Salvatore J. Natoli, "Guidelines for Geographic Education and the Fundamental Themes in Geography," *Journal of Geography* 93 (Jan.–Feb. 1994): 2–6.
- U.S. Department of Education. *America 2000: An Education Strategy* (Washington, DC, 1992).

A copy of the final version of the *National Geography Standards* document (1994) may be purchased from the National Geographic Society in Washington, DC.

═══════════════ ENDNOTES ═══════════════

[1] Curricular information on geography in Russia and the former Soviet Union was summarized from a special issue of Soviet Education (February 1987), entitled "The New Soviet Secondary School Geography Curriculum" and "Soviet Education Today: Aspects of Theory," translated into English by Katherine Judelson and published in Moscow by Progress Publishers in 1984.

[2] Dale Heckman, California Postsecondary Education Commission, speaking about his travels in the Soviet Union in 1986.

[3] Gilbert M. Grosvenor, "Superpowers Not So Super in Geography," *National Geographic* 176 (December 1989): 819–821, and the *San Francisco Chronicle* (November 8, 1989): A-3.

[4] *Course of Study for Upper Secondary Schools in Japan* (Tokyo: Minister of Education, Science, and Culture, 1983): 23.

[5] Jedidiah Morse, *The American Geography: A View of the Present Situation of the United States of America* (Elizabethtown, NY: Kollock, 1789), and *The American Universal Geography*.

[6] Jedediah Morse, *Geography Made Easy: An Abridgment of the American Universal Geography* (Boston: Thomas Andrews, 1814).

[7] *New York Teacher* IV (February 1864): 183.

[8] *Journal of Proceedings and Addresses* (St. Paul, MN: National Education Association,1895): 444.

[9] See the *Report on the Committee of Ten on Secondary School Studies, with the Reports of the Conferences Arranged by the Committee* (New York: American Book, 1894).

[10] William D. Walters, Jr., "Voices for Reform in Early American Geographic Education," *Journal of Geography* 86 (July–Aug.

1987): 159.

[11] William Pattison, "The Four Traditions of Geography," *Journal of Geography* 63 (May 1964): 211–216.

[12] Preston James, "Toward a Further Understanding of the Regional Concept," *Annals of the Association of American Geographers* 42 (March 1952): 195–222; and John Fraser Hart, "The Highest Form of the Geographer's Art," *Annals of the Association of American Geographers* (March 1982): 1–29.

[13] Association of American Geographers, "Experience in Inquiry: HSGP (High School Geography Project) and SRSS (Sociological Resources for the Social Studies)" (Boston: Allyn and Bacon, 1974).

[14] Christopher (Kit) Salter, "The Nature and Potential of a Geographic Alliance," *Journal of Geography* 86 (Sept.–Oct. 1987): 212.

[15] Mark H. Bockenhauer, "The National Geographic Society's Teaching Geography Project," *Journal of Geography* 92 (May–June 1993): 121–124.

[16] Jan O. M. Broek, *Geography: Its Scope and Spirit* (Columbus, OH: Merrill, 1965).

[17] Rhoads Murphey, *The Scope of Geography* (New York and London: Methuen, 1982).

[18] In 1984, the Joint Committee on Geographic Education published *Guidelines for Geographic Education*, detailing the five fundamental themes in geography for use in kindergarten through twelfth-grade classrooms in the United States.

[19] *Geography for Life: The National Geography Standards* (Washington: DC: National Geographic Society, 1994): 1.

3 THE THEME OF LOCATION

We all have been lost at one time or another in unfamiliar surroundings, with no idea of how to get to our destination. Some of us may have felt panic; others, a sense of adventure. In any case, by learning our location we cease to be lost. Likewise, the task of going to a new destination requires a knowledge of one's old and new locations. Such a simple and obvious fact may imply that geographic proficiency is simply a skill in locating places. Unfortunately, this is all too often the limit of geographic concepts and skills taught in schools. Map reading, placename memorizing, and using grid coordinates should not be viewed as the entirety of geographic study, but rather as an entry into modes of geographic thinking that are necessary for more complete understanding of Earth as our home.

The first fundamental theme of geography, Location, is expressed in two of the *National Geography Standards* as the following:

Standard 1: *The geographically informed person knows and understands how to use maps and other geographic representations, tools, and technologies to acquire, process, and report information from a spatial perspective.*

Standard 3: *The geographically informed person knows and understands how to analyze the spatial organization of people, places, and environments on Earth's surface.*

What the telescope is to the astronomer, the map is to the geographer. Maps are the instruments of understanding geographic concepts. It is sometimes said that if the topic in question can be mapped, it is geographic. The same general rule holds for the idea of location. If the topic involves the location, movement, or distribution of things or people on the planet's surface, then it is of interest to geographers. In fact, the concept of *location* and the use of maps are inseparable, and they are therefore the two primary topics of this chapter.

Location is the first of the five fundamental themes in geography. Without location, it is impossible to understand the other four themes. The geographer's first question is, "*Where* is it?" Subsequent questions are, "*Why* is it there?" "*What* is there?" "*What processes* caused this locational situation?" and "*What are the future prospects* for this phenomenon?" Map reading is the most commonly taught geographic skill in schools, so this discussion of the location theme begins with what we believe every informed citizen should know about maps and their use.

MAPS AND MAP INTERPRETATION

The map is the most fundamental tool of geography. Visualizing and interpreting patterns and distributions in cartographic form is one of the skills necessary in understanding geographical concepts. Its necessity in the school curriculum cannot be overstated. In fact, if there is one aspect of Earth study that can be claimed to be the exclusive domain of the geographer, it most certainly is the design and construction of maps. This subfield of geography, cartography, can be traced into prehistory.

Early Cartography

The use of maps is perhaps as old as human curiosity. Travelers in most cultures variously have recorded their journeys in words, or scratched on rocks, or in assemblies of reeds and sticks. The oldest surviving map is a clay tablet showing land holdings in ancient Babylon. We know that the valley of the Nile River was mapped and are almost certain that wherever ownership of land was important there were maps to record the boundaries. Tax collectors no doubt used them for record keeping.

The father of classical mapmaking was the Egyptian Claudius Ptolemy, who lived in the Roman Empire of the second century A.D. He gathered geographic information from earlier Greek documents and tackled the problem of depicting a spherical Earth on a flat surface. Many maps used during the Age of Exploration, including the one used by Christopher Columbus, were copies of Ptolemy's earlier work.

As explorers brought back new information, particularly about the coastlines of distant places, data were added to an ever-growing store of cartographic information. Sometimes mapmakers' imaginations filled all the empty places labeled **Terra Incognita**, but for the most part their accuracy was remarkable considering the conditions under which they worked. With the invention of the chronometer in 1765, longitudinal position (east-west) on Earth's surface could be computed even beyond the sight of landmarks. Thereafter, global navigation became commonplace.

Large-scale mapping of topography and land ownership became more accurate and more common in the nineteenth century. By the beginning of the twentieth century almost all nations had a "mapping bureau." As technology in printing improved, the quality and number of available maps increased.

The need for maps became even more important during the world wars. In the 1930s the availability of aerial photography made mapping more accurate and faster. The geographic specialty called *photogrammetry* emerged, using air photos with ground surveys to produce accurate maps of all parts of Earth (see the box on page 32).

The map is as distinct a form of communication as is writing, spoken language, photography, or computer graphics. It expresses conceptual information about locations, distributions, and distances that can hardly be expressed in other ways. An informal list of maps in everyday use includes drivers' road maps, weather maps, bus route maps, newspaper location maps, zip code maps, children's treasure maps, or scratch maps to a friend's home. Indirect uses of maps that affect our lives include navigation maps for travel and shipping, geologic maps for mineral exploration, zoning maps for land-development decisions and government planning, and topographic maps used by the military, land-management agencies, and recreationists. The thematic maps in this book offer a cartographic potpourri to those who wish to declare themselves map-literate.

A map is commonly thought of as a graphic representation of the surface of Earth that is drawn to scale on a flat surface, such as paper, vellum, or plastic. Exceptions quickly become apparent for we have Moon maps, bathymetric (undersea) maps, maps on television or computer screens, and "mental maps."

However, all true maps have (or should have) four elements: title, scale, directional indicator, and a legend. There are many **cartograms** such as rapid transit guides that are not to scale and therefore technically are not maps.

The most common types of maps are those that show locations of things by means of points, lines, areas, and symbols. These maps get us from where we are to where we want to go. All have (or should have) a title, so we can quickly discern the purpose of the map and its areal coverage. The map should aid us in understanding what symbolization is used, should be drawn to a known scale so that

THE LIFE AND WORK OF THE CARTOGRAPHER

The tradition of cartographer-as-artist goes back hundreds of years to the beautiful maps of the sixteenth century, which often are seen in museums and libraries. Most cartographers today will claim, if modestly, some artistic sensibility in their craft. However, freehand or mechanical drafting is only a basic skill necessary for successful mapmaking today. Most undergraduate courses in cartography now include studies in the scribing of photographic film, map symbolization, design for creating effective messages, and the mechanics of map reproduction through photography or printing. Other courses may include work in computer cartography, surveying, or the transfer of data from air photos or satellite images to maps. Some facility with mathematics and a familiarity with the basics of photogrammetery and remote sensing also are important skills to have.

The cartographer's employer may be one of 40 federal agencies, such as the Department of Defense, the U.S. Geological Survey, or the USDA Forest Service. There are at least 400 different kinds of maps and charts made, with the distribution of hundreds of millions of copies of maps. Private-sector employers of cartographers particularly include resource-oriented companies, such as petroleum, forestry, or mining firms. Commercially produced maps enter homes and workplaces at the rate of about 4 billion per year or more.

The actual workday of today's cartographer is not likely to include design and construction of an entire map or set of maps. Depending upon the size and mission of the employer, it is more likely that the cartographer will specialize in data acquisition, map composition, data processing, reproduction, field validation, or some other technical task. All members of the team, however, will share the pride and satisfaction of seeing the final product put to use by its intended audience.

we may determine distances, and should be easy to orient to a known direction (usually north). Over and above these basics, a map may present much more information, depending upon its intended use.

Dot maps are usually used to show numbers, locations, and patterns of selected data, such as population, events, or commodities (Figures 3-1 and 3-2). *Graduated dot maps* use dots of varying size to show the amount or significance of whatever the dots represent, such as the relative sizes of cities. Dot maps also can show the pattern or arrangement of items, the density of what is being mapped, or the dispersion of a phenomenon from a central core to the outlying fringes. The most common form of dot map is a population map.

Isopleth maps are used to show lines of equal value. A topographic map such as Figure 3-14 shows isolines, or contours, of equal elevation. A weather map shows lines of equal barometric pressure as **isobars** (Figure 3-3). Lines of equal cost, such as transportation cost, are known as **isotims.**

The closeness of lines, such as **contour** lines on a topo-

graphic map, indicates the steepness of the gradient or slope of the phenomenon mapped. By placing known values as points on the map, the cartographer can interpolate between the known values and construct the isolines. The more known value points there are, the more accurate will be the constructed map.

Choropleth patterns are useful for showing areas on a map that are compared with others of a different value. Examples in Figure 3-4 show how a choropleth map may be used to show the number of telephones available in each country. Choropleth maps can be used to compile a great deal of complex information into a simple and readable format. Microcomputer programs exist that calculate the intervals between categories of the data to be mapped and then plot or print the finished map.

There are, of course, other kinds of maps, some very specialized and some very "artistic" at the expense of accuracy. It is interesting to give special attention to maps used in popular publications and on television. Critical examination of television maps, for example, will show limita-

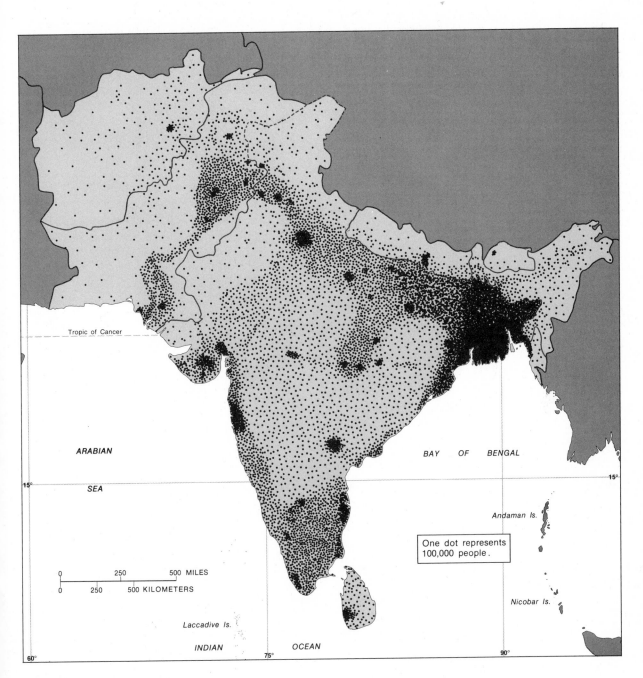

Tropic of Cancer

ARABIAN

SEA

15°

BAY   OF   BENGAL

15°

Andaman Is.

One dot represents
100,000 people.

0   250   500 MILES
0   250   500 KILOMETERS

Nicobar Is.

Laccadive Is.

INDIAN   OCEAN

60°   75°   90°

FIGURE 3-1

*Dot maps are ideal for showing gradations in density and are easy to compare with other factors
such as physical features. (From James S. Fisher,* Geography and Development: A World Regional
Approach, *3rd ed., Columbus, OH: Merrill Publishing Co., 1989, Fig. 25-13, p. 574. Reprinted with
permission of Merrill Publishing Co.)*

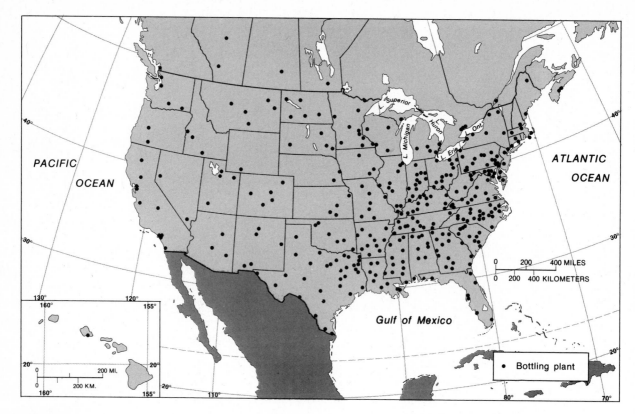

FIGURE 3-2
Dot map of Coca-Cola bottling plants in the United States. Where precise location is important, a dot map is valuable. This illustration shows the relation between soft drink plant locations and major population centers. (From James M. Rubenstein, The Cultural Landscape: An Introduction to Human Geography, *Columbus, OH: Merrill Publishing Co., 1989, Fig. 10-7, p. 362. Reprinted with permission of Merrill Publishing Co.)*

tions in resolution resulting in the need for very large lettering and bright, contrasting colors. Computer-generated maps generally do not have the detail of drafted or scribed maps, but they have the advantage of being faster to produce. Computer-automated drafting, however, usually has the advantages of speed, accuracy, and ease of revision when compared with hand-drafted map production (see the box on page 37). Most of the large mapping firms and government agencies are computerized. The box on page 38 summarizes the life and work of a cartographer at the National Geographic Society.

EARTH'S GRID: A LOCATIONAL SYSTEM

By convention, location is usually identified on a map by use of a grid system. The grid may consist of a row of let-

ters across the top margin with numbers along the side margins, or something more elaborate, depending upon the intended use. A classroom seating chart is a locational system, often laid out in a grid with rows and seat numbers. The ultimate challenge to cartographers through the ages has been to place a locational grid on the spherical Earth and then transfer it to a flat map with a minimum of distortion.

The locational grid has a vertical component—lines of **longitude**—and a horizontal component—lines of **latitude** (see Figure 3-5).

For longitude, the reference points are the South and North Poles and the axis that runs between them. An infinite number of imaginary lines running from pole to pole along Earth's surface make references to east and west locations possible, usually expressed as being so many degrees east or west of the prime meridian (zero degrees or 0°). By agreement among nations, the prime meridian—the 0° line of longitude—is located at Greenwich, England.

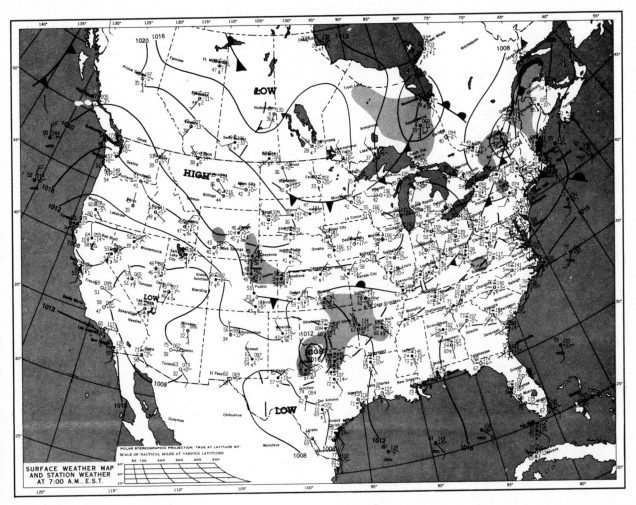

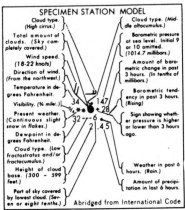

FIGURE 3-3

United States weather map. This system of symbols is used to best convey, at a selected moment, the dynamic state of the atmosphere with several variables—wind direction and velocity, temperature, air pressure, and precipitation. (National Oceanic and Atmospheric Administration.)

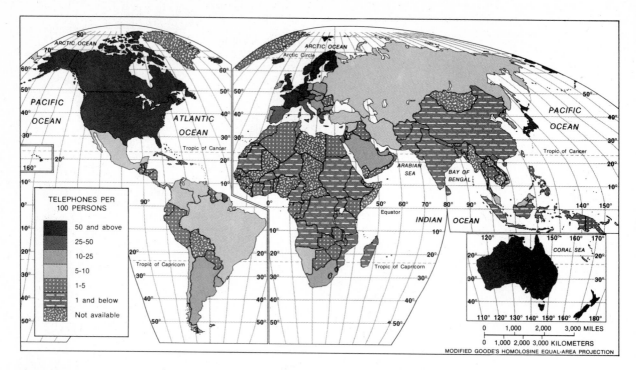

FIGURE 3-4

Telephones per 100 persons. A choropleth map is used to simplify ordinal data into categories for comparison, usually by political units. This map could be compared to another, such as gross national product for evidence of relationships between data sets. (From James M. Rubenstein, The Cultural Landscape: An Introduction to Human Geography, *2nd ed., Columbus, OH: Merrill Publishing Co., 1989, Fig. 8-4, p. 324. Reprinted with permission of Merrill Publishing Co.)*

On the opposite side of the globe the 180° meridian is used as the International Date Line. Here at every midnight a new calendar day begins.

Precise longitudinal location is possible by subdividing the two 180° hemispheres east and west of the prime meridian into degrees, minutes, and seconds. Seconds are used only when it is necessary to establish very precise locations. For most purposes, only degrees and minutes are used; for example, Slippery Rock, Pennsylvania, is located at 80 degrees and 3 minutes west longitude (commonly written 80°3′ W).

To meaningfully locate a place, however, it is necessary to have the other set of grid lines, the lines of latitude. The two sets of lines intersect, creating a coordinate system. Latitude lines (or "parallels") are all parallel to the Equator and to one another. They measure location north or south of the Equator.

Using the latitude-longitude coordinate system, the location of Paris, France, is 2.2° east longitude (2.2° east of the prime meridian running through Greenwich, England), and 48.5° north latitude (48.5° north of the Equator). The location is commonly written 2.2°E, 48.5° N (refer to Figure 3-5).

Much is made of Earth's **grid system** in elementary, secondary, college, and university textbooks. It is one of geography's most basic concepts. Nevertheless, a clear understanding of the usefulness of this locational grid system has escaped enough of the general population to compel us to include a bit more about it.

Superimposed on the grid of Earth used for locating places is a set of "special" parallels used to better understand the relationship between Earth and the sun and the significance of that relationship in our daily lives. Figure 5-4, a diagram of positions taken by Earth and sun at various times of the year, shows that the seasons are a result of the annual cycle of these relationships.

If the planet's axis (from North Pole to South Pole) were perpendicular to the plane of its orbit around the sun, there would be no seasons. The sun would always be directly overhead at the Equator. Earth's axis, however, is 23.5° off from being perpendicular to the plane of orbit. This angle creates an unequal distribution of solar energy

TIPS FROM THE CARTOGRAPHERS

You do not have to be a professional artist to draw accurate and pleasing personal maps. Here are a few simple steps to drawing maps of publishable quality with basic desk tools and a photocopying machine. There are several drawing and drafting programs for home computers that will also do the job.

Your first task is to locate a base map on which you can place your own information. Auto clubs, state highway departments, city or county planning departments, and government mapping agencies such as geological surveys are good sources for base maps. Photocopy the base map, reducing or enlarging it as needed. (If your map will be published, you may need to acquire permission to use a copyrighted base map; however, maps from state and U.S. government agencies generally are not copyrighted and no permission is needed.)

Extraneous information may be opaqued, or the needed information may be traced from the base map. Especially if your map will be published, your base map should be significantly larger than the size you want for your finished product. (The reason is that, when it is reduced in size for printing, most of your semi-professional rough lines will smooth out.)

The next step is to add your information to the map. Get into the habit of always including a title, a legend wherein your symbols are defined, a north arrow, and a bar scale. Also be sure to credit your data sources. Draw your information, usually as points or lines. (Shaded patterns can be tricky, so you may want to work closely with a printer who can help you.)

Use a sharp pencil and straight edge. Dots or circles may be drawn from an inexpensive template purchased at your local stationery or art-supply store. If an area is to be shaded, use a dark No. 2 pencil. These functions are done for you on command if you are using a computer drawing or drawing program. It is best not to use color on originals or copies. Good color mapping usually is more trouble than it is worth for personal use; however, this situation will change as color copiers become more commonplace in the future.

Lettering can be typed directly on the map paper or typed onto clear or white adhesive-backed labels, which can then be cut and stuck down in the right places. You can also use a wordprocessor with different font sizes for this task. Be sure your lettering does not cross other symbols or lines and that it can be easily read after reduction to your finished size. Cartographic conventions are to italicize the names of freshwater and saltwater features using initial capitals and lower case letters. Large waterbodies are entirely in italicized capitals. Mountain range names usually are "all caps," using roman type. Most landform features and placenames are lettered with initial capitals and lower case letters in roman type. Boldface usually designates larger features, such as major cities. Standard type indicates smaller ones, such as towns.

If unique symbols are used, a legend must define them. Use a border if appropriate, leaving reasonable margins. The final step is to return to the photocopying machine and reproduce a clean copy of your map. Shadow-lines from cut-and-paste work can be whited out on your first copy and that, in turn, can be used as a master for more copies.

MEET THE GEOGRAPHER
A Geographer's Geographer at the National Geographic Society

Alice Rechlin

As students of an eclectic discipline, some geographers are unsure of what career direction to take. The range of possibilities may seem confusing or overwhelming. However, this dilemma can present intriguing opportunities for diversity in various occupations while remaining in the discipline of geography.

After completing a doctorate in geography at the University of Michigan, Alice Rechlin went to Valparaiso University to share her enjoyment of the subject with her students. When elected chair of the geography department, she moved into the realm of administration, also a rewarding experience. Later a compelling career change was in the offing, and she joined the staff at the National Geographic Society, eventually settling into the Cartographic Division.

Today she holds the title "The Geographer." The Geographer's position carries with it a very exciting job description. It is her responsibility to keep track of the changing world so that the lines and names on maps reflect the world's reality when publications of the National Geographic Society occur. With the rapid changes continuing to take place on the Eurasian continent in recent years, the maps of these areas are continually being redrawn. Alice says that building a worldwide network of professional contacts is the key to staying abreast of the ever-shifting information we see on maps. "In keeping with the Society's mission to increase and diffuse geographic knowledge, we try to produce maps with the most current information possible."

So while Alice started her career in the academic area of the profession as a cultural geographer with some cartographic background, it was possible to switch her career direction in midstream. "It was the right thing to do. There is nothing more stimulating than change, and I continue to get new lessons daily. I am never bored, and I am learning to expect the unexpected. I have also learned to be a team player. The university is a place for independent thought and research. At NGS, the emphasis is on team effort. NGS is the most frustratingly exciting place I have ever encountered, and at this juncture of my life I wouldn't trade it for any other."

throughout the year, depending upon where the rays of the sun strike Earth at a more vertical angle. For example, the most direct rays of the sun are focused upon the northern hemisphere during May through August, and so people living there have their summer at that time. On the other-hand, solar radiation entering Earth's atmosphere at a greater angle away from perpendicular must travel through more atmosphere before reaching Earth's surface, thereby losing more heat and light by absorption and reflection. The part of Earth that experiences this will have their winter at this time.

The perpendicular rays of the sun are therefore important to track and have been so tracked for most of recorded history. On some globes you will find an *analemma* (which looks like a figure 8) which shows the latitude where the sun is overhead at noon on any day of the year.

To understand the workings of the seasons, as explained in Chapter 5, it is important to know the location where the sun's rays are most vertical at various times of the year. The most northerly position of the overhead (vertical) sun is at the Tropic of Cancer (23.5° north latitude), and the southernmost position of the overhead sun is at the Tropic of Capricorn (23.5° south latitude). The sun's northernmost position in the northern hemisphere (highest point in the sky) occurs about June 21st (the summer solstice), and it is lowest in the sky about December 21st (the winter solstice). The reverse is true in the southern hemisphere, where the warmest season occurs from November through February.

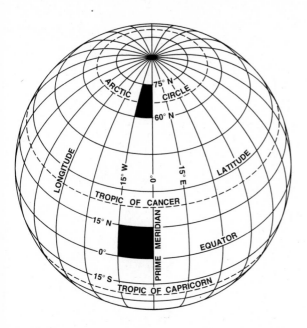

FIGURE 3-5
Earth's grid is made up of locational reference lines known as meridians (longitude) and parallels (latitude) that are numbered as degrees of an arc. All points on Earth can be located by reference to degrees north or south of the Equator and east or west of the prime meridian. (From T. D. Rabenhorst and P. D. McDermott, Applied Cartography: Source Materials for Mapmaking, *Columbus, OH: Merrill Publishing Co., 1989, p. 5. Reprinted with permission of Merrill Publishing Co.)*

As illustrated in Figure 5-4 in Chapter 5, the Arctic Circle marks the latitude north of which the sun cannot be seen at noon on about December 21st. The Antarctic Circle marks the latitude south of which the sun cannot be seen at noon on about June 21st. These circles mark the latitudes tangent to the Circle of Illumination (the line between day and night on Earth).

TIME ZONES

Meridians of longitude on the globe not only mark east-west references for **absolute location.** They are also used as established references for an agreed-upon time system. Earth completes one rotation on its axis every day, and a complete circle of rotation is 360°. Divide 360° by 24 hours in a day, and you can see that Earth rotates 15° each hour. It was agreed in 1884 at an international conference that every 15° would mark a one-hour time zone on Earth, dividing the planet into 24 one-hour time zones.

Each new day begins at midnight at the International Date Line (180° longitude). At that instant it is 12:00 noon on the prime meridian at Greenwich, England (0° longitude). By moving around Earth in units of 15° we may subtract or add hours from Greenwich time to determine local time anywhere on Earth.

Well, almost anywhere. It seems that such a system, no matter how regular and logical, warrants a few exceptions and variances. Figure 3-6 shows that the **time zone** system has not been generally accepted by all nation-states of the world. Some parts of the world prefer to set their clocks to fractions of the hour, such as in the central time zone in Australia. Because of this, it is 4:30 in Darwin when it is 5:00 in Sydney on the east coast of Australia, which is less than 15° away. Other nation-states use their own time system, such as Saudi Arabia's Arabic time, which is based on the times of sunrise and sunset. Still other countries recognize a uniform time in all their regions, thereby eliminating time zones altogether. Examples are the People's Republic of China, India, and Iran. Some flexibility in time-zone designation is necessary on most land areas to avoid bisecting populated regions.

MAP PROJECTIONS

One of the greatest problems in early exploration and navigation was how to accurately represent the spherical surface of Earth on a flat map, because no flat map can be exactly accurate. The term given to the technique of transferring locational information from the globe to a map is **projection.** All flat maps originate from a map projection that is selected to provide the least distortion of the area to be shown. It is physically impossible to do this with complete accuracy, so map projections demonstrate interesting and sometimes amusing attempts to compromise the presentation of the basic properties of maps—area, shape, distance, and direction. Various types of map projections are shown on Figure 3-7.

On those maps where *areas* are accurate throughout, the shapes of the areas shown may be distorted. An example of such an "equal area" or **equivalent area projection** is shown in Figure 3-7.

Maps that show true *shapes* of features on the globe are said to be **conformal.** They also may be drawn to show true direction and are therefore used in navigation.

Mercator first introduced his map in 1569 to aid compass navigation, because it shows a rhumb line (a line of

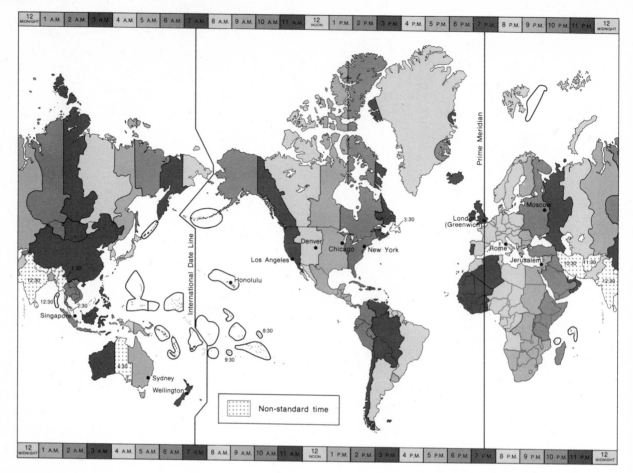

FIGURE 3-6
Time zones of the world. (From James M. Rubenstein, The Cultural Landscape: An Introduction to Human Geography, *2nd ed., Columbus, OH: Merrill Publishing Co., 1989, Fig. 1-4, p. 13. Reprinted with permission of Merrill Publishing Co.)*

constant compass direction) as a straight line (see Figure 3-8). The *shortest* distance between two points on a globe, however, is a great circle, which has a constantly varying compass heading.

The meridians on the Mercator projection are all vertical and perpendicular to the east-west parallels. Therefore, all lines such as travel routes and various isopleths are shown in their correct directions and locations. This is why this type of projection is often used to show data such as climate patterns or ocean currents. However, the price one pays when using the Mercator projection is the sacrifice of a constant or uniform scale. Map scale increases rapidly at the higher latitudes, so that, for example, Greenland appears to be as large as South America when it is, in fact, about the size of Mexico. The scale varies as one moves away from the Equator toward the poles, and the size of the areas shown nearer the poles is deceptively large.

Another commonly used map projection is the Goode's equal area projection. As its name implies it was designed by Dr. J. Paul Goode to show all areas on the map at the same scale. It is therefore useful in showing the areal extent of selected data, such as soil types or dominant languages. The major disadvantages of the projection are the distortions of shapes of places, and the map is necessarily split into sections that center on land areas. Direction and distance between places are difficult to estimate on a Goode's projection.

The most universally accepted global map projection used in schools in the 1990s, and the one we recommend, is the Robinson projection (Figure 3-9). It was designed for the mapmaking firm of Rand McNally & Company by Dr. Arthur Robinson in 1963, and it has been adopted by other important mapmakers and publishers such as the National Geography Society. Its advantages are that it is

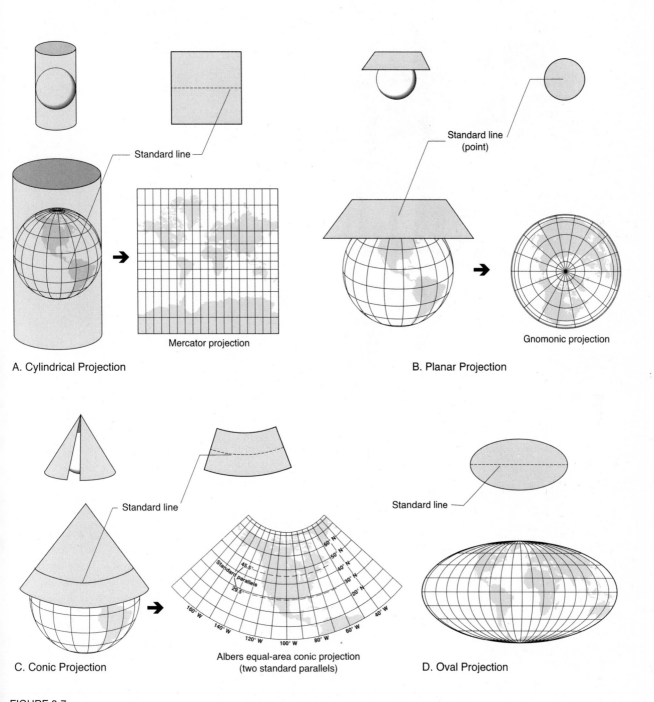

A. Cylindrical Projection

Standard line

Mercator projection

B. Planar Projection

Standard line
(point)

Gnomonic projection

C. Conic Projection

Standard line

Albers equal-area conic projection
(two standard parallels)

D. Oval Projection

Standard line

FIGURE 3-7
General classes and perspectives of map projections. (From Robert W. Christopherson,
Geosystems: An Introduction to Physical Geography, *New York: Macmillan, 1994, Fig. 1-19, p. 24.*
Reprinted with permission of Macmillan Publishing Co.)

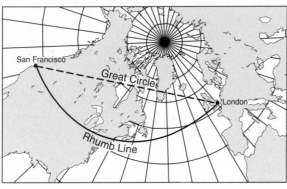

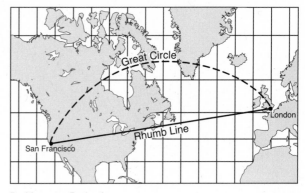

A. Gnomonic Projection

B. Mercator Projection

FIGURE 3-8

Mercator and Gnomonic projections. Comparison of rhumb lines and Great Circle routes from San Francisco to London on Mercator and gnomonic projections. All Great Circle routes are the shortest distance between two points on Earth. (From Robert W. Christopherson, Geosystems: An Introduction to Physical Geography, *New York: Macmillan, 1994, Fig. 1-20, p. 26. Reprinted with permission of Macmillan Publishing Co.)*

FIGURE 3-9

Robinson Projection developed by Arthur H. Robinson, 1963. (From Robert W. Christopherson, Geosystems: An Introduction to Physical Geography, *New York: Macmillan, 1994, Fig. 1-22, p. 27. Reprinted with permission of Macmillan Publishing Co.)*

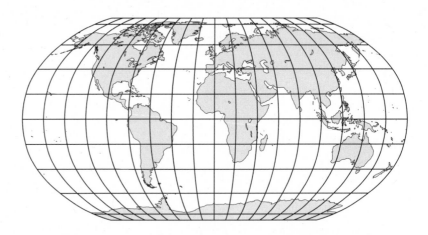

uninterrupted and displays features on Earth and their relationships to one another as accurately as possible with as little distortion as possible.[1] Nowhere, however, is it completely accurate!

OTHER LOCATIONAL TOOLS OF THE GEOGRAPHER

Air Photos and Satellite Imagery (Remote Sensing)

Viewing Earth from above became a reality with consistent and dependable airplane flight early in the twentieth century. Today high-altitude aircraft and orbiting satellites give us detailed information about Earth's surface at incredible

levels of resolution. Both satellite imagery and aerial photography can be vertical or oblique. The use of vertical photographs requires more experience in their interpretation, but they provide a high level of accuracy. Oblique views, because they provide more information on shape, readily enable the identification of landscape features. (Compare Figure 3-10, an oblique view of Mount St. Helens, with Figure 3-11, a vertical view of the same mountain with contour lines and placenames added.)

Pairs of overlapping aerial photographs can be viewed through a stereoscope to obtain a three-dimensional image. This image can control a plotter, making it possible to depict elevation lines, or contours, on a map. Most government topographic maps are made by assembling mosaics of air photos, verifying their scales, and using them as a database for map projection.

FIGURE 3-10
Mount St. Helens in Washington State may be seen from an area of several hundred square miles in the Pacific Northwest and is easily recognizable from the ground. This photo was taken during its 1980 eruption. (Photo by Jim Hughes of the USDA Forest Service; provided courtesy of Geo-Graphics, Portland, OR.)

Photography need not use just the visible light spectrum. Use of infrared and microwave bands of the spectrum enable selective views of landscape features, based on the research needs of the analyst. Infrared photography has proven particularly valuable in surveying agricultural areas for plant diseases, soil moisture content, and other selected features. Earth information in the near-infrared portion of the spectrum is not visible to the eye but can penetrate atmospheric haze and cloud cover to produce color-infrared photographs in "false color." In these images, growing healthy vegetation has the "false color" of red; clear water is black; sediment-laden water is light blue; and urban areas are blue-gray.

Radar is also used in gathering Earth images. Unlike other methods, which rely upon simply receiving energy that is radiated or reflected from Earth, radar generates its own energy, which is reflected back to the radar sensor to create images. An advantage of radar is that it can "see" through cloud cover very well. This type of remote sensing can record surface relief features with remarkable accuracy. Remote sensing technology has been used in government intelligence gathering and in planning and land use studies, resource exploration, climate and weather prediction, flood control, and crop inventories.

Orbiting satellites scan segments of Earth along lines or by covering small areas with televisionlike cameras, and the information gathered is digitized. These digital picture elements, or *pixels,* are transmitted back to Earth where computers reassemble them. They then can be displayed on a television screen or can be used to generate a photograph. The **resolution** (level of obeservable detail) available to the users of satellite images is very fine, as evidenced by the use of such images by government intelligence agencies. These images are now frequently shown on television newscasts and are available to anyone through government agencies and some European private vendors.[2] An infrared satellite view of Washington, DC, is shown in Figure 3-12.

FIGURE 3-11
Those familiar with Mount St. Helens as shown in Figure 3-10 would not necessarily recognize it from this vertical view. However, this air photo provides more specific information than the view in Figure 3-10. When topographic information is overlaid on the photo as shown here, even more information is provided. (Photo courtesy of Geo-Graphics, Portland, OR.)

COMPUTERS AND GEOGRAPHIC INFORMATION SYSTEMS

Computer Use by Geographers

Computers have been used by geographers to manipulate statistical data since the first machines became available in the 1950s. They are very important for the storage of large quantities of data and make possible the presentation of that data in a variety of formats such as spreadsheets, graphs, and charts. Computers are also used to process remotely sensed data and to draw maps (computer cartography). Computers also have facilitated the rapid application of mathematical **models** for site-location analysis, diffusion modeling, and simulation models.

Location models are used in several types of location decision making where several variables are considered. Simulation models are used in many research situations because they can recreate the actions of climatic phenomena, stream flows, or energy transfers. Pollution studies often are aided with these techniques.

The most dramatic technological change in geography has been the creation and use of computer-generated maps. Computer mapping has evolved from over-the-counter graphics packages to highly sophisticated geographic information systems (GISs). Computer automated design (CAD) programs, which allow drafting and mapping with a computer, provide the accuracy required by architects and planners and have the additional advantages of providing greater ease in making revisions and in storing data.

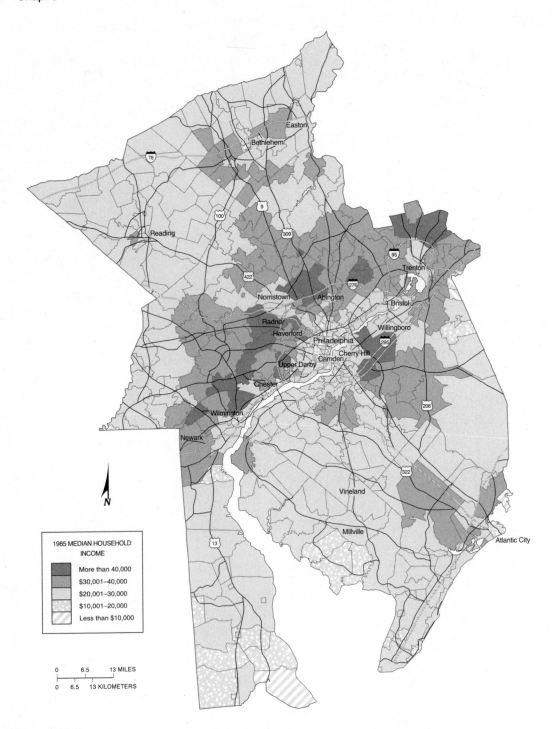

1985 MEDIAN HOUSEHOLD INCOME

More than 40,000
$30,001–40,000
$20,001–30,000
$10,001–20,000
Less than $10,000

0   6.5   13 MILES

0   6.5   13 KILOMETERS

FIGURE 3-13
Location of upscale gyms in Philadelphia. Market area analysis and planners for firms nationwide frequently want to know the median household income of census tracts in a metropolitan area such as Philadelphia. High rent, medium rent, and low rent neighborhoods can then be targeted for various products. This figure illustrates how census data can be used to create maps. (From Anthony R. de Souza and Frederick P. Stutz, The World Economy, *New York: Macmillan, 1994, Fig. 9-34, p. 396. Reprinted with permission of Macmillian Publishing Co.)*

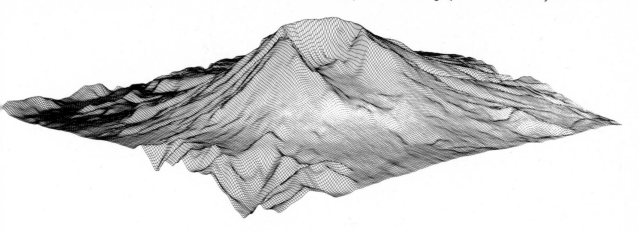

FIGURE 3-14
Computer-drawn terrain model of Mount St. Helens in Washington State. Known elevation values are entered into a database and the computer interpolates "empty" (details) areas to construct the surface as shown here. (Courtesy of Raven Maps and Images, Medford, OR.)

A computer mapping system now available to automobile drivers produces street maps on a screen and locates the driver and the destination at a choice of scales. It also computes driving distance and fuel mileage. This is accomplished with a digitized map database on a tape cassette linked to a computer in the car. (Soon, perhaps, gas station attendants will no longer be interrupted by lost motorists!)

A convenient way to plot distributions is to present the data on computerized outline base maps. For example, an outline map can be generated on the computer screen, and census data on housing, income, and population can be superimposed upon it. Original boundary files may be created by the user, or packaged files can be obtained for these base maps. Likewise, published data can be utilized or one's own data can be entered into the system. Figure 3-13 is a computer-generated map of the location of upscale gyms in Philadelphia. Data are from the 1980 census. Now that a wide variety of data from the 1990 census (for example, data on income, ethnicity, age, and educational levels) is available to educators, it can be entered into the database, and a map much like the one shown in Figure 3-13 can be generated with relative ease.

Geographic Information Systems

Computers can be used to generate a variety of map projections, and they can be used to create trend surface maps or terrain models. A terrain model such as that shown in Figure 3-14 requires the entry into the computer of three-dimensional coordinates (length-width-height). The image is then drawn as a simulated three-dimensional block diagram.

Geographic information systems (GISs) are perhaps the most comprehensive and complex computer-based analysis and mapping tools available to the geographer. Applications range from taking environmental inventories of regions to planning alternatives for cities. GISs utilize a variety of data sources such as census statistics, remote sensing imagery, and traditional maps to create data inventories. They have the ability to manipulate the data into map or graphic form. Figure 3-15 conceptually illustrates the variety of data sources that may be used in GISs. Figure 3-16 presents some displays from a variety of GIS and other computer-mapping applications.

Geographic information systems are usually housed within fairly large computer systems. Data are entered by keyboard or by digitizing with a mouse or scanner. The encoded data are then edited and stored. Upon demand, a choice of data presentations is available: a display on a monitor, a printout, a transmission elsewhere by telephone, or a drawing by a plotter.

Advantages of GIS-based working environments are the ability to continually add new data, to edit previous work, to manipulate information with statistical or simulation models, and to present almost any number of variables or overlays as required. Disadvantages relate to the newness of the technology, such as high initial cost of the system and the need for trained operators.

An example of the map-overlay concept commonly used in GISs is shown in Figure 3-17. In this case, physical and cultural geographic data are combined to produce a composite site analysis for a proposed development project.

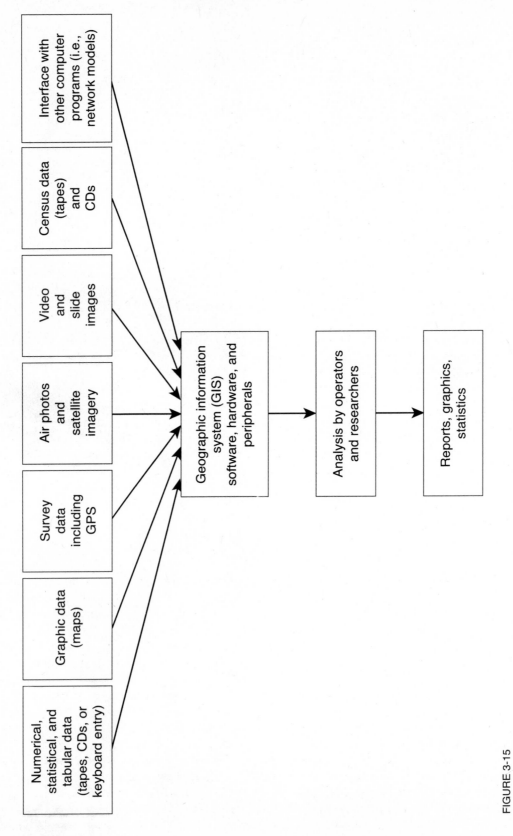

FIGURE 3-15
Data sources and outputs most commonly used in geographic information systems (GISs) for geographic analysis.

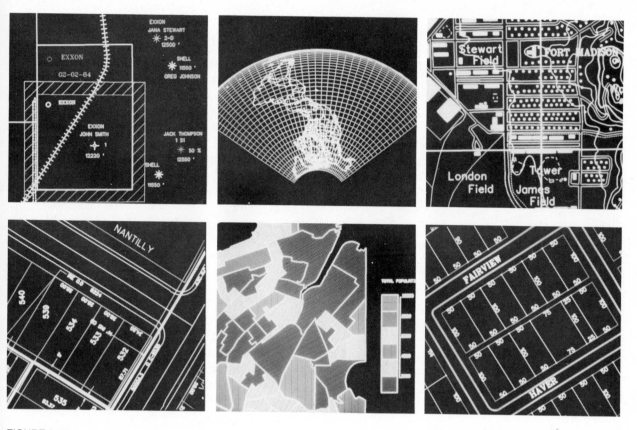

FIGURE 3-16
Six representative GIS applications. (Courtesy of Synercom Technology, Inc.)

LOCATION CONCEPTS USING GEOGRAPHIC TOOLS

For the purposes of this section, "geographic tools" are maps, air photos, space imagery, computers, mathematical functions, and—most important—human observation and analysis. Knowledge of the following concepts will lead to answers to geographic questions.

Scale

Reducing reality to a workable size, or **scale,** is of fundamental importance in understanding Earth's surface. The nature of the problem determines the scale of inquiry. Geographic studies may be large scale, covering an urban neighborhood, or small scale, covering a global region. (Chapter 13 illustrates several geographic problems at different scales.) Figure 3-18 shows the same location at decreasing scales, each with more area covered but in less detail.

Map scale is simply the relationship between distance shown on the map and the corresponding real distance on the surface of Earth. Map scale can be expressed in three different ways: as a "verbal scale," as a bar scale in graphic form, or as a ratio or "representative fraction."

An example of a verbal scale that explains distances on a map would be the words "one inch represents one mile." The graphic scale is most often shown as a bar or line with markings on it representing specific distances in feet, miles, meters, or kilometers. Such scales are used on maps throughout this book. The advantage of a bar scale in our age of photocopying is that it remains accurate when the map is enlarged or reduced, whereas verbal and ratio scales do not.

The ratio scale or representative fraction is the third type of map scale. On many U.S. Geological Survey topographic maps the representative fraction scale is 1/24,000 (or as a ratio scale, 1:24,000), which means that one inch on the map represents 24,000 inches on the surface of Earth. It also means that one foot on the map represents 24,000 feet on Earth, or that one centimeter represents

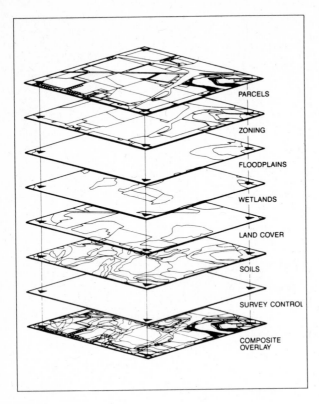

PARCELS

ZONING

FLOODPLAINS

WETLANDS

LAND COVER

SOILS

SURVEY CONTROL

COMPOSITE
OVERLAY

FIGURE 3-17
A computer-based GIS can combine many types of mapped information at the same scale. Analysis of several variables that have been electronically combined may indicate solutions to resource management problems or answer questions of optimum location. (From U.S. Geological Survey.)

24,000 centimeters on Earth, and so on. This type of map scale may be expressed using any choice of measurement units in any country because the "1" on the left side of the ratio or fraction is always in the same measurement units as the number on the right side.

An important point about map scale is that smaller-scale maps show larger areas in less detail than large-scale maps. For example, a 1:1,000,000 (small-scale) map would show only the locations of entire cities, but a 1:100 (large-scale) map could show details of streets and buildings. Figure 3-18 shows a small-scale, large-area map and a large-scale, small-area map, which encompasses less land surface.

A useful fact about air photos and remotely sensed images is that their scales cannot be completely accurate. This is because only the focal point of the camera is directly vertical over Earth's surface. Other points on the image are viewed at a slight angle, which distorts the scale.

Absolute and Relative Location

Knowing your location is helpful only if you know the context of that location. Our **absolute location** at any one time can be at the corner of Market and Broadway Streets, at our home address, or at a specific latitude and longitude. Other absolute locational systems include the Universal Transverse Mercator system, a metric grid used mostly by the military and some scientists; the U.S. State Plane Coordinate system; and various land survey systems, such as the U.S. Township and Range survey system. All French government maps use a locational system based on the "grad," which is 1/100 of a right angle (one degree equals 0.9 grad). Its zero meridian goes through Paris and its zero parallel is the Equator.

Relative location may be your location relative to events in your life. Most often, though, it refers to the geographic context of the relationships among the locations of people, events, and places. Recall for example, a real estate salesperson's description of the ideal home: "close to schools, shopping, and the interstate highway." The combinations of possible locational relationships are almost infinite, but they offer geographers the major challenge of our profession. We want to know not only where things are but why they are there, and how they affect and are affected by other things, people, or events.

Site and Situation

The selection of a site for a nuclear power plant or a hazardous waste disposal facility depends on local and regional factors. Local or **site** factors may include such matters as soil conditions, wind direction, security, land value, and visibility. **Situational** factors, which relate to a broader field of decision making, include considerations such as transportation costs, political climate, influences of special interest groups, user locations, and the characteristics of competing sites.

Most successful cities throughout history have enjoyed some site advantages in their initial growth stages, and almost all carried situational advantages as they grew. Competing cities with fewer situational advantages, such as accessibility or centrality, often did not do as well in population or economic growth.

Resolution

The concept of resolution refers to the level of detail that it is possible to display on a map or to understand in a real landscape. A map, air photo, or remotely sensed image with high resolution will show clear detail. Images with the high-

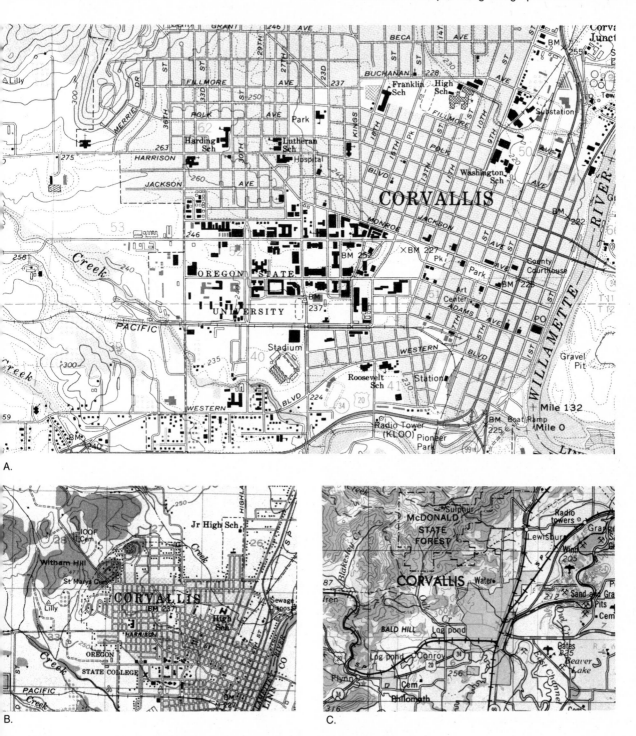

FIGURE 3-18
*Corvallis, Oregon, mapped at the three most commonly used scales produced by the U.S.
Geological Survey. Scale of Map A is 1:24,000; Map B is 1:62,500; Map C is 1:250,000.
(Reproduced from U.S. Geological Survey topographic quadrangle maps.)*

est resolution and greatest capacity for showing detail on the surface of Earth may show objects as small as a basketball. Such high-resolution images may be valuable to security agencies but are generally too detailed and expensive for such tasks as regional land use mapping. The task of the map and image user, then, is to match the level of resolution needed with the task at hand. A television or computer screen showing graphics may also have high or low resolution, depending on how many pixels (picture elements) it displays.

Pattern, Density, and Dispersion

Pattern, density, and **dispersion** of the dots on a map are terms that describe the arrangement of these dots into a meaningful message. On the population dot map of Egypt shown in Figure 3-19 each dot represents 50,000 persons. The *pattern* of dots is such that they are arranged in linear form, straddling the banks of the Nile River. The *dispersion* of dots in an elongated pattern is noticeable, with more near the river and fewer away from it. This implies that the

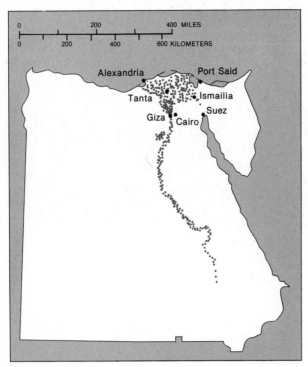

FIGURE 3-19
The population of Egypt shown by dots. Each dot represents 50,000 persons. The extent of Egypt's land area is irrelevant, because the population is concentrated along the Nile River and its delta.

number of persons per square mile, the *density*, decreases as distance from the river increases.

Finally, if map users wanted to take the trouble, they could count the dots, multiply by 50,000, and derive an estimate of Egypt's total population (to the nearest 50,000 people) for the period or year represented by the map.

Location Models and Distance Decay

A **model** is an abstract representation of reality. A model may be a physical object constructed to scale, or it may be a mathematical formula. A three-dimensional map showing both vertical and horizontal scales is often called a *terrain model.*

One of the first—and still valid—mathematical location models was proposed by W. J. Reilly in 1931 and applied to determining market area boundaries by P. D. Converse.[3] Generally, this "law of retail gravitation" or **gravity model** states that the breaking point between the influence of two market centers ("commercial service boundary") was a function of the population of place A, the population of place B, and the distance between them. The formula is

$$Db = \frac{D}{1 + \sqrt{\dfrac{pa}{pb}}}$$

where

Db = distance in air miles from B to the market area boundary

D = distance in road miles between the two centers A and B

pa = population in thousands of market center A

pb = population in thousands of market center B

For example, if large market center A has 100,000 people, small market center B has 20,000 people, and the road distance between them is 26 miles, then

$$Db = \frac{26}{1 + \sqrt{\dfrac{100}{20}}} = \frac{26}{1 + \sqrt{5}}$$

$$= \frac{26}{1 + 2.24} = 8 \text{ miles}$$

If market center A has 100,000 people, B has 50,000 people, and 31 miles separate these market centers, the commercial service area boundary from B is

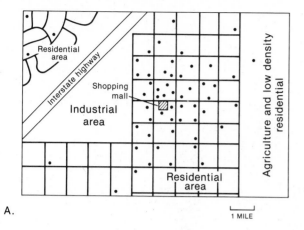

A.

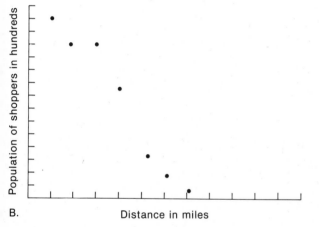

B.

FIGURE 3-20

Two of several ways to display spatial data. Part A utilizes a conventional map and part B uses a scattergram to show the relation between number of customers and distance to a shopping center.

$$Db = \frac{31}{1 + \sqrt{\frac{100}{50}}} = \frac{31}{1 + \sqrt{2}}$$

$$= \frac{31}{1 + 1.41} = 12.9 \text{ miles}$$

If you use this formula for each of the smaller cities surrounding A, and connect the break points of "gravitation," then the commercial service area for market center A is delimited.

Figure 3-20 illustrates two ways of representing the same spatial data. The number of patrons at a shopping center are shown on the map with increasing density as their locations are closer to the center. The graph shows the same relationship with the number of shoppers on the y-axis and the number of miles from the market center on the x-axis. Notice the effect of the interstate highway barrier in each illustration. Both figures are generalizations or abstractions of reality that show patterns of customer density in most North American shopping centers. This characteristic of shopping-center market areas is used by site location analysts to help determine how many customers they may "capture" from competing centers.[4]

A related concept shown in Figure 3-20 is **distance decay.** This is simply the observation that (in most cases) where density is greatest around a node (such as a city), the farther from the center, the lower the density. This distance decay concept, usually based on land use intensity, is used to explain the locations of land uses, the costs for goods and services, and the complexity of locations observed.

THE IMPORTANCE OF LOCATION

As mentioned earlier, the theme of location is a critical underpinning to any understanding of geography. In our present world situation, where almost all newsworthy events have some universal significance, it is dangerous if not inexcusable that so many people are indifferent to locational geography. Some examples of the significance of location in areas of the globe that are often in the news follow.

The Importance of Location in the News

We see reference to locations in Russia and the other republics of the former Soviet Union frequently. Matters of location that have been important to an understanding of Soviet policy and actions over the past 50 years have included:

1. Russia is located only 3 miles (5 km) from the United States at its closest point and borders Canada across the Arctic Ocean. The countries making up the former Soviet Union border many other nation-states, most of which were generally perceived as hostile to the Soviet Union in the last half of the twentieth century. Figure 3-21 offers a (now historic!) Soviet view of Earth centered on the North Pole.

2. A large percentage of Russia is located north of the 49th parallel (the line which also marks the boundary between the United States and Canada). This far-northerly location, as well as distance from the moderating in-

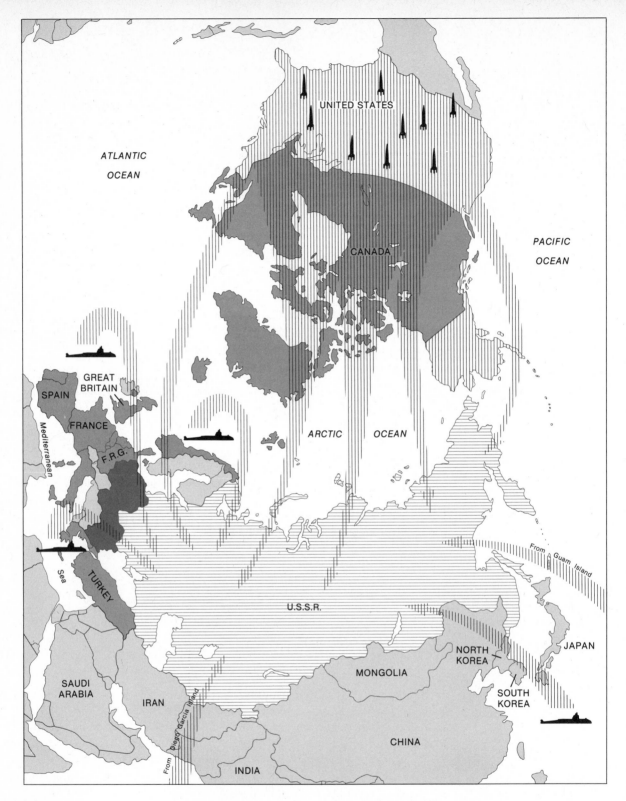

FIGURE 3-21
A Soviet view of Earth taken from sources in the former Soviet Union and redrawn for North American students of political geography. (From M. I. Glassner and H. J. de Blij, Systematic Political Geography, *4th ed., New York: John Wiley & Sons, 1989, p. 44. Reprinted by permission of John Wiley & Sons.)*

fluence of the oceans, limits the number, volume, and consistency of crops grown in Russia. Indeed, this locational disadvantage has contributed to severe food shortages, requiring the government to buy grain and other foodstuffs from other areas of the world. This situation, in turn, has created dependencies that affect world trade, military postures, and diplomatic relations with the rest of the world.

3. The former USSR's **landlocked** location affected its policy toward its neighbors. A search for warm-water ports and the need to protect them has, at least in part, accounted for Soviet occupation of the Kuril Islands, part of Finland, the Baltic republics of Estonia, Latvia, and Lithuania, and for continued heavy Russian and Ukrainian presence in the Black Sea area.

News stories about location appear and reappear in reports on the Middle East as well. The problems of oil transport, the locations of Iraq, Somalia, Haiti, and Cuba, and the locations of military strong points, all highlight the importance of location in understanding world affairs.

Other locational factors that bear examination in world news stories include the locations of **resources** such as rain forests, water, or minerals; the locations of critical trade passages such as the Strait of Malacca or the Panama Canal; and the locations of sensitive borders or **boundaries** such as those between Iraq and Iran or between North and South Korea. Certainly the importance of the locations of Cuba, Israel, Croatia, Hong Kong, and Rwanda are self-evident in the events that transpire in those places.

The Role of Location in History

The importance of location in almost any historical event is worth examination. Knowing the location of the battles at Hastings, Waterloo, or Gettysburg is only the beginning; important locational matters appear in other ways. Here are a few examples of the importance of location in the North American experience, taken from the many identified by geographer Ralph Brown in *Historical Geography of the United States:*[5]

- New England colonial harbors were not chosen for their depth or size. Rather, shelter from wind and defensibility were the primary locational criteria.
- Choice of the location for Jamestown was deliberate, having been selected after an 11-day and 40-mile (64-km) search. Although safe from Indian attack, the choice was a poor one because its swampy location meant that it had poor soil, little protection from weather, and no easy access to the outside world.

- California Spanish mission locations were based on availability of water, soil suitablity for agriculture, and an adequate number of Indians to convert to Christianity. They also were deliberately located one day's travel from one another.
- The chosen location of the 49th parallel boundary between the United States and Canada had little justification on the ground, even though the line on the map looked reasonable enough. In 1810, potentially antagonistic populations were located within a short distance of this boundary, which had not yet been surveyed and marked.
- The locations of swamps in the South Carolina low country were influential in locating rice plantations. Inland swamps with level land and freshwater were chosen as sites, whereas brackish or saltwater swamps were avoided for environmental and perceived health reasons.
- In the early nineteenth-century, wagon routes across the Appalachian Mountains followed animal and Indian trails, which also were the locations of gaps and salt licks. The three major points of focus for the transmontane routes were the forks of the Ohio River (Pittsburgh and vicinity), the Cumberland Gap, and the Valley of East Tennessee.

It is important not to confuse the effect of location in historical decision making with the effect of environmental conditions, even though such a distinction was probably not made by the actors in the time and place under study. Many social or cultural factors were often as important as physical conditions in location decision making. It also is not a good idea to ascribe locational decision making to determinism. For example, the idea that the location of gold and oil in Alaska "dictated" the dramatic migrations and settlement there may not be true, since the means to extract the resources for the benefit of workers and investors also had to be there.

The Role of Location in Urban and Environmental Planning

On a larger scale than the examples noted above, circumstances of historical location have affected the site and form of cities. Our capital city of Washington, DC, was built according to preconceived plans on a site carefully selected by virtue of a central position according to population, a location near the falls of the Potomac River, and with attention to slight topographic undulations that afford sites for monuments. London and Quebec City were chosen for their **defensibility**, Salt Lake City for its isolation, St.

Petersburg, Florida, for its harbor, and Warsaw for its **centrality.** Each has its own interesting story that details and expands the factors affecting place location.

The patterns of city growth included locational factors in decision making as well. In the twentieth century, where the process of planning imposes upon the evolution of city form, matters of location continue to be critical. Urban planners in public service concern themselves with locations of and **accessibility** to transit. The locations of incompatible land uses are important. The locations of governmental, commercial, or other high-traffic or employment **nodes** are critical to the proper functioning of a city. All location decisions in a city, whether they be group or individual decisions, leave imprints on the landscape. The sequence of maps shown in Figure 3-22 reflects the rather arbitrary location of streets in early Tokyo. The location of streets on grids has influenced the subsequent layout of streets of the city since World War II.

The land use planner continually seeks the optimum use of land to benefit the most people and to meet the needs of owners, neighbors, and the rest of society. While not always successful, the planner is to be commended for bringing organization to a complex set of variables that continually create change in the human environment.

The Role of Location in the Site Selection of Economic Activities

Locational decision making has become very important to businesses, the success of which depends upon transportation availability and cost, and upon accessibility to customers. Location-related factors critical to a new manufacturing plant might be land cost, transportation cost of raw materials, labor supply, pollution abatement potential, taxes, market location, and locations of competitors. Locational factors related to the placement of a new fast-food franchise may include some of those cited above, plus traffic flows and the social and ethnic makeup of the area, including their purchasing power, and the visibility of the restaurant from a distance.

Retail-store site selection may be facilitated by use of gravity models, by a checklist approach, or by an analog method (which finds the circumstances of other similar and successful projects). Many decision makers use all three methods. Most computerized site-location models used in marketing today are considered trade secrets, but works by D. D. Achabal and colleagues, and J. L. Goldstucker and colleagues demonstrate the nature of such models.[6]

A retail-store decision checklist might include details on the retailer's characteristics and needs, the nature of the market area under consideration, and evaluations of the competition.

Personal location need not always be the address at which you reside or work. Telephone area codes, postal zip codes, and census tract numbers are also used as location devices. Zip codes, because they are particularly suited to mail advertising, are used in the marketing of consumer products. Market researcher M. Weiss has compiled the lifestyle characteristics (such as income, entertainment preferences, reading habits, and political attitudes) of the U.S. population into 40 clusters or "types."[7] He contends that everyone fits into one of the lifestyle types and that the majority population of every zip code also fits into one of the types or clusters. We are our zip code.

The Role of Location in Social and Political Issues

Social geographers (and some unsocial ones!) view many problems and issues from a locational perspective. The locations of participants and decision makers usually have a great deal to do with issues focusing upon segregation, lifestyle, quality of life, crime rates, voting patterns, or other socially significant concerns.

Feminist geographers have taken steps to ensure that the concerns of women in the population are considered in locational decisions. Exemplifying this concern are the many sessions on women's issues held at the annual meetings of the Association of American Geographers and other professional geography organizations in many parts of the world, and the numerous publications addressing issues of special concern to women worldwide.

One example of this type of geographic inquiry was accomplished in Great Britain in the 1980s: *Geography and Gender,* a recent study by the Women in Geography Study Group of the annual conference of British Geographers, focused its attention upon women's role in the location patterns of industry and regional development. A distinct regional distribution of "jobs for women" has emerged in recent years because there are more and more women in the labor force in the 1990s. This in turn has affected regional development and industrial location patterns in England.

Women traditionally are found mainly in the service sector of the economy (as teachers, nurses, and office workers) or in the lowest-paying skilled labor jobs. As women have entered the work force in ever-greater numbers, changes in the organization and location of industries have generated a new demand for female labor. This study emphasizes that too many aspects of women's employment have been left out of the geographic literature in the past. A number of other significant research studies have been

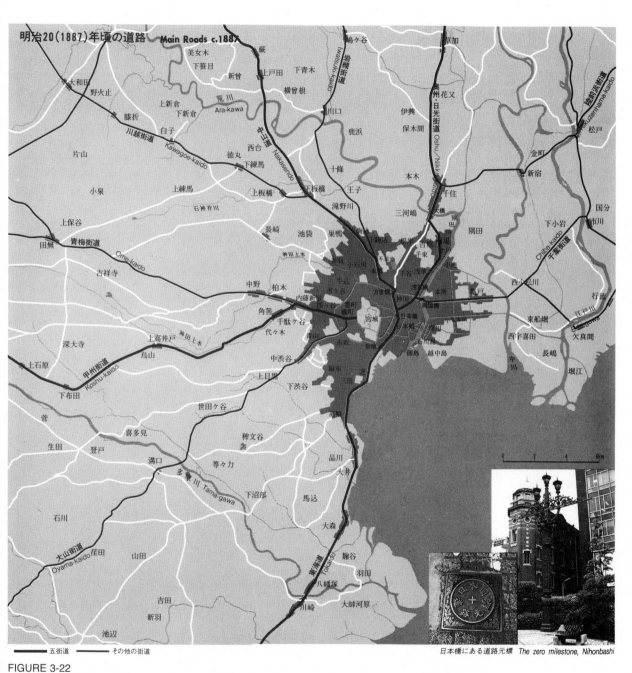

明治20（1887）年頃の道路   **Main Roads c.1887**

五街道 ——— その他の街道

日本橋にある道路元標 *The zero milestone, Nihonbashi*

FIGURE 3-22
The street layout of Tokyo in 1887 and 1930 was very similar to the pattern of today's main roads. Relics of historical location decisions often are part of the contemporary landscape. (From Atlas Tokyo, *Tokyo: Yasuo Publishers, 1986, pp. 50, 54.)*

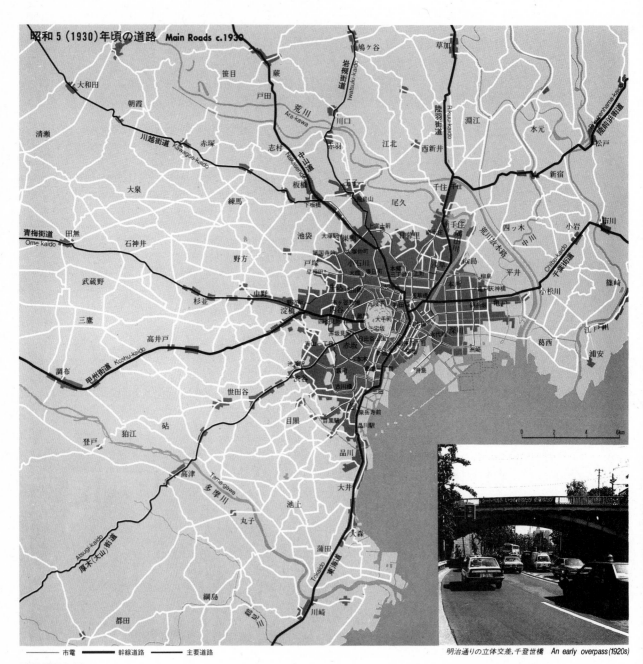

昭和5（1930）年頃の道路　Main Roads c.1930

明治通りの立体交差,千登世橋　An early overpass (1920s)

市電　　幹線道路　　主要道路

FIGURE 3-22
(continued)

現在の道路 **Tokyo's Main Roads**

━━━ 国道   ━━━ 主要道路   ━━━ 環状道路   ══ 有料道路

FIGURE 3-22
(continued)

FIGURE 3-23
Locational solution to an environmental noise problem involves the geographer's ultimate question: Where?

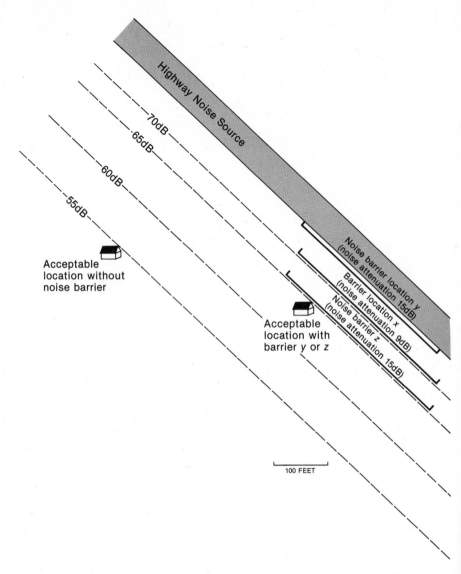

conducted by geographers that investigate women's lives as they relate to locational considerations around the globe.[8]

Other locational issues in a political geographic framework are addressed in Kevin Cox's *Location and Public Problems: A Political Geography of the Contemporary World.*[9] In it he considers several topics of importance to governments and individuals at the metropolitan, international, and intranational levels, such as residential quality, income, competition, urban size, policymaking, and development, all from a locational viewpoint.

The Role of Location in Environmental Issues

Almost all environmental problems or issues have a locational dimension. Chapter 10 discusses the locational as-

pects of many of these issues. Locations of polluters, the location of receptors, and the location of persons subject to the environmental hazards are all obvious and quite mappable. A frequent solution to local environmental problems, although not always the best solution, is to locate the pollutant source away from populated areas. Of course, the best solution would be to eliminate the pollutant source altogether.

The reduction of exterior noise from transportation sources is just one example of a locational problem. With each doubling of the distance between the noise source and the receptor, the sound level at the receptor is reduced six decibels. If the sound level 50 feet (15m) from the source is 70 dB, it will be 64 dB at 100 feet (30m), 58 dB at 200 feet (60m), and so on. The solution, therefore, is to move the receptor back to an acceptable sound level, usually 55 dB.

If this is not possible, a noise barrier may be constructed. This decision creates a new problem. If the barrier is at location *x* as in Figure 3-23, the noise, level will be less attenuated than if it were placed at *y* or *z*. The reasons for this difference are found in the principles of physics, but the optimal location problem is a geographical decision.

In the process of site **planning** the development team may work with location decisions on a daily basis. The planner of a residential subdivision, for example, must first amass environmental information about geology, wildlife, soils, storm water runoff, slope, sun orientation, and vegetation in the area to be developed. All of these are usually mapped for their locational significance to the goals of the project. The human and technical considerations are then added to the plan. The locations of municipal services, access roads, and electricity, water, and sewer lines are considered. Only then can the planners begin to locate their lots, streets, and boundaries on the subdivision map. For large projects the optimum location of parks, schools, and community centers must also be added. The best site plan almost always is the best arrangement of location for various land uses on the map.

Location decision making is of such importance that it has become a specialty in many geographers' training and provides employment opportunities for many university graduates. The above examples are but a few of the many applications that may be found in this subfield of geography.

We stated earlier that the geographer's first question is "Where is it?" The foregoing discussion of location and its allied concepts sets the foundation for the geographer's next question, "Why is it there?" Geographic analysis and geographic literacy among all citizens are based on recognizing patterns and processes of location decision making in human activity. The following chapters, which discuss the themes of Place, Human-Environment Interaction, Movement, and Region, all begin with this locational component.

=========== LEARNING OUTCOMES ===========

Learning outcomes in *Guidelines for Geographic Education* for the theme of Location are listed here. Students will:

1. Be able to locate places using a system of mathematical coordinates in an arbitrary grid system (absolute location).

2. Be able to describe locations in terms of relationships with other locations (relative location).

3. Understand that location is a significant aspect of every activity, event, person, place, and physical and cultural feature on Earth.

4. Be able to explain how location influences activities and processes that occur in different places.

5. Be able to describe how the physical and cultural attributes of one location interact with the attributes of other locations.

6. Be aware that the significance and importance of locations change as cultures change their interactions with each other and with the physical environment.

7. Recognize the differential rates of change in interactions between places.

8. Realize that knowledge of locations and their characteristics are key factors in understanding human interdependence.

9. Know the locations of major water bodies and land masses.

10. Be able to identify and locate a large number of important places and features in many parts of the world.

11. Be able to use maps to ask questions about the distributions of Earth features.

12. Be able to discuss maps as a primary geographic tool.

13. Be able to distinguish among various map projections and discuss how map projections distort perceptions of relationships on Earth.

The *Model Curriculum Standards* for grades nine through twelve in California lists specific teaching outcomes or understandings that all students should master. They are fairly representative of the social studies curricular goals in most states. In U.S. history-geography courses, the locational aspects of the following topics are particularly identified as being important: landforms, physical features, resources, major events, significant foreign powers, economic partners, agricultural land, culture and economic regions, the "West," migration routes, cities, transportation modes and routes, environmentally sensitive areas, ethnic differences, environmental hazards, and differing lifestyles. The world history and geography courses have even more emphasis upon location, with particular attention to world culture regions. The standard most specific to locational knowledge and skills reads:[10]

> A course should develop the student's knowledge of physical and placename geography and encourage relation of that knowledge to specific historical or contemporary events and conditions. Students should be encouraged to identify the major mountain ranges, arid regions, rivers, seas, islands, straits, etc. to show how they have affected the course of history and development in various periods. Other locational aspects of other topics in the list of standards include: religions, economic systems, culture groups, cities, governmental forms, historic regions, and events.

KEY CONCEPTS

Absolute location	Geographic Information	Legend	Resolution
Analemma	System (GIS)	Longitude	Robinson projection
Cartogram	Globe	Map projection	Scale
Cartography	Grid system	Mercator projection	Time zone
Conformal projection	Isobars	Model	
Contour lines	Isotims	Relative location	
Equal area projection	Latitude	Remote sensing	

FURTHER READING

Joel Makower and Laura Bergheim, eds., *The Map Catalog* (New York: Vintage Books, 1986), claim to list "every kind of map and chart on Earth and even some above it." Map reading is made easy and entertaining in Mark Monmonier's *How To Lie with Maps* (Chicago: University of Chicago Press, 1991).

A guidebook for the maximum use of maps is *Map Use and Analysis* by John Campbell (Dubuque, IA: William C. Brown, 1991). Another well-illustrated introduction to the mapping and remote-sensing work of the U.S. Geological Survey is Morris M. Thompson's *Maps for America* (Washington, DC: U.S. Government Printing Office, 1979).

A comprehensive survey of applied geography is found in John W. Frazier, ed., *Applied Geography: Selected Perspectives* (Englewood Cliffs, NJ: Prentice-Hall, 1982). Included are the employment of geographic concepts and geographers in planning, health care, environmental management, marketing, and several other topics.

Two highly recommended reference atlases are *Goode's World Atlas,* edited by Edward Espenshade, Jr., 18th ed. (Chicago: Rand McNally, 1991) and *Hammond Atlas of the World: Concise Edition* (Maplewood, NJ: Hammond, 1993). Their thematic maps are particularly informative.

Traditional mapmaking design and techniques are comprehensively presented in John Campbell, *Introductory Cartography* (Englewood Cliffs, NJ: Prentice-Hall, 1984). See also Thomas Avery and Gradon Berlin's *Fundamentals of Remote Sensing and Air Photo Interpretation* (New York: Macmillan, 1992). A commonly used reference for air photo interpretation by the same authors is *Interpretation of Aerial Photographs* (Minneapolis, MN: Burgess, 1985).

William E. Barrows, *Deep Black: Space Espionage and National Security* (New York: Random House, 1987), is an account of America's high-technology system for conducting surveillance from space over all portions of the globe during the Cold War.

Two highly recommended works in the series *Applied Cartography* are by Thomas D. Rabenhorst and Paul D. McDermott, *Introduction to Remote Sensing* and *Source Materials for Map Making,* both pubblished by Merrill (Columbus, OH, 1989).

Applied Remote Sensing (New York: Longman, 1986) includes chapters on this technology's use in studies of population, atmosphere, geology, geomorphology, hydrology, biota, agriculture, soils, and land use. Recommended also is *Introduction to Digital Image Processing* by J. R. Jensen (Englewood Cliffs, NJ: Prentice-Hall, 1986).

The location and form of cities from a historical perspective is discussed in *An Introduction to Urban Historical Geography* by Harold Carter (London: Arnold, 1983); John Borchert, "American Metropolitan Evolution," *Geographical Review* 57 (July 1967): 301–332; and L. Benevelo's *The History of the City* (Cambridge, MA: MIT Press, 1991).

Computer books are almost always out of date by the time they are published, but two that explain what can be done by those interested in the topic are David J. Maguire's *Computers in Geography* (New York: Halsted Press, 1989) and *Computer Applications in Geography* by Paul M. Mather (New York and Chester, NY: John Wiley, 1991). Another handy reference on computer mapping is B. T. Dent, *Principles of Thematic Map Design* (Reading, MA: AddisonWesley, 1985).

For an overview of the value of geographic information systems in research and applied work see David J. Maguire, Michael F. Goodchild, and David W. Rhind, eds., *Geographical Information Systems: Principles and Applications* (New York: John

Wiley, 1991). Another survey of this exciting research application in geography is *Introductory Readings in Geographic Information Systems,* edited by Donna J. Peuquet and Duane F. Marble (London: Taylor and Francis, 1990).

An instant course in understanding and using geographic location models is *Gravity and Spatial Interaction Models* by Kingsley E. Haynes and A. Stewart Fotheringham (Beverly Hills, CA: Sage, 1984). Also recommended on this topic are Kenneth Jones and James Simmons, *Location, Location, Location: Analyzing the Retail Environment* (New York: Methuen, 1987), and *Making Business Location Decisions* by R. Schmenner (Englewood Cliffs, NJ: Prentice Hall, 1992).

Publications on the locational aspects of gender are increasing in recent years. A good starter book is *Geography and Gender: An Introduction to Feminist Geography* by the Women in Geography Study Group of the Institute of British Geographers (London and Dover, NH: Hutchinson, in association with the Explorations in Feminism Collective). Other excellent books on gender and geography include Janet Henshall Momsen and Janet Townsend, eds., *Geography of Gender* (London: Century Hutchinson, 1987), and *Full Circles: Geographies of Women Over the Life Course,* by Cindi Katz and Janice Monk (London and New York: Routledge, 1993).

ENDNOTES

[1] John B. Garver, "New Perspective on the World," *National Geographic* 174 (December 1988): 911–913.

[2] Catalogs of available imagery may be obtained from (1) SPOT Image Corporation, 1897 Preston White Drive, Reston, VA 22091; (2) EROS Data Center, U.S. Geological Survey, Sioux Falls, SD 57198; and (3) EOSAT, 4300 Forbes Boulevard, Lanham, MD 20706.

[3] W. J. Reilly, *The Law of Retail Gravitation,* 2nd ed. (New York: Pillsbury, 1953), and Paul D. Converse, "Development of Marketing: Fifty Years of Progress," in H. G. Wales, ed., *Changing Perspectives in Marketing* (Urbana: University of Illinois, 1951).

[4] The most common model (it has many variations) used to locate stores and shopping centers was introduced by D. L. Huff in 1963. Huff argued that the probability of a consumer visiting a selected store is equal to the ratio of the utility of that store to the sum of the utilities of all the alternative stores considered. Expressed as $P(C_{ij})$, the probability that shopping place j will be visited by a consumer is

$$P(C_{ij}) = \frac{\dfrac{S_j}{T_{ij}^{\beta}}}{\displaystyle\sum_{j=1}^{n} \dfrac{S_j}{T_{ij}^{\beta}}}$$

where

$P(C_{ij})$ = probability that shopping place j will be visited by a consumer

S_j = square footage of selling space or "utility" of the shopping place

T_{ij} = travel time from home to the store

β = distance exponent

Gravity model formulations in geography may vary depending upon what is being measured, such as population and distance. Each application may be unique, but the basic formula is

$$I_{ij} = \frac{KP_i P_j}{d_{ij}^{h}}$$

where

I_{ij} = interaction between the two places i and j

K = constant of proportionality

P_i = population of place i

P_j = population of place j

d_{ij} = distance between i and j

h = power to which distance is to be raised

These elements may be found in most predictive gravity models used in location research or regional planning. See D. L. Huff, "A Probabilistic Analysis of Shopping Center Trade Areas," Land Economics 39 (1963): 81–90.

[5] Ralph H. Brown, *Historical Geography of the United States* (New York: Harcourt, Brace & World, 1948).

[6] D. D. Achabal, W. L. Gorr, and V. Mahajan, "Multiloc: A Multiple Store Location Decision Model," *Journal of Retailing* 58 (1982): 5–25, and J. L. Goldstucker, D. Bellenger, T. Stanley, R. Otte, *New Developments in Retail Trading Area Analysis and Site Selection* (Atlanta: Georgia State University, 1978).

[7] Michael Weiss, *The Clustering of America* (New York: Harper & Row, 1988).

[8] On gender location issues, see Joni Seager and Ann Olson, *Women in the World Atlas* (New York: Simon & Schuster, 1986). For other sources, consult the excellent recent publication by Michael Levasseur entitled *Finding a Way: Annotated Bibliography on Underrepresented Groups in Geography* (Indiana, PA: National Council for Geographic Education, 1993).

[9] K. R. Cox, *Location and Public Problems: A Political Geography of the Contemporary World* (Chicago: Maaroufa Press, 1979).

[10] California State Department of Education, *Model Curriculum Standards in History/Social Science* (Sacramento, 1985).

4 INTRODUCTION TO THE THEME OF PLACE

Florida's Epcot Center reflects a future sense of place. (Photo by author.)

This chapter introduces various geographic concepts central to understanding the theme of Place. The following *National Geography Standards* are addressed in this overview of the theme of Place:

> Standard 2: *The geographically informed person knows and understands how to use mental maps to organize information about people, places, and environments in a spatial context.*
>
> Standard 4: *The geographically informed person knows and understands the physical and human characteristics of place.*
>
> Standard 6: *The geographically informed person knows and understands how culture and experience influence people's perception of places and regions.*

In previous chapters you were introduced to the structure of geography, its areas of emphasis, and the perspectives from which it is viewed. The first of the five fundamental themes, Location, was introduced in Chapter 3, along with an assemblage of concepts relative to that theme. In the next five chapters we will study the second fundamental theme, Place. Once you understand the major concepts related to the theme of Place you will be prepared for further analysis of geographic thinking in the remaining three fundamental themes—Human-Environment Interaction, Movement, and Region.

Places may be real or fictional. They may be as small as your armchair or larger than the solar system. A place often is completely unique, although many places have certain characteristics in common. Knowledge of specific places was the core of geographic inquiry for centuries, but in current practice, geographers also study the relations among places, including their associations, patterns, similarities, differences, and connections.

EARLY REFERENCES TO PLACE

What the earliest travelers brought home from places they had visited were narrations of their adventures, descriptions of exotic environments, interesting artifacts, and sometimes maps. This information most often was based on limited fact, blended with conjecture and fantasy. The motive for **exploration** was not so much to fill in a blank map as it was to verify accepted thinking about distant places.

Geographer John Allen's essay on imagination and geographical exploration reminds us that exploration is a process rather than a series of distinct events. According to this geographer: "The process of exploration began only after objectives were established, based on the perceived or imagined nature of the lands to be explored."[1]

At times, journeys undertaken by explorers were based on misinformation. Examples include the Greeks' search for the "Tin Islands" of Great Britain, the Spanish exploration to find the Seven Cities of Cibola in the American Southwest, and the journey by American explorers Lewis and Clark in search of a fictitious all-water route across North America.

Trade also was a motive for ancient explorer-mariners. The Phoenicians built an empire based on navigation and trade in the Mediterranean Basin, and the Portuguese, Spanish, and others would do the same around the world's oceans fifteen centuries later.

The Age of Exploration in Europe after 1492 expanded the not only the influence of Christianity but also filled in some blank areas on the map that often were labeled **Terra Incognita**. John K. Wright, in his presidential address to the Association of American Geographers in 1946, eloquently described the fascination of places unknown as

> *Terra Incognita:* These words stir the imagination. Through the ages, men have been drawn to unknown regions by Siren voices, echoes of which ring in our ears today when on modern maps we see spaces labeled "unexplored," rivers shown by broken lines, islands marked "existence doubtful."[2]

In Figure 4-1, we see three maps of the world reproduced from Martin and James's book *All Possible Worlds.*[3] It is interesting to note the similarities and differences among these conceptualizations of place. It is obvious from these maps that different people saw places in the world in different ways, depending on time and space.

Places are continually being "discovered" by modern-day tourists. The major purpose of most popular geographical societies and national tourist bureaus continues to be personalizing the experience of place.

There are many ways to experience a particular place. We can read about it, photograph its landscapes, conduct a scientific inventory of its features, or visit it as a tourist. Places have both physical and human characteristics. Experiencing place also may be more subjective or phenomenological, focusing on the storing of impressions and memories. Place impressions also may be experienced vicariously through visual and print media, expanding our horizons to the limits of Earth.

THE MEANING OF PLACE IN EVERYDAY LIFE

Try to describe the earliest place you recall. Is it a place where you felt happy or content? Perhaps you remember a fearful place. Chances are very good that your experience of that place was emotionally intense for it to be imprinted so clearly on your consciousness at such an early age. Places often evoke strong feelings. Think of the importance of place in literature, in film, and in real-life experiences. We could not understand Tom Sawyer or the Brothers Karamazov as characters in a narrative unless we also understood the context of their lives. It has been said that "place is to geography what time is to history." It is the stage upon which the drama of life on Earth takes place.

Categories of place are limitless, but we suggest beginning with attention to scale. Children easily learn small, immediate, and familiar places first; then their horizons expand to include more complex places, often called regions.

Personal Space

Personal space may be thought of as an invisible bubble or aura that surrounds us. We begin to evaluate our surroundings in infancy, and throughout our lives we continually attempt to personalize the space around us.

Personal space and **spatial behavior** vary with environmental constraints. Both physical settings (such as the size of a room, the nearness of other people, and the number of occupants) and cultural constraints (such as the personality traits of the individuals, their ages, ethnicities, or genders) may affect one's perception of personal space. Examples of these interactions may be observed in a crowded elevator or on an open beach. It is obvious that personal space requirements vary greatly between these two extremes.

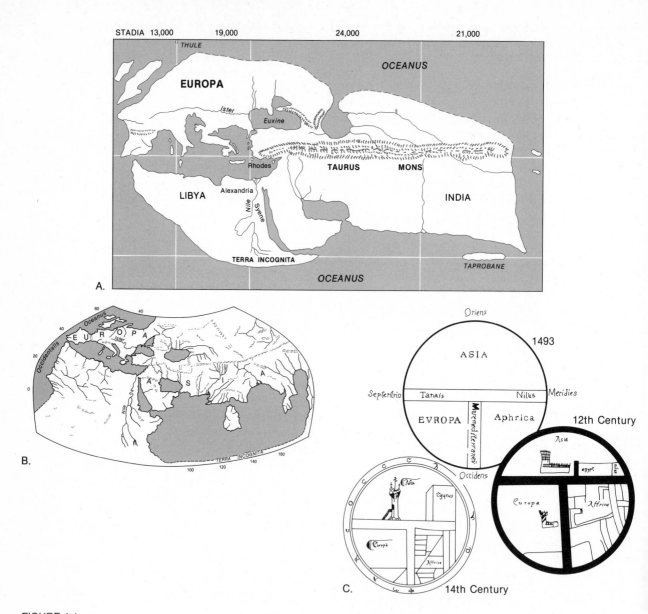

FIGURE 4-1
The world according to: A. Eratosthenes (third century B.C.*).   B. Ptolemy (second century* A.D.*).
C. Medieval theologians. (From Geoffrey J. Martin and Preston E. James,* All Possible Worlds: A
History of Geographical Ideas, *New York: John Wiley & Sons, 1993, Fig. 2, p. 20. Reprinted with
permission.)*

Anthropologist Edward Hall, who uses the term *prox-mics* for his studies of personal space, has found that misunderstandings among people of different cultures often can be attributed to differences in what they consider proper distances for communication.[4] What may be considered conversational space by people of one culture may be viewed as spatial invasion by those of another. Americans who have visited the Middle East have experienced the close proximity that is customary during conversations among people of that region. Hall emphasizes cross-cultural knowledge of personal and social consultive distances as an important safeguard against violating the norms of others.

Home

The attachment to hearth and **home** often is our most significant concept of place. Home is a place that, hopefully, we can create and modify to our own liking. It is sometimes a refuge or sometimes an operation center for our personal interactions with surrounding places. Home is where we begin our interaction with the rest of the world. When introduced to a new map or atlas, is not one of our first actions to locate our home town?

Geographer Yi-Fu Tuan extended the notion of home to include *homeland*. Tuan suggested that cultural groups everywhere tend to view their home space as the center of the universe. He found that this profound attachment to homeland appears to be worldwide and cross-cultural. According to Tuan:

> It is not limited to any particular culture or economy. It is known to literate and nonliterate peoples, hunter-gatherers, and sedentary farmers, as well as city dwellers. Place is an archive of fond memories and splendid achievements that inspire the present; place is permanent and hence reassuring to man who sees frailty in himself and chance and flux.[5]

The concepts of patriotism and nationalism often considered in political geography are extensions of the personal nature of one's homeland and the importance of place as home.

Neighborhood

Neighborhoods, by definition and as a matter of practice, are also usually personal places. Geographic studies of neighborhoods usually consider their limits or boundaries, their house types or street patterns, their unifying social characteristics, or perhaps their political significance. Neighborhood as place is, in reality, many things to many people. A crowded residential neighborhood in Calcutta may be viewed as a dangerous environment to tourists and other outsiders but is perceived as a safe refuge from the city's stresses by its local residents. Likewise, different players in the neighborhood scene have their own definitions and perceptions of the place. Taxi drivers, school children, garbage collectors, church members, newcomers, and long-time residents all have varying images of the "same" locale. Place as multiple reality is explained by urban geographer David Ley as follows:

> Places are constructed, constructed not simply in the physical engineering sense, but more profoundly that they are objects given meaning by a subject and that

their reality is thereby socially constructed and socially contingent. It follows from this that place is an idea as well as an object and likely to have a multiple reality to groups with varied concerns.[6]

The intermediate-scale personal place here termed "neighborhood" most often is mappable but more often is understood as a mental construct by people involved in it. The neighborhood may have a center such as a school or a transit station. It may possess a hierarchy of activities such as children's play, business transactions, social interactions, or isolation. The neighborhood serves as an intermediate link from home to community. Communities are distinct places and may be of almost any size. A hometown also is a place.

On a longer but still personal scale is one's feeling toward nation as a special place. Nation-states throughout the world are perhaps the largest differentiation of places as separate entities. In the United States, most state curricular frameworks guide teachers through a variety of scales of places from home, to neighborhood, to community, to nations. This "expanding universe" model of curriculum development builds on students' knowledge of home space and, eventually, global space.

All places have a specific areal extent and are somehow differentiated from other places. The differentiation may be observed as boundaries or perhaps as transition zones. These limits may be quite dramatic, or they may be very subtle. Places delimit territory, and territoriality, at whatever scale, delimit the range of selected human activities. The concepts of territoriality, transition zones, and boundaries are discussed in Chapter 7.

MAJOR CONCEPTS IN UNDERSTANDING THE THEME OF PLACE

The Time-Space Continuum

All events in the human experience occur in a place at a time. The **time-space continuum** may be thought of as a vertical and horizontal reference grid whereon events are pinpointed. In some cases the time component is more critical, and those who study history are better qualified to analyze the significance of the event. If place or environment are more significant, however, it is appropriate that geographers be involved. In reality, there is a great deal of overlap

FIGURE 4-2
*Perceptual maps of Los Angeles
drawn by school children.
A. Perceived by upper middle class
white residents of the suburb of
Westwood.   B. Perceived by
African-American residents of
Avalon.   C. Perceived by Hispanic
residents of Boyle Heights. (From
P. Orleans, "Differential Cognition of
Residents," in Roger M. Downs and
David Stea, eds.,* Image and
Environment, *Chicago: Aldine
Publishing Co., 1973, Figs. 7-3 and
7-4. Reprinted with permission of
Roger M. Downs.)*

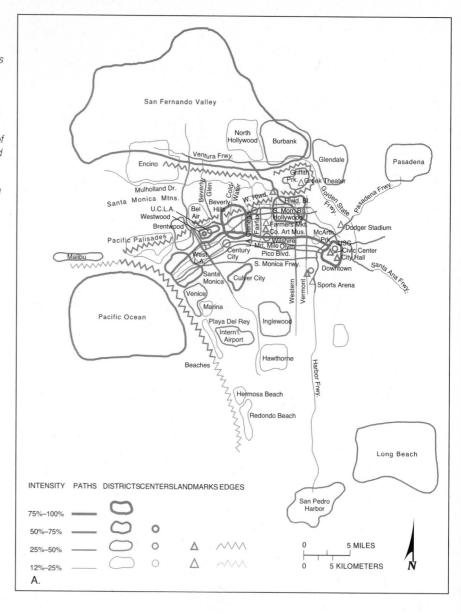

in understanding the human story through time on the surface of Earth.

This concept was illuminated by geographer Jan Broek:

> Whatever the kind of place, we must always keep in mind its position in time as well as its location in space. A place cannot be understood by merely observing the interaction of present day forces. Knowing the legacy of the past and sensing the persistence of change are essential qualities of the geographic mind.[7]

Mental Maps

As we learned in Chapter 3, maps are symbolic representations of the surface of Earth. **Mental maps** may be considered one step removed in that they are mental images and memories, rather than tangible documents. Mental maps often are remarkably accurate. Others may be inaccurate but still contain information useful in geographic analysis. The National Geography Standards contains several references to the importance of students' developing solid and detailed mental maps of their world.

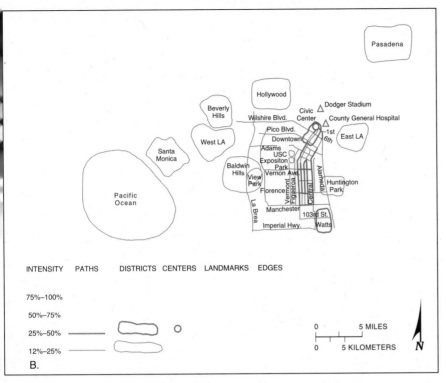

B.

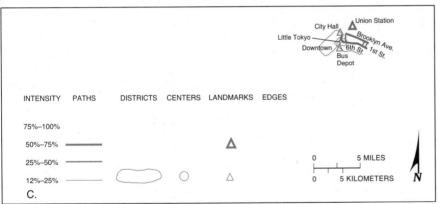

C.

The maps shown in Figure 4-2 are mental maps drawn by school children in downtown Los Angeles. These maps constructed by different ethnic groups reveal different images of the same neighborhood and are a reflection not only of ethnicity but mobility. Hispanic children (Map C) had a somewhat clearly defined but rather spatially limited image of their home area, while African-Americans included more information on the map (Map B) and expanded the areal extent of their neighborhood. White children had the largest and most detailed maps of the neighborhood (Map A). Their maps also contained fewer barriers and boundaries.

Mental maps teach us how different people think about the same place. They sometimes reveal prejudices or mind sets about places that may not be grounded in truth. Mental maps are indicators of one's experiences in travel, in observation, and in spatial awareness. In general, the closer a place is to one's home, the more accurate is its mental map. Faraway places tend to be more distorted or under-represented. Compare the "New Yorker's Idea of the United States" shown in Figure 4-3 with an accurate map of the nation. Why is this perceptual map drawn to evoke humor? What makes it humorous?

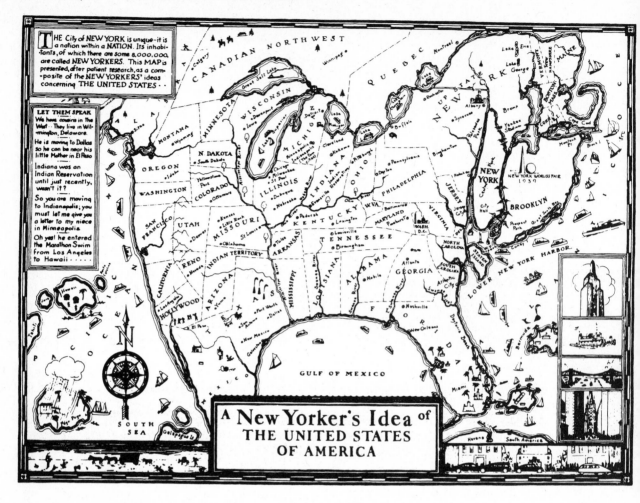

FIGURE 4-3
A New Yorker's idea of the United States of America. (From The Saturday Review of Literature *15, no. 5, November 28, 1936, p. 4. Reprinted with permission.)*

Sense of Place

Another important concept useful in understanding the second geographic theme is **sense of place.** Here we are referring to the subjective and sometimes emotional connections to our immediate environment. These connections may be feelings of attachment, or they may be more negative feelings. The sense of place may include memories of past events and an anticipation of the future. Much of our sense of place is built on the uniqueness of that particular place in our lives.

An analysis of particular urban ethnic neighborhoods provides the geographer with an excellent opportunity to observe sense of place in action. Oftentimes the profound sentiment that attaches the new residents to their original homeland continues to be a strong human emotion in the

new land. Loyalty to a group's homeland, socialized into the individual during childhood, remains important even when the ethnic group is transplanted to a new place. This new place, in turn, creates special bonds and feelings based in part on former experiences. Examples are the Chinatowns that are found in many large cities of the world.

One of the reasons that certain tourist destinations are successful is that they convey a special sense of place. The designers of places such as EuroDisney went to great lengths to create a memorable sense of place that can be imprinted on postcards, souvenirs, and visitors' memories. A sense of place for tourists may be based on both human and physical characteristics. Even spectacular natural features such as Ayers Rock in the Uluru National Park in Australia convey the essence of a unique location that we have here labeled "sense of place."

Art and music also convey a strong sense of place to students. Reggae music links African and Caribbean sounds, providing a rhythmic case study for sensing place. Likewise, no unit on folk culture would be complete without a demonstration of the similarity between Appalachian Mountains bluegrass music and its Scotch-Irish roots. Addition of the African banjo to traditional European string instruments created a unique "Appalachian" country sound. Distinctive settings such as these may be illustrated in the classroom with music, art, stories, and legends about place.

Clearly, places affect people, and people create or change places. The landscape character of a place does have a power of its own, and according to Yi-Fu Tuan, "Place is the center of meaning constructed by experience."[8]

Behavioral Response to Place

Some places create behavior patterns and human reactions. These places may be specifically designed to accomplish a behavioral reaction, such as amusement parks, dance halls, or shopping centers. Other places may evoke behavior patterns based on a need for security or protection. These may be responses to the natural hazards of a place. Others may be socially caused, such as the high-crime districts of the inner city. Behavioral responses to places may be both positive and negative. Calm, reverent behavior may be evoked from a visit to a familiar church or cathedral. A unique physical setting also may trigger certain behavior patterns in visitors; consider the commonest response to a first-time view of the Grand Canyon of the Colorado River: "Wow!"

Territoriality

Personal space may contain a possessive element. Children's games and the arrangement of furniture in a classroom often exhibit territorial motives. The use of space in a single-family home or yard arrangement may reflect territoriality. For example, the front yard is usually semipublic, while the back yard remains more private. The side yard may become storage space or a boundary defining ownership and responsibility for maintenance.

Territoriality may be observed in almost all social settings. Consider the school classroom as an example. Robert Sommer studied the successfulness of students in a classroom, compared with their choice of seating in the room.[9] As shown in Figure 4-4, Sommer found that those who participated more—the more successful students—sat in the front and center of the class, while those who partici-

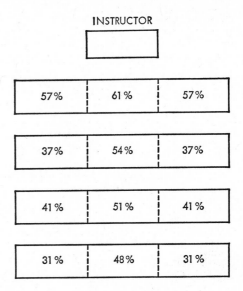

FIGURE 4-4
The ecology of participation in straight-row classrooms. (From Robert Sommer, Personal Space: The Behavioral Basis of Design, *Englewood Cliffs, NJ: Prentice-Hall, 1969, Fig. 5, p. 118. Reprinted with permission of Robert Sommer.)*

pated least—the least successful—occupied the rear and sides of the room. Eye contact and active participation were stronger among the students in the front-center of the classroom. In this and almost all other studies of furniture arrangement and personal space, territory is staked out very quickly and adhered to almost universally. Studies of spatial invasion show that when one's personal territory is invaded, flight or conflict usually results.

The importance of territoriality may be seen clearly in urban neighborhood settings, particularly along their **boundaries.** Examples of these places are often marked by urban graffiti, incidents of crime, and other social indicators of cultural stress. A more positive use of the territorial concept has been developed by site planners and is termed "defensible space." Territorial markers such as hedges, window views, and pavement changes help define units of personal space as separated from public space in housing complexes.

PLACE PLANNING

Planning is a process by which people change the natural environment to suit their needs and desires. Urban planning, if done wisely, creates safe, comfortable, attractive cities. Planning at any scale involves melding governmen-

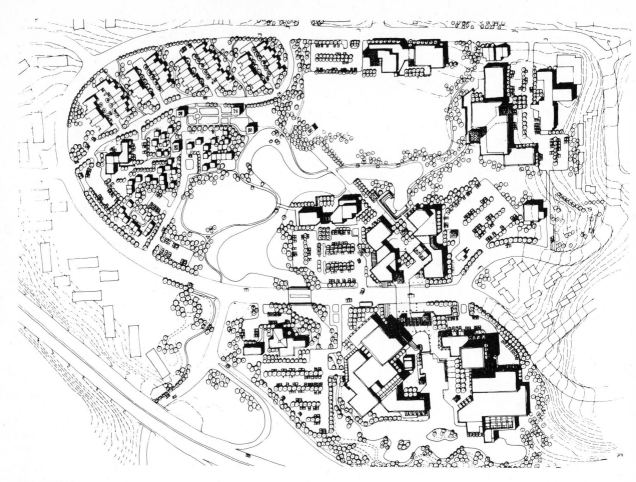

FIGURE 4-5
*A representative site plan for proposed development: Marin City master plan, Marin County
Redevelopment Agency, California. (From David Gates Associates brochure, 1985. Reprinted with
permission of David Gates.)*

tal regulations with an owner's needs and the ultimate functions and activities of users. Planning is thus advanced decision making about human behavior in selected places.

Site planning is the process of analyzing a specific place with a mind toward changing its form and use. According to Kevin Lynch and Gary Hack in their widely used text, site planning is defined as "the art of arranging structures on the land and shaping the spaces between."[10]

The site-planning process usually begins with an environmental inventory of what is presently there. Factors mapped and analyzed may include topography, geology, soils, natural vegetation, hydrology, climate, sun angles, wildlife, and surrounding land uses. The inventory is usually in the form of a map accompanied by written text or diagrams as necessary. After taking this inventory of the site's natural environment, a similar process may be undertaken for similar socioeconomic conditions such as vehicle and pedestrian traffic, noise, economic viability, municipal services, and known problems.

The *program* for the project is then defined. Ultimate uses of the site are listed, and exact activities of the users are detailed. This is where design begins. Creation of the site plan on paper or by computer now becomes a process of combining the goals of the client with the current physical attributes of the site and the program of user activities (Figure 4-5). The intent is usually efficient use of space combined with aesthetic and artistic goals. Naturally, many other factors are included such as administrative regulations, cost, and economic return. A successful project will eliminate previous problems associated with the site, enhance its natural attributes, and improve the quality of life for its visitors and users. (See the box on page 73.)

GEOGRAPHY AT WORK
The Role of the Land Development Planner in Urban Design

Site planner Joanna Callenback at work. (Photo by Joanna Callenback.)

All too often the so-called process of land "development" has been at the expense of the natural and historic elements of a place. Often the architect, engineer, and real estate agent have worked toward their own goals with only limited communication among themselves, usually at the direction of the developer or government official. The team was competent and committed but lacked a quarterback.

Land developer Joanna Callenback credits her success to the integration of the characteristics of place into an aesthetic as well as a functional use of the land. She first became interested in geography upon taking a cultural geography survey course at the University of California—Berkeley. She remembers learning that "places reflect the culture of their occupants" and that it would be fun and interesting, to "make" places that people would enjoy. She earned a master's degree in landscape architecture at Harvard University, then worked for both global and local design firms. After gaining experience by working on a variety of large projects, she became a project manager with YCS Investments, a land development firm in San Francisco.

This site-planner's opinion of an ideal project is a situation in which she can successfully coordinate the efforts of the market analyst, the engineer, the architect, and the environmental planner. She states, however, that the process is really more "hands-on," and "I often *am* the market analyst, the site-planner, and the environmental monitor, etc." After an inventory of natural features and constraints is undertaken, she will do a dozen or more preliminary plans for a project. The process of inventory and of analysis by working through these preliminary plans turns into a synthesis of environmental, political, engineering, and marketing conditions.

Joanna notes that the most difficult aspect of her work is dealing with government regulations and bureaucratic constraints in the process of getting permits to build the places she has created on paper or computer screen. There are usually several agencies to deal with and, of couse, political support for any new idea is critical. "Planners and developers must have patience and a knowledge of everyone's role in the process." After completion the new place should be environmentally sound, aesthetically pleasing, functional, and it should bring a financial return to investors.

FICTIONAL PLACE

The previous section described the process of creating new places for living, work, and recreation. This section demonstrates the value of fictional places in understanding geographic descriptions of Earth. Radio, television, and film can create a clear sense of place. Radio programs such as the popular *Prairie Home Companion* made fictional Lake Wobegon vivid to everyone who tuned in. We all know of a favorite movie that gives us a sense of being in a region or a specific place.

The use of literature in geography focuses upon one of two themes: understanding a region through the literature written in and about it, or understanding of landscapes for specific places through literature.

Understanding a Region through Its Literature

Fiction can identify places with physical features, cultural attributes, or economic activities that may have taken place there or are important in the contemporary significance of that area. Regional fiction in the United States is exemplified in the works of Faulkner, Steinbeck, Cather, Hawthorne, Rölvaag, Frost, Thoreau, Twain, L'Amour, and George R. Stewart. There are many others, including Jack London, whom we quote below.

Most of us who have never been in the far north can begin to understand the significant elements of an Arctic landscape by reading Jack London's succinct and powerful descriptions of them. Consider the vividness of a redwood forest in this passage from London's *The Valley of the Moon:*

> Never, in all their travels, had Saxon seen so lovely a vista as the one that greeted them when they emerged. The dim trail lay like a rambling red shadow cast on the soft forest floor by the great redwoods and over-arching oaks. It seemed as if all local varieties of trees and vines had conspired to weave the leafy roof–maples, bug madronos and laurels, and lofty tan-bark oaks, scaled and wrapped and interwound with wild grape and flaming poison oak. Saxon drew Billy's eyes to a mossy bank of five finger ferns. All slopes seemed to meet to form this basin and colossal forest bower. Underfoot the floor was spongy with water. An invisible streamlet whispered under broad-fronded brakes. On every side had opened tiny vistas of enchantment, where young redwoods grouped still and stately about fallen giants, shoulder-high to the horses, moss-covered and dissolving into mold.[11]

Another American writer, Willa Cather, wrote a vivid narrative about her sense of place in the American West in *O Pioneers:*

> The Divide is now thickly populated. The rich soil yields heavy harvests; the dry bracing climate and the smoothness of the land make labor easy for men and beasts. There are few scenes more gratifying than a spring plowing in that country, where the furrows of a single field often lie a mile in length, and the brown earth, with such a strong, clean smell, and such a power of growth and fertility in it, yields itself eagerly to the plow; rolls away from the shear, not even dimming the brightness of the metal, with a soft, deep sigh of happiness.[12]

Understanding Landscapes through Literature

The author in many cases will know a landscape from his or her own experience and will be able to isolate important elements from it that best give setting to a story. In reading the following paragraph by Charles Dickens describing nineteenth-century London, we may understand the power of place in affecting his characters. In fact, the place is a

part of the character, while the character is a part of the place:

> To reach this place, the visitor has to penetrate through a maze of close, narrow, and muddy streets, thronged by the roughest and poorest of water-side people, and devoted to the traffic they may be supposed to occasion. The cheapest and least delicate provisions are heaped in the shops; the coarsest and commonest articles of wearing apparel dangle at the salesman's door, and stream from the house-parapet and windows. Jostling with unemployed laborers of the lowest class, ballast-heavers, coal-whippers, brazen women, ragged children, and the very raff and refuse of the river, he makes his way with difficulty along, assailed by offensive sights and smells from the narrow alleys which branch off on the right and left, and deafened by the clash of ponderous wagons that bear great piles of merchandise that rise from every corner. Arriving, at length, in streets remoter and less-frequented than those through which he has passed, he walks beneath tottering house-fronts projecting over the pavement, dismantled walls that seem to totter as he passes, chimneys half crushed half hesitating to fall, windows guarded by rusty iron bars that time and dirt have almost eaten away, and every imaginable sign of desolation and neglect.[13]

Many stories become believable when a literary sense of place is created effectively. For example, one of the best ways for a westerner to understand prerevolutionary Chinese agricultural landscapes is to read Pearl Buck's *The Good Earth*.[14] The sense of place of rural China lingers long after one has finished reading her book. A fairly well-read student can assemble a collection of geographic images for places throughout the world with some well-chosen reading.

A best selling Chinese-American writer, Amy Tan, describes her mother's memories of a favorite Chinese landscape in her book *The Joy Luck Club:*

> "I dreamed about Kweilin before I ever saw it," my mother began, speaking Chinese. "I dreamed of jagged peaks lining a curving river, with magic moss greening the banks. At the tops of these peaks were white mists. And if you could float down this river and eat the moss for food, you would be strong enough to climb the peak. If you slipped, you would only fall into a bed of soft moss and laugh. And once you reached the top, you would be able to see everything and feel such happiness it would be enough to never have worries in your life ever again.[15]

Another geographic concept often exemplified in fiction is **sequent occupance.** This term, discussed in detail in Chapter 11, refers to the sequence of peoples who occupy and interact on a landscape in one locale over time. This process may be created in semifictitious places by skilled authors, as particularly exemplified in the works of James Michener. In most of Michener's stories fictional characters are intertwined with historical persons and events through time.[16]

Human-environmental conflict and environmental attitudes also may be effectively expressed in literature. Edward Abbey's *The Monkey Wrench Gang* is a fictional sermon about environmental degradation.[17] John Steinbeck's short story "Flight" and George Stewart's *Earth Abides* and *Storm* all illustrate the overwhelming power of environmental processes and forces that affect human activity on Earth.[18]

In addition to environmental attitudes, a limitless number of characters in fiction exemplify perceptions or attitudes about their personal physical environment. *Landscape perception,* discussed earlier in this chapter, often is a means by which authors can instill motives or the rationale for other kinds of decision making in their characters, as in the works about Native Americans by Tony Hillerman and in the decisions and experiences of a variety of Arizona residents by Barbara Kingsolver.

A creative and enterprising author may create imaginary places that have never existed. The science fiction writer must be particularly adept at this. Austin Tappan Wright's *Islandia* was a very thorough creation of an entire continent, complete with maps.[19] A similar work was written by Jim Crace, entitled *Continent.*[20] In 1915 another American writer, Charlotte Perkins Gilman, wrote *Herland,* a story about a fictitious place dominated by strong women.[21]

Leonard Lutwack illustrates how writers have created and changed places in literature for various purposes.[22] He notes that the current concern for place in literature is a result of the recent awareness of environment and ecologically related topics. In Lutwack's work the properties and uses of place are surveyed, and extensive examples of real and symbolic places are offered, particularly as they are described in rational literature. A useful list of commercial films that depict places exceptionally well awaits compilation.

Some real places are the inspiration for fictitious ones. Disneyland, Disney World, and clone entertainment centers or theme parks are based on imaginary works. Other tourist attractions, such as Cannery Row in Monterey, California, were at one time functional "real places." Now they are tourist destinations, typified by converted buildings that now house book stores, gourmet restaurants, and souvenir shops. Cannery Row's sardine packing plants and rowdy fishermen, upon which Steinbeck's book was based, were last seen working in the 1950s (see the box on page 76).

A satisfying study of place requires systematic analysis blended with a dose of imagination and, perhaps, inspiration. As Yi-Fu Tuan put it, "enclosed and humanized space is place. Compared to space, place is a calm center of established values" (see the box on page 78).[23]

HOW SYMBOLIC LANDSCAPES CREATE A SENSE OF PLACE

Certain landscape images convey strong meanings and shared perceptions of place. According to geographer Donald Meinig, every mature nation has its symbolic landscapes. Places in the United States such as the New England village, "Main Street" of Middle America, and a California suburb all convey distinct images of place. Meinig suggests posing the following six geographic questions to guide the study and observation of these symbolic landscapes:[24]

- What were the landscapes which have served as the bases for these symbols really like?
- How do actual landscapes become symbolic landscapes?
- How can we assess the impact, the power of the symbol?
- How do we define and assess the significance of the difference between the ideal and the real?
- What does this threshold set of symbols tell us about America?
- What is happening? Is any new pattern discernible in the landscapes of American communities?

This list of questions may serve as an inspiration for reading the landscape and interpreting the culture of the United States (or any other nation), and for learning more about the characteristics of place on the surface of Earth.

CHARACTERISTICS OF PLACE

The meaning and importance of place has been addressed in this chapter as background for the use of the place

CANNERY ROW AND THE PRESERVATION OF LITERARY LANDSCAPES

Cannery Row, Monterey, California: an example of a literary landscape. (Photo by author.)

Over 15 million people a year visit the Monterey Peninsula on the central California coast. The quaint Victorian homes of Pacific Grove, the natural splendor of Point Lobos, the paths-of-history walk in Old Monterey, the tranquility of Carmel Mission, or the restaurants and art galleries of Fisherman's Wharf are favorite places in the area. Others may prefer the tennis courts of Carmel Valley or the wildness of Big Sur. The largest single attraction in terms of attendance, however, is Cannery Row, with over 3 million visitors a year.

The physical nature of Cannery Row is described only in three short passages in Steinbeck's 1945 book, but they are vivid and memorable:[*]

Cannery Row in Monterey is a poem, a stink, a grating noise, a quality of light, a tone, a habit, a nostalgia, a dream. Cannery Row is the gathered and scattered, tin and iron and junk heaps, sardine canneries of corrugated iron, honky tonks, restaurants and whore houses, and little crowded groceries and laboratories and flophouses. In the morning when the sardine fleet has made a catch, the purse-seiners waddle heavily into the bay blowing their whistles. The deep-laden boats pull in against the coast where the canneries dip their tails into the bay. The figure is advisedly chosen, for if the canneries dipped their mouths into the bay, the canned sardines which emerge from the other end would be, metaphorically at least, even more horrifying.

Then cannery whistles scream and all over town men and women scramble into their clothes and come running down to the Row and go to work. They come running to clean and cut and can the fish. The whole street rumbles and groans and screams and rattles while the silver rivers of fish pour out of the boats and the boats rise higher out of the water until they are empty. The canneries rumble and rattle and squeak until the last fish is cleaned and cut and cooked and canned and then the whistles scream again and the dripping, smelly, tired men and women straggle out and droop their way up the hill into town and Cannery Row becomes itself again—quiet and magical.

Fish canning began in Monterey at the turn of the century with the opening of the Booth Cannery, which employed immigrant Italian and Norwegian fishermen and packers. At its peak in the 1940s, the Row contained 19 canneries and 20 reduction plants, employing 5000 people and processing 235,000 tons of sardines per year. Unfortunately, the industry's demand for sardines was greater than the fish's reproductive rate. They were fished out in the 1950s, when Steinbeck's sequel *Sweet Thursday* was published, and the last cannery closed in 1962.

Immediately after the release of Steinbeck's books about marine biologist Doc Ricketts and his friends, the tourists began to appear. The "real" Ed Ricketts told of an "invasion" into his laboratory-apartment where he physically had to

concept in cataloging the landscape features of Earth. Specific characteristics of place, then, are its inventory of the natural environment with its overlay of human features. In Chapter 5 we begin this place-based inventory of the physical environment by considering atmospheric and hydrospheric systems. Chapter 6 continues our discussion of the physical environment with an examination of systems in the lithosphere and biosphere.

In Chapter 7 we turn our attention to human patterns as we focus upon human imprints on places. These imprints may be belief systems, technological innovations, lan-

throw people out. The first two restaurants appeared in 1957, and by 1961 Steinbeck, after living in New York for several years, wrote in *Travels with Charley:*[†]

> The canneries which put up a stinking stench are gone, their places filled with restaurants and antique shops and the like. They fish for tourists now, not pilchards, and that species they are not likely to wipe out.

Under the protection of a permissive City General Plan and consolidated ownership of most properties, Cannery Row now claims a world-famous public aquarium, two dozen major restaurants, a historic carousel, two major hotels, and various shops. To Monterey's credit, the aquarium, a biological supply house, and an aquaculture enterprise do help maintain a marine atmosphere in the area.

The conflict of priorities among businesses in the visitor industry, preservationists, planners, and local citizens revolves around the nature of the place that was originally described in two works of fiction, as compared with the nature of the place today and what it might be like in the future. This dilemma about what a place *ought* to be like is familiar to those dealing with the preservation of historic places such as Williamsburg, Virginia, or Charleston, South Carolina. An additional consideration in the case of literary preservation is whether to focus on the author, the fictitious characters, or the landscape described. Decision makers in places like Mark Twain's Hannibal, Missouri, must face these issues.

Steinbeck's opinion about the development of Cannery Row might be worth pondering by those who enjoy the sense of place that can be created in our minds by good writers:[‡]

> My own suggestion will get me exiled from the Peninsula. Young and fearless and creative architects are evolving in America. They are, in fact, some of our very best artists in addition to knowing the sciences and the materials of our period. I suggest that these creators be allowed to look at the lovely coastline and to design something new in the world, but something that will add to the existing beauty rather than cancel it out. The tourists would not come to see a celebration of history that never happened, an imitation of limitations, but rather, a speculation on the future...

---

[*] *John Steinbeck,* Cannery Row *(New York: Viking Press, 1945), pp. 1–2. Reprinted with permission.*
[†] *John Steinbeck,* Travels with Charley *(New York: Viking Press, 1962), p. 182. Reprinted with permission.*
[‡] *John Steinbeck, quoted in the* Monterey Peninsula Herald, *March 8, 1957.*

guage patterns, or other cultural and political expressions. Chapter 8 also deals with human characteristics of place but from an economic perspective. Goods and services and their distributions on Earth form the heart of this discussion of economic patterns.

Chapter 9 considers the urban characteristics of place as a composite of constructs that make up a city. These may include built forms, residential patterns and commercial nodes, and the overall energy flows that link cities together. Taken together, these five chapters on the theme of place will strengthen your understanding of *this* place, which we call Earth.

GEOGRAPHER AT WORK:
A Poet Speaks as a Geographer

Geography and bioregionalism as poetry: poet Gary Snyder. (Photo by Gary Snyder.)

Pulitzer Prize–winning poet Gary Snyder grew up in western Washington, knowing that he was not "from a place" but was a *part* of place. With an education consisting of mountaineering, travel, university study, and Zen Buddhist training, he practices his beliefs in the foothills of the Sierra Nevada Mountains on a mountain farmstead. Along with his neighbors he lives by a combination of nineteenth- and twentieth-century technologies: wood stoves for heat, photovoltaic cells for electricity.

Snyder has published fourteen books of poetry and prose and is probably the only writer in the country who is self-supporting on his poetry. His book *Turtle Island* won the Pulitzer Prize in poetry in 1975. He has been the recipient of a Bollingen grant, was a Guggenheim Fellow, and is a member of the American Academy and Institute of Arts and Letters.

Snyder puts his belief in community culture and the need for sustainable economic skills into practice by living and working closely with the land, his neighbors, and his two sons. As noted in correspondence with the authors, he is a long advocate of *bioregionalism* (see Chapter 10) and its alternative perspective of political geography:

> The political side of bioregionalism, for starters, is recognizing that there are real boundaries in the real world which are far more appropriate than arbitrary political boundaries. And that this is just one step in learning where we really are and how a place works.

In remarks given at the University of California–Davis, Snyder discussed the idea of place:

> One of the most striking things about the last three decades in both capitalist and socialist nations is the relentless advance, short in distance so far, but still

LEARNING OUTCOMES

Learning outcomes in *Guidelines for Geographic Education* for the theme of Place are listed here. Students will:

1. Know that places are distinctive in terms of their physical and human characteristics.

2. Describe and interpret complex physical processes that produce the geological, biological, and other natural features of places.

3. Be able to relate how human activities and culture create a variety of different and similar places.

4. Explain how intensive human activities can dramatically alter the physical characteristics of places.

5. Describe ways in which people define, build, and name places and develop a sense of place.

6. Discuss why places are important to individual human identity and as symbols for unifying a society.

7. Understand how different groups in society may view places differently—depending on their life stage, gender, class, ethnicity, or values and belief systems.

8. Give examples of how humans view a single place from many perspectives—for example, as a cultural center, source of an important resource, political trouble spot, or origin point of a desired product.

9. Give examples of how the significance and meaning of places change over time.

10. Realize that places can be damaged, destroyed, or improved through human actions or natural processes.

coming, of a newer sensibility in regard to nature. This sensibility would see the human enterprise against the scale of biological history which is the matrix of our being, cultivating a morality of respect for all species and extending our sense of membership into the vast inorganic territories, even beyond that.

Human beings, being finite creatures, are still basically local, and are members of specific webs of exchange. A child bonds to a landscape. Even the farthest ranging migratory birds follow known paths. Our relationship to nature is a relationship to place. Yet the idea of "membership in place" is ignored, scorned, and even actively opposed by both capitalist and socialist managers and thinkers as economically and socially backward.

This kind of thinking and writing (about personal commitments to places) can actualize a concern that leaps over to include the whole planet, and of course, the human populations included in that. But it does not necessarily map its territories by the boundaries of national states—it may visualize the world as a mosaic of natural regions, the zones of different lives.

A literature of place can also help jog the occasional recollection—for each of us at any moment—that the whole universe is right here, right now.

In a message to readers of this book, Snyder states:

Each of us is a small contingent creature with ancestors, family, community, and place. The place is really a part of the larger community—it is a watershed, a big family of plants, birds, and animals, a configuration of flats or slopes—it is the territory in and on which we live. Landscape is one meaning of, an extension of, the mandala of being. If we think of ourselves as separate from it, we delude ourselves.

One of the ways to celebrate, and also make clear the nature of a place and our being in it, can be seen from songs, stories, and poems in the world. They speak a double language—the flowers or birds that are evoked are real flowers or birds, but they may also be a metaphor for a state of mind, a symbol of the spiritual condition. These ancient strategies of literature are not irrational or inaccurate, but are grounded on a fabric of shared existence. It is only a sign of our times that people must now reach for place, must reach for an image of bird or plant. That means that, at this time, the artist plays not only a forward or *avant-garde* role, but a very traditional and archaic role, that of conservator of planetary heritage.

KEY CONCEPTS

Exploration	Perception	Sense of place	Temporal-Spatial continuum
Mental map	Personal space	Sequent occupance	Terra incognita
Neighborhood	Planning	Symbolic landscape	Territoriality

FURTHER READING

The literature about specific places is as unlimited as the nature of places and place itself. The following works are only samples of what is available to those interested in the unique character of the landscapes we create and inhabit.

Descriptions of place by early explorers provide fascinating insights into their images and perceptions of new discoveries. Geoffery J. Martin and Preston E. James, *All Possible Worlds: A History of Geographical Ideas* (New York: John Wiley, 1993)

chronicles the role of early geographers in building a comparative understanding of known and newly discovered places. Their bibliography is particularly useful in locating the actual sources of early descriptions.

The standard work covering geographical exploration from Greeks to modern times is John N. L. Baker, *A History of Geographical Discovery and Exploration* (London, 1931; reprinted New York: Cooper Square, 1967).

Daniel J. Boorstin, in *The Discoverers: A History of Man's Search to Know His World and Himself* (New York: Random House, 1983) places the importance of early geographical discoveries and descriptions into the context of the development of present culture and civilization. *The Mapmakers: The Story of Great Pioneers in Cartography from Antiquity to the Space Age* by John Noble Wilford (New York: Alfred A. Knopf, 1981) demonstrates how early cartographers combined limited information with varying quantities of misinformation, or no information at all, to create maps of how places were thought to be rather than how they actually were.

A clear statement about living in a place and the need for a personal sense of place is Edward C. Relph's *Place and Placelessness* (London: Pion, 1984). Another readable and comprehensive look at the theme of place is Kay Anderson and Fay Gale's *Inventing Places* (Melbourne: Longman Chesire, 1992).

One of the most fascinating recent books discussing the blending of humanistic and quantifiable aspects of place is J. Nicholas Entrikin's *The Betweenness of Place: Toward a Geography of Modernity* (Baltimore, MD: The Johns Hopkins University Press, 1991). An earlier contribution by Entrikin, presenting a framework for analyzing the changed qualities of place, is *The Characterization of Place* (Worcester, MA: Clark University Press, 1991).

The study of place as experience is rich with the contributions of Kevin Lynch, Peter Alexander, and Yi-Fu Tuan. Lynch's two best-known works on this topic are *The Image of the City* (1960) and *What Time Is This Place?* (1972), both published in Cambridge, Massachusetts, by MIT Press. Alexander's most popular work, *A Pattern of Language* (New York: Oxford University Press, 1977), suggests that places and the physical patterns they create should be keyed to the social and biological living patterns of their inhabitants, namely people in communities. Yi-Fu Tuan is best known for relating environment to individual human behaviors and feelings. Particularly recommended is *Topophilia* (Englewood Cliffs, NJ: Prentice-Hall, 1974) and *Space and Place: The Perspective of Experience* (Minneapolis: University of Minnesota, 1977).

The experiencing of place has definite gender, socioeconomic, ethnic, and racial dimensions. Gender-based images of a dramatic regional landscape, the American Southwest, are perhaps nowhere more effectively grasped and portrayed than in Vera Norwood and Janice Monk, eds., *The Desert is No Lady* (New Haven, CT: Yale University Press, 1987). The role of human's place in nature is explored in Gary Snyder's *The Practice of the Wild* (San Francisco: North Point Press, 1990).

Personal space is analyzed in a classic work by Edward T. Hall, *The Hidden Dimension* (Garden City, NY: Doubleday, 1966). Somewhat related is a formerly popular work *The Territorial Imperative: A Personal Inquiry into the Animal Origins of Property and Nations,* by Robert Ardrey (New York: Atheneum, 1966).

Excellent references about mental maps and the behavior that springs from them include David Lowenthal and Martyn J. Bowden, eds., *Geographies of the Mind* (New York: Oxford University Press, 1976); Roger M. Downs and David Stea, *Maps in Minds: Reflections on Cognitive Mapping* (New York: Harper & Row, 1977); and Peter Gould with Rodney White, *Mental Maps* (Harmondsworth, Middlesex: Penguin Books, 1974).

Homes, neighborhoods, and special places are analyzed and described in David Ley's text for urban social geography, *A Social Geography of the City* (New York: Harper & Row, 1983). Other recommended references dealing with space and behavior, particularly in urban environments, are Robert Sommer's *Personal Space: The Behavioral Basis of Design* (Englewood Cliffs, NJ: Prentice-Hall, 1969), and John Jakle, Stanley Brunn, and Curtis C. Roseman, *Human Spatial Behavior* (Prospect Heights, IL: Waveland Press, 1987).

The importance of the theme of place in land use planning is discussed in Thomas F. Saarinen, David Seamon, and James L. Shell, eds., *Environmental Perception and Behavior: An Inventory and Prospect* (Chicago: University of Chicago, Department of Geography, 1984), and Kevin Lynch and Gary Hack, *Site Planning,* 3rd ed. (Cambridge, MA: MIT Press, 1984).

The nature of fictional places in the study of geography may be reviewed by referring to Christopher Salter and William Lloyd, *Landscape and Literature* (Washington, DC: Association of American Geographers, 1976), and two recent publications from Syracuse University Press: William Mallory and Paul Simpson-Housley, eds., *Geography and Literature: A Meeting of the Disciplines* (1987) and Leonard Lutwack, *The Role of Place in Literature* (1984).

Brief descriptions of unique places in North America were compiled for the International Geographical Union meetings in 1993 and published in *Geographical Snapshots of North America,* edited by Donald G. Janelle (New York: Guilford Press, 1992). A second book emphasizing the theme of place in North America is Michael P. Conzen, ed., *The Making of the American Landscape* (New York and London: Routledge, 1990).

═══════════════════ **ENDNOTES** ═══════════════════

[1] John L. Allen, "Lands of Myth, Waters of Wonder: The Place of the Imagination in the History of Geographical Exploration," in David Lowenthal and Martyn Bowden, eds., *Geographies of the Mind* (New York: Oxford University Press, 1975): 43.

[2] John K. Wright, quoted in Martin and James's *All Possible Worlds: A History of Geographical Ideas* (New York: John Wiley, (1993): 63.

[3] Martin and James, *All Possible Worlds: A History of Geographical Ideas* (New York: John Wiley, 1993): 20, 44.

[4] Edward Hall, *The Hidden Dimension* (Garden City, NY: Doubleday, 1966): 126–127.

[5] Yi-Fu Tuan, *Space and Place: The Perspective of Experience* (Minneapolis: University of Minnesota Press, 1977): 154. Reprinted with permission.

[6] David Ley, A *Social Geography of the City* (New York: Harper & Row, 1983): 133. Reprinted with permission.

[7] Jan O. M. Broek, *Geography: Its Scope and Spirit* (Columbus, OH: Merrill, 1965): 6. Reprinted with permission.

[8] Yi-Fu Tuan, "Place—An Experiential Perspective," *Geographical Review* 65 (1975): 151–165.

[9] Robert Sommer, *Personal Space: The Behavioral Basis of Design* (Englewood Cliffs, NJ: Prentice-Hall, 1969): 118.

[10] Kevin Lynch and Gary Hack, *Site Planning* (Cambridge, MA: MIT Press, 1984): 1.

[11] Jack London, *The Valley of the Moon* (Santa Barbara, CA: Peregrine Smith Books, 1978).

[12] Willa Cather, *O Pioneers* (Boston: Houghton Mifflin, 1913): 76. Reprinted with permission.

[13] Charles Dickens, *Oliver Twist* (New York: Dodd, Mead, 1941): 338–339. Reprinted with permission.

[14] Pearl S. Buck, *The Good Earth* (New York: Modern Library, 1944).

[15] Amy Tan, *The Joy Luck Club* (New York: G. P. Putnam's Sons, 1989): 21. Reprinted with permission.

[16] See, for example, James Michener, *Hawaii* (1959); *Centennial* (1974); *Chesapeake* (1978); *Texas* (1985), and *Alaska* (1988); all published by Random House.

[17] Edward Abbey, *The Monkey Wrench Gang* (New York: Avon Books, 1976).

[18] John Steinbeck, *Flight, A Story* (Covelo, CA: Yolla Bolly Press, 1984); George R. Stewart, *Earth Abides* (Los Altos, CA: Hermes, 1974); and *Storm* (New York: Random House, 1941).

[19] Austin Tappan Wright, *Islandia* (New York: Holt, Rinehart & Winston, 1958).

[20] Jim Crace, *Continent* (New York: Harper & Row, 1987).

[21] Charlotte Perkins Gilman, *Herland* (New York: Pantheon Books, 1979).

[22] Leonard Lutwack, *The Role of Place in Literature* (Syracuse, NY: Syracuse University Press, 1984).

[23] Yi-Fu Tuan, *Space and Place: The Perspective of Experience* (Minneapolis: University of Minnesota Press, 1977): 54.

[24] D. W. Meinig, "Symbolic Landscapes," in *The Interpretation of Ordinary Landscapes:* (New York and Oxford: Oxford University Press, 1979): 164–192.

5 THE THEME OF PLACE: PHYSICAL SYSTEMS I

Permafrost sign in northern Alberta, Canada. Physical geography explained to park visitors in northern Alberta, Canada. (Photo by author.)

In the preceding chapter the essence of a place is described from a *humanistic* point of view. It was noted there and is amplified here that place is also established by *physical features and processes*. These are studied by physical geographers in four interrelated realms:

- The atmosphere (weather; climate)
- The hydrosphere (water; the oceans)

- The lithosphere (the solid surface, with its landforms, soils, and geomorphic processes)
- The biosphere (the living planet, with its plant and animal distributions; the environment)

These four "spheres" are naturally occurring but are influenced more and more by human activity. In this chapter we turn our attention to Earth's atmosphere and hydrosphere. The following chapter examines the lithosphere and biosphere. Both chapters on physical geography focus on content contained in the following *National Geography Standard:*

> Standard 7: *The geographically informed person knows and understands the physical processes that shape the features on Earth's surface.*

The features of each place we study are a result of past and present processes. The old idea of an "abiding Earth" is no longer a generally accepted viewpoint of geographic thought. Energy continually flows through Earth environments, ensuring that Earth is dynamic and ever-changing. Understanding these changes at many levels of scale and generalization creates the unique perspective of the geographer.

THE SYSTEMS APPROACH

To study our dynamic planet, a framework is needed in which to examine complex interrelationships, to integrate information, and to craft descriptive models. The concept of **system** provides such a framework. A system is an organized collection of components that function as an integrated whole. It is inherently impossible to understand a system by only examining its constituent parts. The key to systems analysis is establishing the **linkages** among constituent components. The flow of energy in a system is often studied to determine these linkages.

You probably already are familiar with the concept of system. If you drove a car today, you relied on the workings of a propulsion system under the hood of your vehicle. Alternatively, you may have been a passive participant in a "mass transit system" or utilized the controlled paths of a system of streets and sidewalks. Your digestive system regularly reminds you of its status. Although we usually are conscious only of system components that require attention, we participate in systems throughout each day of our lives.

The various component parts of a system function together as a whole. Using the examples listed above, it is obvious that a system's various component parts are interrelated. Inside your car the carburetor works in conjunction with the fuel pump and control signals from the driver. Your digestive system is subject to malfunction when the interlinked organs of the body fail to coordinate operations. Systems analysis seeks to define the limits of system com-

ponents, to determine their internal structure, and to evaluate energy sources and flow.

Types of Systems

To grasp the system concept, visualize a system as having some type of **boundary**. Systems are classified according to their relationship with items outside their boundary. The three classifications are the following:

Isolated systems have no connection beyond their boundaries; that is, the boundaries are impermeable. (Isolated systems are not found in nature and thus are not discussed in our study of physical geography.)

Closed systems exchange matter across their boundaries but do not exchange energy. As there is a fixed amount of energy to operate the system, the results are predictable: closed systems tend to shut down eventually as their energy source becomes exhausted. (Although closed systems are rare on Earth, it is vital to understand the difference between closed and open systems when attempting to understand complex environmental relationships.)

Open systems comprise essentially all systems on Earth. Solar energy passes through the bounding atmospheric envelope of our planet. The Earth system is open with respect to energy, although essentially closed with respect to matter. An automobile is an open system dependent on the merging of energy (fuel) and oxidant (air) in the carburetor. Open systems characteristically involve energy exchange across their boundaries. In many cases, both energy and matter flow across boundaries (Figure 5-1).

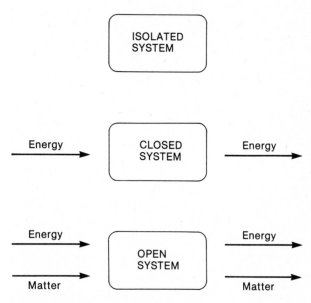

FIGURE 5-1

The three types of systems: isolated, closed, and open.

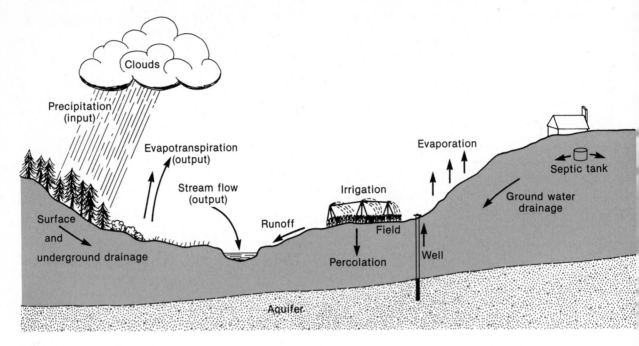

FIGURE 5-2
Simplified model of water movement in a small basin.

Models

Now let us make this systems approach fit the real world. To do so we will use the concept of models and the technique of model building. In Chapter 3 you were introduced to simple models as abstractions of reality. Similar ideas of abstracting and simplifying variables may be applied to physical systems.

Static models—those lacking change—may be simple to develop. Other models are dynamic and therefore more complex; they rely on flow charts, quantitative analysis, or simulation to describe a system's behavior. To be useful, models must correctly predict system behavior. This is particularly important because models are used to project system changes when a variable is altered. What will happen to the other variables? How does the system change?

Figure 5-2 depicts the variables of water movement in a hydrologic basin. Try to visualize how change in one variable progressively ripples through the system (see the box on page 85.)

A geographic system may be based on map layers (layers of spatial data) that combine to establish system boundaries and functions. Major physical systems include climate,

vegetation, soil, topography, and organisms. The purpose of this chapter and Chapter 6 is to demonstrate how interrelated elements of systems on Earth create the physical

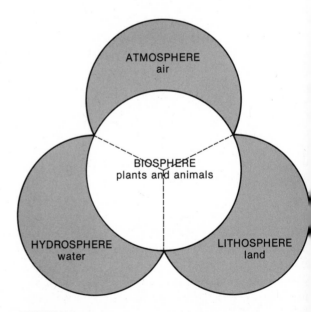

FIGURE 5-3
The four Earth geosystems interconnected.

CLIMATE MODELING

How can human societies prepare for an uncertain climatic future? Present and future climates can be simulated mathematically with the help of high-speed computers. Such mathematical models have been used to simulate present climates and are helping to explain the evolution of past climates, including those of the ice ages. The accuracy of these paleoclimatic simulations, in turn, has built confidence within the workers who employ these same models that predicting future climates will be possible. An example of an extremely simple model is one that calculates the average temperature of Earth as an energy balance arising from Earth's average reflectivity and the average greenhouse properties of the atmosphere. More sophisticated general circulation models predict the evolution not only of temperature but also of humidity, wind speed, wind direction, soil moisture, and other variables.

But even the most complex general circulation model varies sharply in the amount of spatial detail it can resolve. No computer is fast enough to calculate climatic variables everywhere on the Earth's surface and in the atmosphere in a reasonable length of time. In addition, the low spatial density of the calculations is a primary cause of error in the models.

Calculations in a model are executed at widely spaced points that form a three-dimensional grid at and above the surface. To fully simulate a climate, a model must take into account the complex feedback mechanisms that influence it. Snow, for example, has a destabilizing positive feedback effect on temperature. When a cold snap brings a snowfall, the temperature tends to drop even further than might be expected because snow, being highly reflective, absorbs less solar energy than bare ground.

Using a climate model verified by fossil evidence, it is known that in the middle of the Cretaceous Period about 100 million years ago broad-leafed tropical plants grew in the mid-latitudes. Alligators lived near the Arctic Circle, which, like the Antarctic, was free of permanent ice.

Work is presently being conducted to simulate the global warming and other climatic effects induced by the enhanced greenhouse effect (that is, the effects of increased carbon dioxide emissions into the atmosphere). Soon one will be able to state with more confidence how the impacts of rising levels of greenhouse gases might be distributed. Until then, only circumstantial evidence indicates how these impacts are likely to be significant. The Earth already is more than 0.5°C warmer than it was a century ago. In addition, efforts to model the comparatively short-term climatic effects of a nuclear war are now the subject of climatic simulation studies.

landscape. A flowchart or a symbolic model often is used to depict the interrelationships involved.

Earth's systems are subdivided by physical geographers into the study of the water, land, air, and living organisms. These are termed the **hydrosphere,** the **lithosphere,** the **atmosphere,** and the **biosphere** (Figure 5-3). Although these four major geographic systems are intrinsically interrelated, we will examine each one separately.

THE SUN: EARTH'S ENERGY MACHINE

Earth and its four physical subsystems are driven primarily by solar energy. The initial distribution of solar energy on the planet, followed by its transformation and redistribution in the atmosphere, lithosphere, and hydrosphere, create markedly varied physical landscapes over Earth's sur-

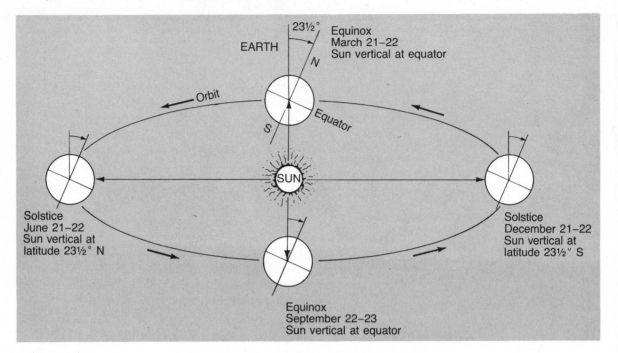

FIGURE 5-4
Earth-sun relationships and seasonal patterns. (From E. Willard Miller, Physical Geography: Earth
Systems and Human Interactions, *Columbus: Merrill Publishing Co., 1985, Fig. 2-3, p. 11.*
Reprinted with permission of Merrill Publishing Co.)

face. As the sun is the primary energy source for Earth, it is
logical to start our investigation of the physical patterns and
processes on Earth by examining its relationship to the sun.

The Energy System

Earth is the third planet of a system of moving bodies that
are held together by the gravitational attraction of the central sun. Two primary motions of Earth affect us all.
Rotation of Earth on its axis results in the diurnal temperature changes of day and night. The **revolution** of Earth
around the sun results in larger-magnitude seasonal changes
during the course of a year.

In Figure 5-4 it is evident that Earth follows a slightly
elliptical orbit around the sun, which causes its distance
from the sun to vary over the year. The changing distance
from the sun, however, results only in minor variations (7
percent) in the amount of energy falling onto Earth. It is
not the elliptical orbit that causes our seasonal changes.

Rotation

Rotation is the spinning of Earth on its own axis, like a top.
The axis is an imaginary line extending from pole to pole.
When viewed from space at a point above the North Pole,

Earth is spinning in a counterclockwise direction. The sun
therefore rises in the east and sets in the west. (Any point
on the surface of Earth is always moving eastward).

The angular velocity of rotation is 360° in a 24-hour
period. This defines the average length of the terrestrial
day. We never feel movement as the planet rotates because
we are moving at the same speed. We know that movement occurs because of the passage of day and night. Earth
is spherical, so half of its surface is always in daylight and
half is in darkness. The dividing line between day and
night defines the *Circle of Illumination.* Actual day lengths
vary over the course of a year because of the tilt of the axis of rotation.

Daily rotation of Earth has profound effects upon the
planet's physical environmental systems. The lives of plants
and animals are governed by the diurnal swing of time.
Night and day bring characteristic levels of light, temperature, humidity, and air motion that regulate processes in
the life layer. The "jet lag" experienced by travelers is evidence of the strong tie of human biorhythms to the diurnal cycle of light.

The distance "traveled" by a point on the surface during a rotation, and the speed of travel, depend upon your
location on Earth. If you are standing on one of the poles,

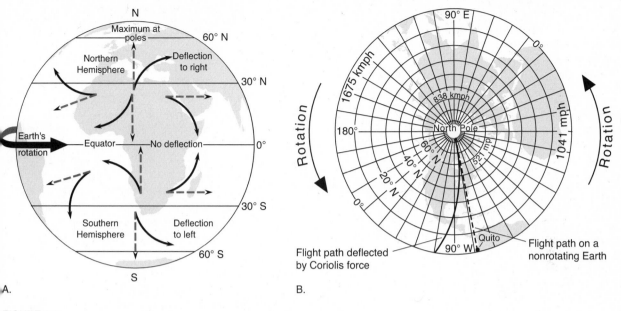

FIGURE 5-5

Distribution of the Coriolis force on Earth: A. apparent deflection to the right of a straight line in the Northern Hemisphere; apparent deflection to the left in the Southern Hemisphere.   B. Coriolis deflection of a flight path between the North Pole and Quito, Ecuador. Note the latitudinal variations in speed of rotation. (From Robert W. Christopherson, Geosystems: An Introduction to Physical Geography, *2nd ed. New York: Macmillan, 1994, Fig. 6-7, p. 153. Used with permission of Macmillan Publishing Co.)*

you are not traveling at all—just being rotated once in 24 hours. However, the closer you are to the Equator, the faster is your rotation speed, and the distance traveled. If you stand on the Equator, your average speed is about 1000 miles an hour (1600 km/hr), and your distance traveled per day is about 25,000 miles (40,000 km)(Figure 5-5).

Again, being fixed to Earth's surface, we do not feel this rotational speed. It does substantially affect winds by deflecting them to the right (clockwise) in the northern hemisphere and to the left (counterclockwise) in the southern hemisphere. This fascinating phenomenon, known as the **Coriolis effect,** is portrayed in Figure 5-5. Rotational influences on movement are discussed in detail later in this chapter as the predictable patterns of winds and ocean currents are outlined.

Revolution

Earth moves—revolves—around the sun on the plane of the ecliptic (Figure 5-4). We make one trip around the sun in approximately 365.25 days, moving through space at a speed of about 18.5 miles per second (29.8 km/sec). This single revolution around the sun determines our calendar year. (The extra quarter of a day beyond our rounded-off 365-day year is accumulated for four years, creating an extra day—February 29th—during "leap year.")

Seasons on Earth

Note that the axis of Earth is not perpendicular ("straight up-and-down") to the plane of the ecliptic but is inclined at an angle of 23.5° from verticality. **Seasons** occur on Earth because its axis of rotation maintains this constant 23.5° tilt in space as our planet revolves around the sun (Figure 5-4). The north end of the axis always points toward the North Star and the south end toward the Southern Cross.

Imagine for a moment that the axis of rotation was changed so as to be perpendicular to the plane of the ecliptic. If this happened, the Equator would always be located exactly on the plane of the ecliptic, the sun's rays would most directly strike Earth on the Equator, and the length of daylight would be unchanging over the course of a year for all latitudes.

Earth's axis, however, is not perpendicular to the plane of the ecliptic, and it always tilts 23.5° from vertical. (Viewed another way, the Earth's axis at all times is tilted 66.5° from the plane of the ecliptic—Figure 5-4). Only on two days out of each year is the sun perfectly overhead at noon on the Equator: about March 21st (the first day of spring in the northern hemisphere) and about September 21st (the first day of autumn).

On June 22nd solstice (the first day of the northern hemisphere's summer), the North Pole is tilted toward the sun. By the December 22nd solstice (the first day of the northern hemisphere's winter), Earth has revolved halfway around the sun, and the North Pole now is tilted away from the sun.

It is this axial tilt that causes seasons on Earth. When the northern hemisphere is tilted toward the sun, solar energy more directly hits the surface (the angle of the sunlight as it impacts upon Earth is closer to 90° at noon), and the length of the daylight period is longer. This more direct and longer period of heating results in warmer temperatures and our summer season.

Consider again the 23.5° tilt from vertical of Earth's axis. At the June **solstice,** when the northern hemisphere is tilted the maximum 23.5° toward the sun, the area where the sun's rays strike Earth perpendicularly has shifted to its most northerly point of the year, the Tropic of Cancer (23.5° N). When Earth moves to the opposite position in the orbit (the December solstice), the rays are most vertical at the most southerly point, the Tropic of Capricorn (23.5° S).

Light falling vertically is the most intense and direct; oblique light heats less effectively. Due to the axial tilt of Earth, light and heat radiation from the sun are vertical only at one latitude at a time. As Earth revolves in its orbit around the sun each year, the location of these vertical rays moves between the Tropic of Cancer and the Tropic of Capricorn. The sun's "vertical" rays never strike any part of Earth poleward of these parallels (north of the Tropic of Cancer or south of the Tropic of Capricorn).

Using Figure 5-4, let us review all of this by examining four critical positions on Earth and their relationship to the sun's rays. First, on June 21st or 22nd, the northern hemisphere directly faces the sun. On this date, vertical rays strike the Tropic of Cancer and summer officially begins in the entire hemisphere. This position and date are known as the *June solstice* (*summer* solstice for the northern hemisphere; *winter* solstice for the southern hemisphere).

Six months later (on December 21st or 22nd) the southern hemisphere is tilted so that it directly faces the sun and its rays vertically strike the Tropic of Capricorn. The southern hemisphere thus begins its *summer* season in December, and the northern hemisphere its *winter.* This position of Earth in relation to the sun is called the *December solstice.*

Halfway between each of the solstice positions in orbit, neither the northern hemisphere nor the southern hemisphere are inclined toward the sun. At these positions neither hemisphere receives the vertical rays of the sun; they fall instead at the Equator. These intermediate positions bring equal periods of day and night to every latitude on Earth and are thus known as **equinoxes** ("equal night"). On September 22nd or 23rd the northern hemisphere experiences the *autumnal equinox* and on March 22nd or 23rd the vernal equinox, and vice versa in the southern hemisphere.

Declination is the angle between the Equator and the latitude at which the sun's rays fall at 90°. Therefore, the declination value varies through the year from +23.5 (June solstice) to −23.5 (December solstice). The specific value for each day is available in the astronomical tables of an almanac.

It was knowledge of the relationship between declination and latitude that allowed early seafarers to determine their latitude at noon if they knew the position for the sun on that date. Measuring the solar altitude at noon and combining this value with the declination allowed determination of latitude. (Longitude determinations are far more complex and awaited the invention of chronometers.)

Altitude and **latitude** are important concepts as we begin to explore the relationship between the sun's energizing rays and the surface of Earth. This relationship provides a foundation for analysis of another system, that of the Earth-atmosphere interface. Analyzing this system will explain energy distribution patterns on Earth.

THE EARTH-ATMOSPHERE SYSTEM

Earlier in this chapter we defined a closed system as one that makes no exchanges across system boundaries. In our analysis of the relationship between Earth's surface and the atmosphere, we shall consider this system closed with respect to matter and open with respect to energy. Much energy moves freely across the boundary in each direction, while the boundary is mostly impermeable with respect to mass. (This is a simplification, because meteorites, spacecraft, and other types of matter do move across the atmospheric boundary between space and Earth. In fact, the difference between energy and matter, according to Einstein's famous $E = mc^2$, is simply a matter of perspective!)

The sun is the major source of energy that drives the

entire Earth system. This energy comes as various waves, including the ultraviolet, visible, and infrared parts of the electromagnetic spectrum. The average amount of this electromagnetic energy received at the outer edge of the atmospheric boundary during a unit of time is known as the *solar constant.* Space satellites have continually recorded the magnitude of the solar constant and have charted minor fluctuations in the sun's energy output. Sensors on board orbiting spacecraft also have photographed much of Earth's surface using film sensitive to both visible and invisible (infrared) reflected light.

Incoming solar radiation (**insolation**) is called *shortwave* radiation. Earth both reflects some shortwave radiation and emits thermal radiation of its own. This emission of thermal radiation is Earth's cooling mechanism. Without it the absorption of shortwave radiation would quickly heat the surface to the melting point! The emitted energy is at longer wavelengths and consequently is referred to as *longwave* radiation.

Earth constantly emits longwave radiation, 24 hours per day. Very little incoming solar energy is absorbed by the atmosphere. Most passes through the transparent atmosphere, is absorbed at the surface, and is reradiated as longwave radiation. It is by absorption of this longwave radiation that the atmosphere is heated. (The atmosphere is heated from below by radiation from Earth, not by the sun, which explains the decrease in air temperature as we rise into the atmosphere). The Earth-atmosphere system therefore experiences a *cascade* of solar energy through transmission, reflection, absorption, and reradiation as it passes into and out of the Earth system (Figure 5-6).

The percentage of shortwaves reflected from the surface of Earth is known as **albedo.** Albedo values vary with surface types. A white, snow-covered surface has a much higher albedo rate (65–90 percent) than a darker, plowed field (5 percent)—see Figure 5-7. The albedo of a water surface is very low during times of overhead sun (2 percent) but is high in the morning and evening—consider the time of day at which glare is worst at a beach. The photos in Figure 5-7 illustrate places having very different albedo rates.

Solar energy absorbed at Earth's surface is available to drive Earth processes. The surface of the planet exchanges this energy along different pathways. As energy is converted into other forms it drives biologic, atmospheric, and geologic processes.

A small portion of emitted longwave radiation, about 7 percent, escapes directly through the atmosphere back into space. The wavelengths at which this loss occurs are known as the "atmospheric window" (Figure 5-6).

The Energy Balance of the Atmospheric Subsystem

It is important to remember that a system has both inputs and outputs. As you balance the budget in your checkbook, you add and subtract to keep the system in balance. The overall temperature of Earth does not vary from year to year; therefore the inputs of shortwave solar radiation are equal to the outputs of longwave radiation.

The Energy Balance of the Earth's Surface Subsystem

As you can surmise, and know from personal experience, the Earth-atmosphere system is in constant flux. As shortwave energy from the sun enters the system, longwave radiation from Earth leaves the system. In the atmosphere, energy is moved by absorption, reflection, scattering, and transmission (of solar energy) to the surface of Earth. At this surface, absorption, reflection, and conversion to longwave emission work to balance the system.

Energy Transfer

There are five ways in which energy can be transferred from one place to another: conduction, convection, latency, advection, and photosynthesis (chemical).

Conduction is the transfer of heat from molecule to molecule; no mass is moved. Conduction actually plays a small part in global heat exchange because molecule-to-molecule exchanges are ineffective in gases (air is a good insulator). Conduction of energy into soil or water reservoirs may be important in local microclimates.

Convection is the process of transferring heat within the entire mass. Convection is of great importance in air and water because this type of movement occurs easily in a fluid (air, though a gas, behaves like a fluid). Temperature-related density differences drive convection currents as warm air rises and is replaced by sinking cooler air. Thunderstorms and showers are excellent examples of the powerful energy transfer accomplished in convectional storms.

Latency is the most important transport mechanism. Water, as it is evaporated, stores the heat used in evaporation as latent energy. (The word *latent* means "hidden.") This latent energy is transferred as the water vapor is moved in global wind systems. When the water vapor is condensed in a cloud, the latent heat of evaporation is released. This latent heat is the primary energy source of cyclonic storms, thunderstorms, and hurricanes. Clouds are a sign that energy has been moved from the ground into the atmosphere.

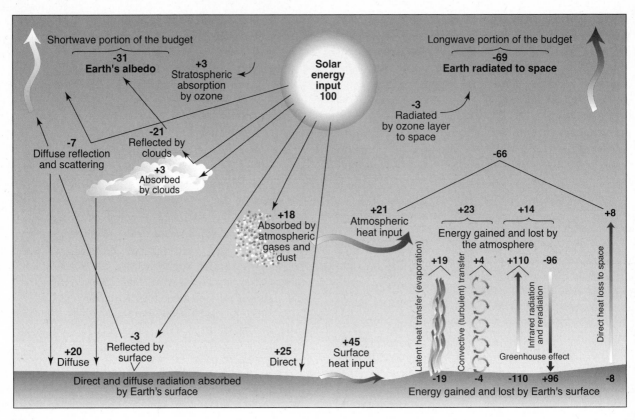

FIGURE 5-6
*Earth-atmosphere energy balance showing the distribution of 100 units of solar energy as it cas-
cades through the system. (From Robert W. Christopherson,* Geosystems: An Introduction to
Physical Geography, *2nd ed., New York: Macmillan, 1994, Fig. 4-5, p. 101. Used with permission
of Macmillan Publishing Co.)*

Advection is another significant method of heat transfer. Advection transfers heat by the horizontal movement of
air (wind) or water (ocean currents). Advection may occur
locally, as in land and sea breezes along your favorite beach,
or may be the movement of tropical air into the middle latitudes. Advection acts on scales from local to global. The
movement of wind is primarily to transport energy from areas of surplus (the Equator) to areas of deficit (the poles).
This is why the atmosphere is most vigorous during the
winter season.

Photosynthesis by plants converts only a small fraction
of incoming energy to chemical energy. Small though this
process may be in global energy terms, it is the foundation
of all life, however. We all appreciate its importance!

We know that an overall energy-budget balance exists
globally between incoming and outgoing energy. However,
during the course of a year a balance may not exist for any
specific location. Seasonal warming and cooling is characteristic of all nontropical climates. As these local differences

develop, heat is transferred from areas of energy surplus
(usually low-latitude equatorial regions) to areas of energy
deficiency (high-latitude regions). This poleward movement of energy is most noticeable in the middle latitudes
and will be examined later in this chapter.

In summary, the varying patterns of weather and climate we experience on Earth are the direct result of energy
transfers within the Earth-atmosphere system. This very
important fact should be kept in mind while reading the
following sections of this chapter.

Composition and Structure of the Atmosphere

Five gases dominate atmospheric processes: nitrogen, oxygen,
water vapor, carbon dioxide, and argon. These gases, along
with particles such as soot, dust, pollen, ice, and water
droplets, constitute the mass of the life-giving atmosphere.

Gases in the atmosphere are divided into two categories, those that are present in constant proportions and

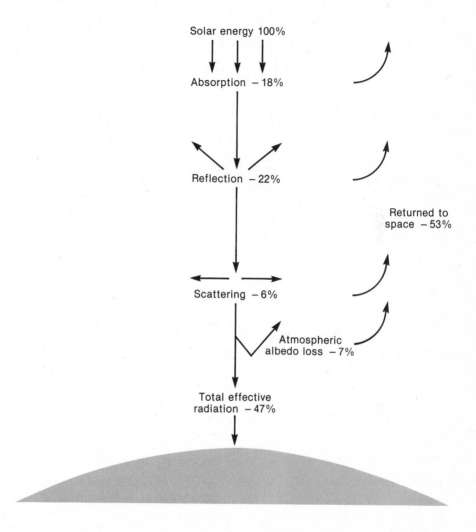

Solar energy 100%

Absorption − 18%

Reflection − 22%

Returned to
space − 53%

Scattering − 6%

Atmospheric
albedo loss − 7%

Total effective
radiation − 47%

A.

B.

C.

FIGURE 5-7

Albedo: a high-latitude forest, glacial surface, and plowed field each reflect/absorb heat at different rates. (Photos by author.)

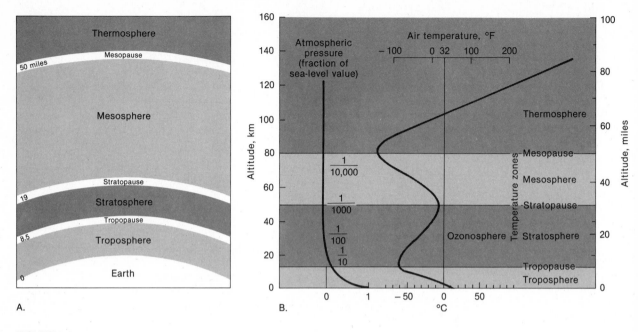

FIGURE 5-8
*A. Layers of Earth's atmosphere. B. Temperature graph of the atmosphere. (From Harm J. de Blij,
The Earth: A Physical and Human Geography, New York: John Wiley & Sons, 1987, Fig. 3-6, p. 98.
Reprinted with permission.)*

those that vary with time and place. Nonvariant gases include nitrogen (78 percent of the atmosphere), oxygen (21 percent), and the inert gas argon (0.93 percent). The most important variable gases are water vapor, carbon dioxide, and ozone.

Although the overall volume of water vapor in the air is small (about 0 to 4 percent), it is important because of its ability to transfer energy and its role in precipitation. Water vapor also selectively absorbs and reradiates longwave radiation, keeping the surface of Earth warm on humid nights.

Carbon dioxide (about 0.003 percent) is a vanishingly small percentage of the volume of air but is a vitally important regulator of life systems on Earth. A primary role of carbon dioxide is to absorb and reradiate longwave radiation (the **greenhouse effect**). Plants extract carbon dioxide from the atmosphere through photosynthesis to create the organic compounds of life and release oxygen as a byproduct.

Since the explosive growth of the population of humankind, carbon dioxide levels have been increasing due to deforestation and the burning of fossil fuels. Some scientists now believe that increased carbon dioxide levels are causing a warming of global climates. What is certain is that carbon dioxide levels in the atmosphere are increasingly a global concern.

Ozone—triatomic oxygen that is concentrated in the stratosphere at an altitude of 15 to 18 miles (24 to 29 km)—is a pollutant in the lower atmosphere that injures lungs, attacks plant leaves, and oxidizes many materials. However, it performs a vital function in the upper atmosphere: ozone is the gas that absorbs dangerous ultraviolet radiation as it passes into the atmospheric system from the sun. More information about the breakdown of the ozone layer is presented in Chapter 10.

The atmosphere can be divided into several layers, based on temperature characteristics (Figure 5-8). Most of us realize that the atmosphere is warmest near Earth's surface and gets cooler away from the surface. You may have heard an airline pilot remark on the cold temperatures at cruising altitudes. The term for the typical cooling with increasing altitude is *normal lapse rate*, which is −3.5 Fahrenheit degrees per 1000 feet gain in altitude. The measured lapse rate at a specific location is called the *environmental lapse rate*.

Temperature measurements using instrumented balloons or kites and satellite sensors show that cooling with height is not consistent. There are actually two zones of the atmosphere where the air is warmest—at the surface and in the ozone layer at about 31 miles elevation (50 km). Temperatures decrease with height in the lower zone, the *troposphere.* Almost all weather is restricted to the troposphere.

In the next zone, the *stratosphere,* ozone is present and the air is heated directly by the sun; the negative lapse rate of the troposphere reverses in the stratosphere. Above the stratosphere lies the *mesosphere,* and above this layer is the *thermosphere.* Still farther out (at about 435 miles or 700 km) is the low density *exosphere.*

Figure 5-8 illustrates the relationship of these layers of the atmosphere and their approximate distances from the surface of Earth. As almost 100 percent of all air lies in the troposphere and stratosphere, both weather and life are restricted to these layers.

Weather and Climate

Understanding the vertical structure of the atmosphere provides a foundation for understanding the workings of the Earth-atmosphere system. Most of us are concerned with the day-to-day events in the atmosphere. These **weather** phenomena describe the state of the atmosphere at a given place and moment, and predict conditions in the near future.

However, physical geographers also are concerned with long-range conditions near Earth's surface. This broader temporal view is known as **climate**. Climatologists have defined physical regions of Earth based on differing climatic parameters obtained from weather data that have been averaged over a long period. Climatologists deal primarily with four basic elements of climate—temperature, atmospheric pressure, wind, and precipitation. Each is discussed subsequently.

Temperature and Climatic Controls

The energy-balance equation for solar energy demands that a long-term overall balance be achieved, not that the energy be distributed evenly on the planet. Unequal heating and cooling is characteristic of Earth's surface and atmosphere. The need to transport energy, thereby balancing energy budgets, is the basic cause of daily changes in the atmosphere (weather) and seasonal variations (climate). When considering weather and climate, *always recall the importance of global energy transfers from surplus to deficit areas.*

In addition to the effect of sun angle on shortwave energy, it also is important to note that the duration of daylight affects daily heating in high latitudes. Other controls on temperature include altitude, the rate of heating of land and water surfaces, and the blocking of air masses by mountain barriers. (The influence of ocean currents on climate is presented elsewhere in this chapter.)

Latitude

Latitude is a primary control on climate, especially temperature. The seasonal pattern of insolation (incoming solar radiation) discussed earlier in this chapter is the heart of any analysis of a climatic system. Latitude influences this seasonal variation, which of course varies the temperature range. Latitude also determines the length of daylight in each 24-hour period.

In general, the farther a location is poleward from the Equator, either to the north or south, the less solar energy is received throughout the year. These areas are cooler than the Equatorial regions of energy surplus and have a larger annual range in temperature. Figure 5-9 shows the distribution of global mean net radiation at ground level on Earth.

Land and Water Differences

While latitude is a major determinant of global temperature zones, more local influences account for other important day-to-day and seasonal changes. After the intensity and duration of solar energy, the second most important climatic control is the type of surface receiving the radiation. Land surfaces have a low specific heat (the amount of energy required to raise the temperature 1 Celsius degree), do not allow much radiation to pass beneath the surface, and only slowly conduct energy. Land surfaces, therefore, both heat and cool rapidly.

If you have traveled in the desert you may have noticed the rapid ground heating after sunrise. Soils characteristically reflect heat quickly from the surface and cool rapidly after sundown. Continental climates, covering large land masses away from coasts, experience great temperature ranges both daily and seasonally when compared with maritime regions at the same latitude.

Water, on the other hand, has a high specific heat, allows light to penetrate beneath the surface, and transports heat by circulation. It therefore takes much longer to heat a water surface, and it takes longer for cooling to occur. Water heats to greater depths than land, and horizontal and vertical motions in the fluid ensure that extreme local heating does not occur.

Acting as a thermal reservoir, water acts as a heat sink during warm summer months and a heat source during cool winters.

This causes nearby land areas to experience reduced temperature ranges on a daily and seasonal basis. Maritime climates tend to be more moderate than continental climates, reflecting the greater constancy of their energy budget.

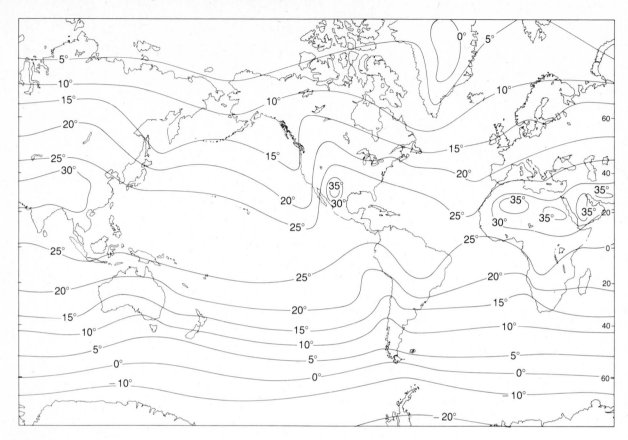

FIGURE 5-9

Distribution of global solar radiation in July. (From Edward J. Tarbuck and Frederick K. Lutgens, Earth Science, 5th ed. Columbus: Merrill Publishing Co., 1988, Fig. 12-15, p. 352. Used with permission of Merrill Publishing Co.)

It is interesting to note the differences between maritime and continental climates by comparing coastal and continental locations at the same latitude. Nice, France (on the Riviera) experiences none of the bitter winters of Mankato, Minnesota (both are near 44° N). Nice, located on the Mediterranean Sea, has a maritime climate. Mankato, located in the middle of a large land mass, has a continental climate. Another example might compare San Francisco, St. Louis, and Washington, DC (all near 38° to 39° N).

Examine the world climatic map shown in Figure 5-22 and compare the climates of Western Europe (maritime) and western Siberia (continental) at the same latitude. Remember that the direction of the prevailing winds is another important factor in determining maritime influences. This map of world climatic zones illustrates the wide variety of climatic patterns on Earth. Our analysis of Earth's circulation patterns later in this chapter may help you understand these patterns more completely.

Altitude and Mountain Barriers

Topographic features are another important control on climate and temperature variations on Earth. The cooling felt as you travel up a mountain reflects the reduction of counter-radiation by the rapidly thinning atmosphere. Another influence of topography is the effect that mountains have as barriers to air movement. This influence on climate is especially noticeable for ranges that run north-south in the middle latitudes, such as the Rocky Mountains in North America and the Andes Mountains in South America. These ranges effectively block the prevailing winds blowing from west to east.

One result is that locations on the windward side will be moister and cooler than corresponding locations on the leeward side of the range. This can be seen by comparing the graphs of annual temperatures at Monterey and Death Valley, California, both at the same latitude but on opposite sides of the Sierra Nevada Mountains (see Figure 5-10).

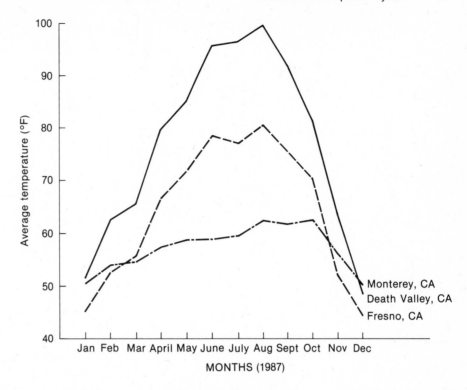

FIGURE 5-10
Average monthly temperatures at Monterey, Fresno, and Death Valley, California—all at approximately the same latitude (about 36.5°N).

The climatic controls described above are important factors in determining local, regional, and global temperature variations. Earth's surface and the atmosphere are parts of an interconnected system, so mechanisms for balancing energy budgets at all scales are necessary. A further mechanism is horizontal air transfer by wind, due to pressure.

To understand the significance of wind, it is necessary to examine its origin: pressure differences resulting from patterns of heating and cooling, or from dynamic need to transport energy poleward (these pressure systems are not related to surface temperature).

Winds and Pressure

The force exerted by air on a unit area of Earth surface is known as **atmospheric pressure.** Pressure differences at the surface reflect whether the air column is slowly rising or descending. These vertical motions often reflect the temperature of the air column and are measured with a *barometer.* Barometric measurements indicate that a "normal atmosphere" at sea level pushes with a force of 14.7 pounds per square inch (1 million dynes/cm²). Air is a compressible fluid; the pull of gravity causes it to be most dense near the ground surface and for pressure to decrease rapidly with altitude. Mountain climbers often experience oxygen deprivation as they attempt to climb in the "thinner" (lower pressure) air of high altitudes.

Small temperature changes cause large density changes in air, resulting in large pressure changes. A hot-air balloon encloses air having an elevated temperature and therefore a low density (low pressure); the balloon rises until the internal pressure of the balloon matches that of the surrounding air. The vertical movement of air columns is relatively slow, however, in comparison with horizontal motions.

Horizontal gradients from high to low pressure at the surface (or at any altitude) result in air moving directly toward the low pressure, seeking to eliminate the pressure gradient. Such a **pressure gradient** is characteristic of large geographic areas and results in air movement from "highs" into "lows" (Figure 5-11). We feel this horizontal air motion as the wind. Winds are measured by an **anemometer.**

Areas of high and low pressure vary over time and space; the resulting wind changes are an important part of our daily weather patterns. However, the overall effect of airflow along pressure gradients is a distinct poleward movement of energy and a balancing of local energy budgets. On a global scale this airflow results in the subsidiary motion of ocean currents.

Pressure is represented on weather maps by lines of equal pressure called **isobars.** Isobars for two days in January and July are shown in Figure 5-12. Seasonal differences in pressure occur as isobars shift north and south

FIGURE 5-11

A pressure gradient is established when air flows from a high to a low. It is felt as wind. (From Robert W. Christopherson, Geosystems: An Introduction to Physical Geography, *2nd ed., New York: Macmillan, 1994, Fig. 6-5, p. 151. Used with permission of Macmillan Publishing Co.)*

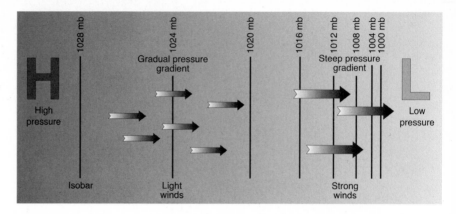

with the apparent migration of the sun from Tropic to Tropic.

Two general types of pressure systems may be identified. Several semipermanent areas of high and low pressure exist that control global climates, global wind patterns, and dominate the troposphere. These *pressure belts* are predictable and can be mapped on a global scale. The other type of pressure system is a *cell,* a more local condition that can be seen on daily weather maps in your local newspaper. These smaller-scale pressure systems are referred to frequently on television weather reports as "lows" or "highs."

If Earth's surface had no topographic barriers, and if Earth were not rotating, it would be easy to draw a map of the world's wind systems simply by connecting high and low pressure belts. This is reflected in one of the first models of atmospheric circulation, proposed in 1735 by George Hadley. Hadley's single-cell model of energy transfer suggested that warm air would rise at the tropics and travel aloft toward the poles. Simultaneously, cold air at the poles would sink and move on the surface toward the Equator (Figure 5-13). The **Hadley cell** was the first attempt at explaining the relation between wind and pressure belts on Earth. The model calls for descending cold air and high pressure at the poles and rising warm air and low pressure at the Equator. Winds would flow along this global pressure gradient.

Earth, however, *is* rotating, and so a single-cell system is not possible. Hadley's model was gradually developed to include a system in which warm air rises at the Equator, descends in the subtropics, rises at the higher mid-latitudes, and descends at the poles, all the while being deflected by the Earth's rotation and land-water temperature differences. Pressure gradients among these cells are the primary cause of the global wind system.

Two forces in addition to the pressure gradient force affect the direction and velocity of Earth's winds. First, the

Coriolis effect deflects the airflow to the right in the northern hemisphere and to the left in the southern hemisphere. The airflow around high-pressure cells in the northern hemisphere is clockwise and around low-pressure cells is counterclockwise. Second, friction between the air and the land surface is a direct result of contact with mountain ranges, urban areas, and other rough surfaces. This friction diverts and slows airflow at low levels.

The combined effect of these three forces—pressure gradient, the Coriolis effect, and friction—is to produce global patterns of airflow directly related to semipermanent zones of high and low pressure. Air moves from areas of high pressure and converges into areas of low pressure, so the atmosphere acts as a great heating system, transferring energy from the Equator to the poles.

Figure 5-13 depicts a simplified model of winds and pressures, summarizing the relations among the major wind systems and pressure belts. Air moving horizontally from the northern hemisphere subtropical high is deflected to the right by the Coriolis effect, resulting in easterly winds at the Equator. Air moving poleward from the southern hemisphere subtropical high is deflected to the left (the Coriolis effect acts in an opposite sense in the southern hemisphere), creating westerlies in the middle latitudes. The general pattern of global winds is best understood as a balance of the pressure gradient force and the Coriolis effect.

A Global Model of Atmospheric Circulation

A close examination of Figure 5-13 shows the relation between prevailing winds and belts of atmospheric pressure. Major global winds include the doldrums, the Horse latitudes, the trade winds, the westerlies, and the polar easterlies.

The doldrums are located in the Intertropical Convergence Zone (ITCZ) and are dominated by low pressure and rising air (centered on the Equator and extending

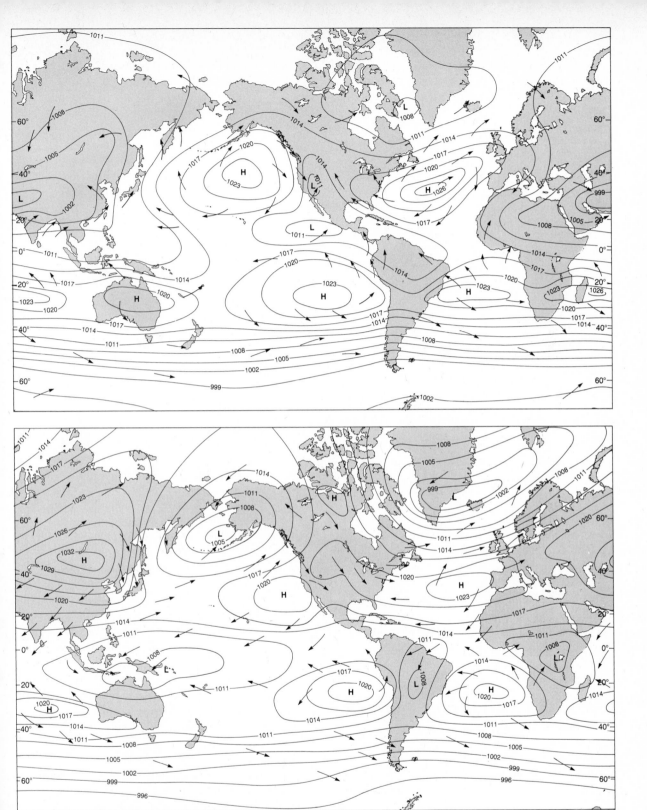

FIGURE 5-12
World map of isobars in July and January. (From Edward J. Tarbuck and Frederick K. Lutgens,
Earth Science, *5th ed., Columbus, OH: Merrill Publishing Co., 1988, Fig. 14-12, p. 393. Reprinted*
with permission of Merrill Publishing Co.)

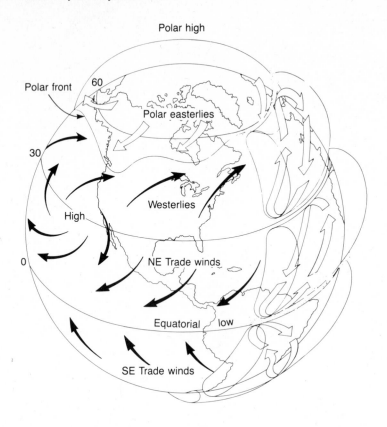

FIGURE 5-13
Idealized global circulation patterns. (From Edward J. Tarbuck and Frederick K. Lutgens, Earth Science, *5th ed., Columbus, OH: Merrill Publishing Co., 1988, Fig. 14-11, p. 392. Reprinted with permission of Merrill Publishing Co.)*

5° to the north and south). Air converges here, so it is forced to rise; as a result there is very little wind. The doldrums got their name from sailors who frequently became stranded in the calm, windless environment (see the box on page 99).

Air blows into the ITCZ from the subtropical high-pressure belts. These high-pressure belts are the origin of both the trade winds and the westerlies. Due to weak horizontal pressure gradients and vertically descending air, the subtropical high is also a region of light, variable winds. Sixteenth-century Spanish sailors carrying heavy loads to colonies in the Americas were said to become so frustrated by the lack of wind power that they threw their horses overboard to lighten the load. This legend has caused the zone of the subtropical high to become known as the Horse latitudes.

Trade winds blow into the equatorial low out of the subtropical high from both the north and south. They are deflected by the Coriolis effect and thus arrive at the Intertropical Convergence Zone from the northeast in the northern hemisphere and from the southeast in the southern hemisphere. The persistence of the trades made them reliable for sailing ships; today we associate them with the warm winds of tropical vacation islands.

Westerly winds are the dominant winds in the middle latitudes. As air moves poleward from the subtropical highs, it is deflected to the right by Earth's rotation (the Coriolis effect) in the northern hemisphere and to the left in the southern hemisphere. This creates a wind that follows roughly a west-to-east direction. Winds are always named for the direction from which they come, so these important winds are known as the westerlies in both hemispheres.

Westerlies are dramatic and changeable because of the large pressure gradient between their origin (the subtropical high) and their destination (the polar low). In the southern hemisphere there is less land friction to slow them down; sailors were often overwhelmed by the sudden show of strength by the westerlies in the South Atlantic and South Pacific. Latitudes along the journey around the tip of South America, in fact, became known as the "Roaring Forties" and the "Furious Fifties."

The westerly winds blow into a zone of low pressure known as the polar low. In the northern hemisphere there is not a continuous low-pressure belt due to interrupting land masses; the low-pressure areas are called the Aleutian low and the Icelandic low. The low pressures are particularly strong in winter when temperatures over the large land masses of Eurasia and North America are colder than those

IN THE DOLDRUMS

While playing the popular game *Trivial Pursuit* you may have answered the question, "What is the belt of low pressure around the Equator called?" The answer, of course, is the doldrums. The word brings to mind boredom, inactivity, and a feeling of having the "blahs" rather than an atmospheric belt.

According to a recent newspaper column by Michael Gartner,* the word *doldrums* is a combination of the words dull and tantrum and was first used by a British reporter in 1811 when he wrote, "I am now in the doldrums, but when I get better... ."

The word is always plural and is always preceded by *the*. In human terms, it describes a state of depression or listlessness. In atmospheric terms, it refers to a low-pressure system immediately north and south of the Equator where the air is calm and stable. The doldrums occur where the trade winds meet from opposite directions and neutralize each other. The doldrums are not a part of the horse latitudes, although both are belts of low or no wind velocity. The horse latitudes lie farther north and south of the Equator, between 30° and 35° latitude.

* Michael Gartner, *"Word Power,"* Sacramento Union *(August 19, 1985), p. 16.*

over the oceans. At this time the need for energy transport is greatest and cyclones are most vigorous.

Polar easterly winds blow out of the polar highs into the subpolar lows. In the southern hemisphere, the continent of Antarctica has a well-developed and somewhat consistent wind pattern dominated by these easterlies (although their velocity is never very strong). In contrast, the northern hemisphere polar easterlies are weakly developed and barely noticeable over the Arctic Ocean.

Air Masses and Cyclones and Anticyclones

Air masses form in regions of relatively stable air and are classified according to the type of surface beneath them. Air masses are named *continental* (meaning "dry") or *maritime* (meaning "wet"), based on moisture characteristics, and are designated *equatorial, tropical,* and *polar,* based on temperature.

Close to Earth's surface, giant cells develop (especially in the mid-latitudes), having a radius of 600–1200 miles (1000–2000 km). These simple circulation cells are the primary means of transporting energy in the mid-latitudes. Cyclones (low-pressure cells) and anticyclones (high-pressure cells) rotate as they carry air from the poles and equatorial regions into the mid-latitudes (Figure 5-14).

Cyclones usually are formed over oceans in high latitudes and subtropical areas of the world and are based on maritime polar and maritime tropical air masses. Anticyclones most often develop over land from continental polar or tropical air masses.

Cyclones and anticyclones are the reason for references on local weather reports to "highs" and "lows." Cold fronts and warm fronts that we hear so much about on weather forecasts are the result of air flows within low-pressure cells. In the winter months, cyclones move quickly across the mid-latitudes. They may travel from one coast of North America to the other in three to five days, although their actual net transfer of air is very low. Both cyclones and anticyclones move less often in summer, and as a result summer weather usually is less changeable and dramatic than winter weather. Clouds and possibly precipitation are associated with cyclonic disturbances. Anticyclones, on the other hand, bring clear skies, low winds, and pleasant conditions.

A front is a zone of contact between two different air masses in a low-pressure cell. A cold front is generally more dramatic than a warm front. Rain showers or thunderstorms may occur as the rapidly moving cold air mass overtakes the slower warm air mass and pushes it upward (Figure 5-15). In both cold and warm fronts, unstable air and precipitation occur because the two bodies of air coming into contact are of different temperatures, forcing the warmer air to rise turbulently.

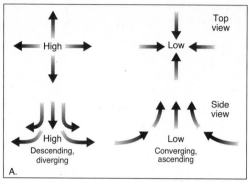

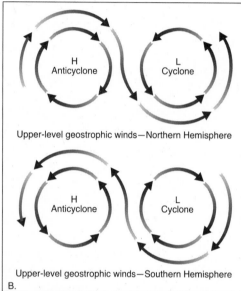

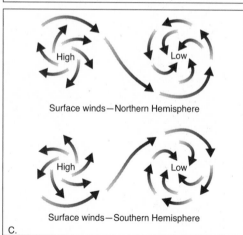

FIGURE 5-14
Cyclones and anticyclones in the northern and southern hemi-spheres. (From Robert W. Christopherson, Geosystems: An Introduction to Physical Geography, *2nd ed., New York: Macmillan, 1994, Fig. 6-6, p. 152. Used with permission of Macmillan Publishing Co.)*

Jet Stream

Another globally significant wind system is the rapidly moving jet stream, named after its discovery by jet bomber pilots flying above 30,000 feet (9150 m) after World War II. These winds extend around Earth between 30° and 60° latitude on both sides of the Equator. Jet streams are associated with the underlying pressure cells; like the cells, they move daily and seasonally in response to temperature and pressure differences.

Average speeds of this wind system vary from about 100 miles per hour (160 km/hr) in summer to about 200 miles per hour (320 km/hr) in winter. Airlines routinely move flight routes and change altitudes to avoid jet streams (when flying west) and encounter them (when flying east). Jet streams rarely affect local weather at Earth's surface but seem to direct the movement of large air masses.

Monsoons

Surface wind systems also are influenced by land and water differences on Earth. The huge landmass of the continent of Asia alters normal heating and cooling patterns and creates its own wind system, the **monsoon** (Figure 5-16). In winter, southern Asia's land surface has lower temperatures and higher pressure than the adjacent Indian Ocean; air moves from the land toward the sea. This causes a dry wind and a winter monsoon, bringing a few months of fair weather. Reversing this pattern in summer, the land is warmer than the water. Low pressure develops over the land causing moist air to blow from the Indian Ocean northward and northwestward onto India, Indonesia, and China. This warm, humid air creates the summer monsoon that brings extremely heavy rainfall to Asia.

Local Winds

Mountain/valley winds and land/sea breezes are local disturbances related to diurnal fluctuations in temperature. As an area warms, the pressure of the overlying air is lowered; this air then rises and is replaced by new air blowing in at the surface. With the heating of mountain slopes during the day air moves upward and a surface breeze up from the valley develops. At night, cool mountain wind blows downslope, as cool, dense air moves downward into the valley.

A similar pattern occurs near water bodies. The rapidly heated land is warmer than the water during the day and an onshore "sea breeze" develops in late afternoon and evening. Nocturnal cooling of the land surface reverses the wind flow and a dry offshore "land breeze" is present by the early morning (see Figure 5-17).

FIGURE 5-15

A. Typical middle-latitude cyclone in the northern hemisphere. Shown are isobaric arrangement, fronts, wind movements, and precipitation areas.   B. Profile of a cold front. Cold air, being denser, forces its way under the warm air, moving it upward along the front, causing rapid cooling and condensation which frequently creates a line of violent weather (squall line). C. Profile of a warm front. Warm air flows smoothly upward over the denser mass of cold air at the warm front. This gentle action creates a large area of stratus clouds, often with an extended precipitation area. (From E. Willard Miller, Physical Geography: Earth Systems and Human Interactions, Columbus, OH: Merrill Publishing Co., 1985, Figs. 7-4, 7-5, 7-6, p. 103. Reprinted with permission of Merrill Publishing Co.)

Northern hemisphere cyclone

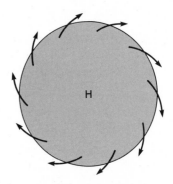

Northern hemisphere anticyclone

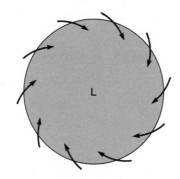

Southern hemisphere cyclone

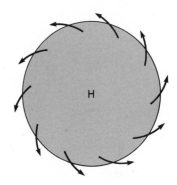

Southern hemisphere anticyclone

Other local winds that are of environmental importance include the *foehn* and *chinook*. These are warm, dry regional winds that descend from high to low elevations over a mountain barrier or a high plateau. Antifoehn medication for chapped lips and skin, and pills for moody foehn behavior, are available in shops at the foot of the Alps in Switzerland and Italy. Similarly, the chinook winds in the eastern Rocky Mountains have a reputation for rapidly melting snow cover and for creating domestic problems and violent behavior among Denver residents.

Descending warm air also is perceived as a negative environmental factor in southern California, when the dry Santa Ana wind blows out of the desert. It has been documented recently that homicide and assault rates increase when this fierce wind blows. Numerous wildfires and losses of residential homes in the mountains rimming the Los Angeles basin also are associated with this wind.

The Santa Ana wind is caused by high pressure building up over the desert to the east of Los Angeles; the winds are further accelerated by funneling through mountain passes, thus increasing their velocity. Late summer fires in the chaparral-covered areas of Malibu, the Santa Monica Mountains, and other trendy places along the southern California coast are a natural hazard of the region.

As an exciting summary of this section of Chapter 5 on weather and climate, sometimes unbelievable world weather extremes are presented in the box on page 104–105. A physical geographer using geographic information system (GIS) technology is featured in the box on page 112.

THE OCEAN CIRCULATION SYSTEM

The circulation of the ocean is closely tied to the circulation of the atmosphere. The ocean's circulation system is influenced by the major global winds—Equatorial currents

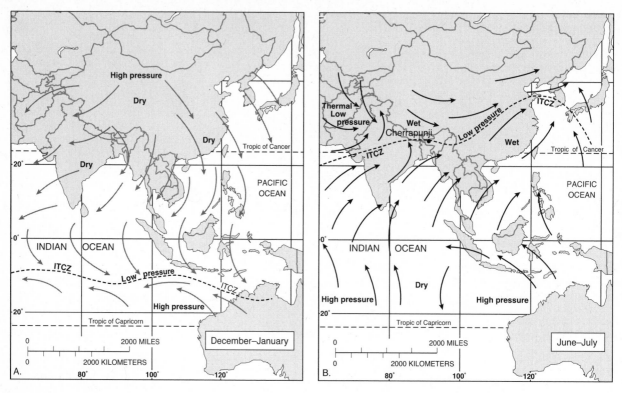

FIGURE 5-16
Monsoon system of southern Asia. A dominant feature of the Asian monsoon is a 180° reversal of wind direction between winter (A) and summer (B) seasons. Note greater migration of the intertropical convergence between the two seasons. (From Joseph E. Van Riper, Man's Physical World, *New York: McGraw-Hill Book Co.,1971, Fig. 7-15, p. 219. Reprinted with permission of McGraw-Hill, Inc.)*

that run from east to west (tropical easterlies) while mid-latitude currents flow from west to east (the westerlies). The Coriolis effect acts on ocean currents in a manner similar to airflow. Currents are deflected clockwise from their path in the northern hemisphere and counterclockwise in the southern hemisphere.

Study the global oceanic circulation in Figure 5-18 and you will notice that the Coriolis effect influences the direction of most major ocean currents. Energy is transported toward the poles by huge, slow-moving, circular whorls of water known as *gyres*. The gyre located off the east coast of North America includes the Gulf Stream, the North Atlantic current, and the north equatorial current. Western North America is influenced by the huge gyre that includes the California current, the North Pacific current, and the north equatorial current. In general, the west side of an ocean basin has warm currents and the east side has cold currents. This reflects their respective source waters in the Equator and the poles.

WATER IN THE EARTH-ATMOSPHERE SYSTEM

Our discussion of ocean currents has illustrated the important links between the atmosphere and hydrosphere. The hydrosphere, the second of the circulation systems of the Earth system, is made up of water in all its forms and movements. This "water world" is interconnected to both the atmosphere ("air world") and Earth's surface systems ("land world" and "life world"). Most important, it is the presence of water that makes life possible on Earth.

No water is exchanged between the atmosphere and outer space, so we can define its movement as a closed system. Water is not readily synthesized in the laboratory, and no immense reservoir is present deep in Earth. Water is a finite resource on our planet. The flow of water in the hydrosphere is called the **hydrologic cycle**.

FIGURE 5-17
*Land and sea breezes. (From
Edward J. Tarbuck and Frederick
K. Lutgens,* Earth Science, *5th ed.,
Columbus, OH: Merrill Publishing
Co., 1988, Fig. 14-14, p. 396.
Reprinted with permission of Merrill
Publishing Co.)*

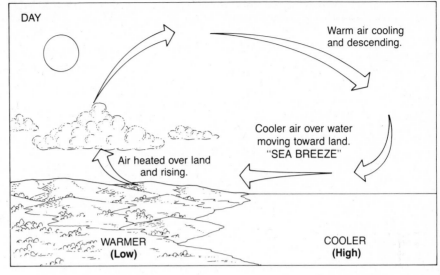

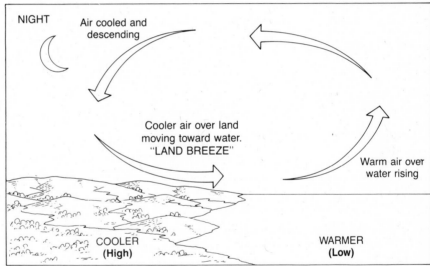

The Hydrologic Cycle

Water present in the oceans, atmosphere, and land is transferred within these system reservoirs via the hydrologic cycle. Every year approximately equal amounts of water are evaporated and fall as precipitation. Within this closed system, water may be stored in several places—in the oceans, in ice caps and glaciers, within the land (in rivers, as groundwater, in lakes, in soil), and in living things. The smallest reservoir of water in the system is the atmosphere, and the largest is the oceans. Water moves through the cycle from one storage area to another through various processes, including evaporation, transpiration, runoff, melting, freezing, sublimation, condensation, and precipitation (Figure 5-19). The most important processes are discussed below.

Water: The Incredible Substance

Before we discuss the processes of the hydrologic cycle, it is important to understand and appreciate the special characteristics of water. Water is an amazing substance. It exists in all three phases in the Earth's system—solid (ice), liquid (water), and gas (water vapor). The passage from one form to another occurs naturally and easily within the system. As each phase change occurs both mass and energy are exchanged.

In addition, water has a high specific heat, which al-

WORLD WEATHER EXTREMES

While patterns of the average conditions of the atmosphere (the climate) are most important to understanding landforms, vegetation, and land use, it is the extreme weather event that makes the prime-time news. Extreme weather events may inflict damage or loss of life, but in most cases they are fairly benign because they generally occur in sparsely populated areas. Nevertheless, extreme events in our atmosphere provide examples of how diverse and interesting our dynamic planet can be.

Rainfall

The world's highest average annual precipitation was recorded on the Hawaiian island of Kauai on nearly mile-high Mt. Waialeale, at 460 inches or about 38 feet (1168 cm). This almost daily rain is caused by the movement of saturated trade winds against the mountain, causing orographic precipitation. Some travel atlases appropriately label this mountain as the wettest place in the world!

The world's greatest 12-month rainfall occurred at Cherrapunji in eastern India, where 1042 inches or nearly 87 feet (2647 cm) fell during 1860–1861. This precipitation was due to heavy monsoon activity.

The rainiest day recorded anywhere on Earth, that is, the greatest 24-hour measured rainfall, was 74 inches (188 cm) or just over 6 feet at Cilaos on the French island of Réunion, in the Indian Ocean east of Madagascar.

The lowest recorded rainfall on Earth occurred at Arica, Chile, in the Atacama Desert, where it has not rained in 20 years! Average annual precipitation here is 0.03 inch (0.08 cm). Cold offshore currents and westerly winds keep the area foggy much of the year, but no rain falls because there are no coastal hills to create orographic lifting of moist air.

lows it to act as an energy-storage medium. Water is chemically active, dissolving rocks and delivering nutrients to plants. Water vapor in the air creates a warm thermal-blanket effect as it intercepts longwave radiation from Earth. Ice, the solid form, has the unusual property of having a lower density than the liquid form, which is why ice cubes float. Water is a truly unique substance.

Evaporation and Transpiration

Water exists in Earth's atmosphere in all three states—solid, liquid, and gas—just as it does at Earth's surface. In fact, water vapor is one of the most important gases in the atmosphere because it absorbs longwave heat energy and is the vehicle for the latent transfer of heat. Water vapor in the air, even though it exists in relatively small amounts at any given time (no more than 4 percent by volume), is also an important factor in the comfort of humans on Earth. When the atmospheric water vapor content (humidity) and temperature are high, human discomfort occurs because perspiration becomes an ineffective cooling mechanism.

Water vapor content is generally reported as *relative humidity* (the actual amount of water vapor in the air, as compared with the maximum amount that saturated air can hold at that temperature). **Absolute humidity** is the exact amount (weight) of water vapor in a given volume of air. Weather reports almost always use the term *relative humidity* because it is much more meaningful in terms of human comfort.

Water vapor enters the air through the Earth-atmosphere system by the process of *evaporation*. Evaporation is the process of liquid water changing to gaseous water (water vapor). This process, like all other movements through the system, involves energy. The energy that is used to transform liquid water to a gas becomes trapped in each water vapor molecule as the "latent heat of vaporization." This energy is later released in the atmosphere as the "latent heat of condensation." Thus, *evaporation is a cooling process* (heat

Snow

There are many places on Earth where it has never snowed in recorded history and some places, such as Greenland, where the snowpack has not melted in centuries. The highest average snowfall recorded in one season was at Rainier Paradise Ranger Station in western Washington State where 1122 inches (2850 cm) or nearly 94 feet was measured in 1971–1972. The highest snowfall in one storm was 189 inches (480 cm) or nearly 16 feet at Mount Shasta in northern California. The record for deepest snow on the ground, excluding the Arctic and Antarctic ice caps, was 451 inches (1145.5 cm) or nearly 38 feet in the Sierra Nevada Mountains in eastern California.

Temperature

The highest recorded temperature on Earth was at El Azizia along the Mediterranean coast of Libya, on September 13, 1922, when the temperature reached 136° F (58° C). The world's lowest temperature was recorded on July 21, 1983 at Vostok, a Russian research station in Antarctica, at −129° F (−89° C).

Wind

The highest wind speed recorded over a 24-hour period was at Mount Washington, New Hampshire, on April 11 and 12, 1934. Winds blew at velocities of 128 mph (206 km/hr), with the peak gust recorded at 231mph (372 km/hr).

Source: Compiled by the Geographic Sciences Laboratory, U.S. Army Corps of Engineers Topographic Laboratories, Fort Belvoir, VA (1985).

---

is absorbed), and *condensation is a warming process* (heat is given off).

The rate and amount of evaporation occurring anywhere depends on the nature of the surface, the wind speed, the humidity of the air, and the temperature. Plants diffuse water molecules through pores (stomata) into the atmosphere. This transfer of water into the atmosphere via plants is known as *transpiration*. Due to the close interaction between evaporation and transpiration at Earth's surface, these two processes usually are referred to with one term, **evapotranspiration.** Wind-induced turbulence greatly increases the rate of evaporation (this is why clothes are mechanically tumbled through the air in a dryer).

Soil moisture, wind speed, and atmospheric humidity may act as limiting factors that increase or decrease the rate of evapotranspiration. The evapotranspiration (ET) concept is used frequently in water and irrigation studies; it is a vital part of water modeling and water-budget analyses discussed later in this chapter. At the most basic level of analysis, E (evaporation) plus T (transpiration) equals ET (evapotranspiration).

Condensation

The opposite of the process of evaporation is the process of condensation. When water vapor is converted from vapor to a liquid through the process of condensation, energy is released. The "latent heat of condensation" is released into the atmosphere and warming occurs. Air must reach the *dew point* (saturation point), and then be further cooled for moisture to be condensed out of the air. As an air mass cools its capacity to hold water is reduced until it reaches the dew point. If the temperature continues to drop, condensation occurs and liquid cloud droplets form.

A number of processes may cool the atmosphere or the surface of Earth until condensation occurs. The rapid nocturnal cooling of solid ground surfaces cools the immediately overlying air and results in condensation. This may

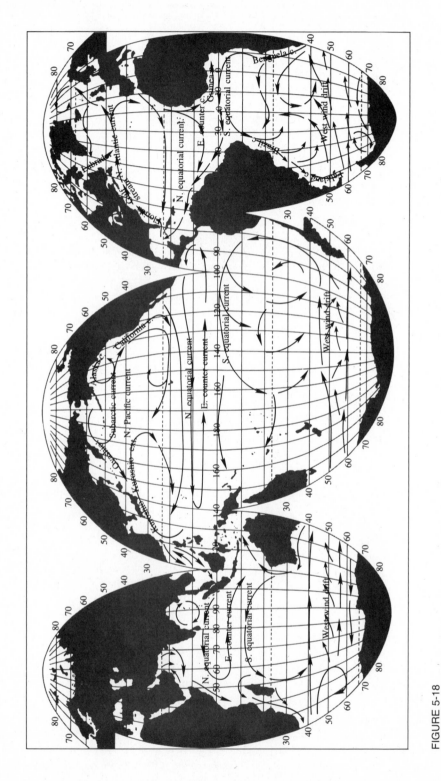

FIGURE 5-18
Global oceanic circulation is affected by rotation of Earth, by local conditions, and by deflection from the continental land masses. Map shows wind-driven surface currents in February and March. (From Harold V. Thurman, Introductory Oceanography, 5th ed., Columbus, OH: Merrill Publishing Co., 1988, Fig. 8-2, p. 193. Reprinted with permission of Merrill Publishing Co.)

FIGURE 5-19

Hydrologic cycle. The oceans are the major source of water in the atmosphere. Winds carry the moisture onto the continents, where part of it falls to Earth as precipitation. Water returns to the ocean by running off the land as surface water, by movement in the ground as groundwater, and indirectly by evapotranspiration. (From E. Willard Miller, Physical Geography: Earth Systems and Human Interactions, *Columbus, OH: Merrill Publishing Co., 1985, Fig. 6-4, p. 86. Reprinted with permission of Merrill Publishing Co.)*

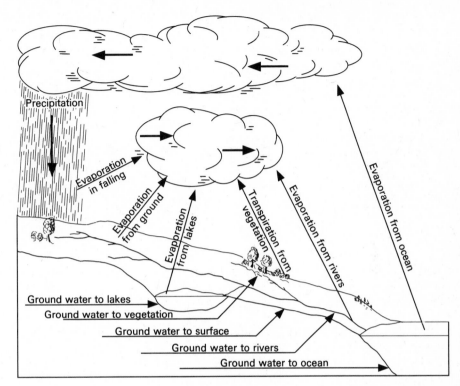

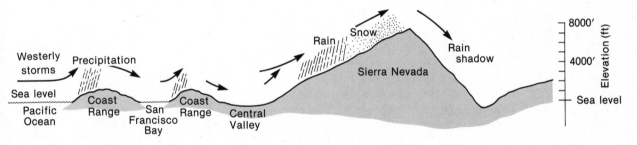

FIGURE 5-20

Orographic precipitation and the rainshadow effect in California.

cause early morning dew on the lawn or a low ground fog that dissipates as the sun rises.

Advectional cooling (the movement of warm air over a cool surface) is also an important process leading to condensation. This creates the fogs associated with Newfoundland or San Francisco. Most important, dynamic cooling associated with the expansion of rising air results in the dew point being reached high in the free atmosphere, causing the flat bottoms of cumulus and stratus clouds. (*Dynamic cooling,* in which either warm air or cool air rises, occurs in low-pressure cells, as air is forced up along fronts and as air is lifted over orographic barriers.)

Precipitation in the Atmosphere

Precipitation is the direct descent of liquid water from the air to the surface of Earth. Water drops form around nuclei in the air (ice or dust particles) and are pulled by gravity toward Earth's surface. Precipitation begins within the minute liquid droplets in the humid atmosphere of clouds. Most clouds fail to produce rain because the cloud droplets are too small to fall under the influence of gravity. However, under special conditions the cloud droplets are able to coalesce. When they grow too heavy to be supported by the rising air, gravity pulls them to the surface. It nev-

FIGURE 5-21
*Global precipitation patterns.
(From James S. Fisher,
Geography and
Development: A World
Regional Approach, 3rd ed.,
Columbus, OH: Merrill
Publishing Co., 1989, Fig.
3-1, pp. 46-47. Reprinted
with permission of Merrill
Publishing Co.)*

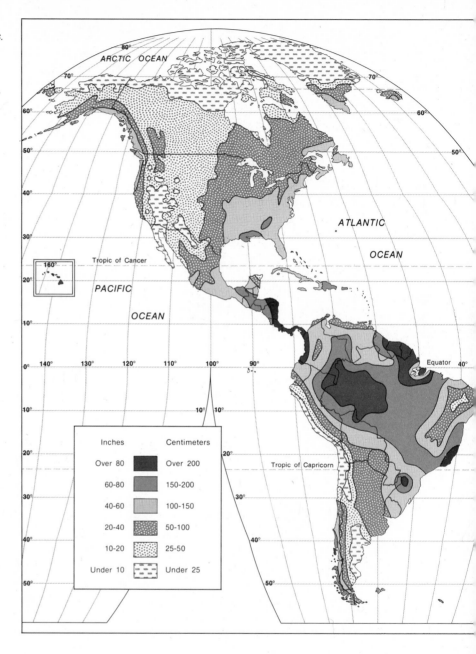

er rains beneath 99 percent of Earth's cloud cover! Precipitation is an under-appreciated phenomenon.

Many different cloud forms occur in the atmosphere. Clouds are usually classified according to their altitude (into low-, middle-, and high-level clouds) and according to their storm potential. The type of precipitation is named for whichever mechanism lifts and cools the air and usually is classified as frontal, orographic, or convectional in origin.

Frontal precipitation is very common; you have often heard the term **front** on the weather news. Recall that a front is the point of contact between two differing air masses. When the warmer (and usually moister) air mass is forced to rise above cooler air, a discontinuity of surface temperature and air mass is created. Along such a front, precipitation, is particularly common. As the warm air is pushed off the ground, cold temperatures and gusty winds mark the passage of the front.

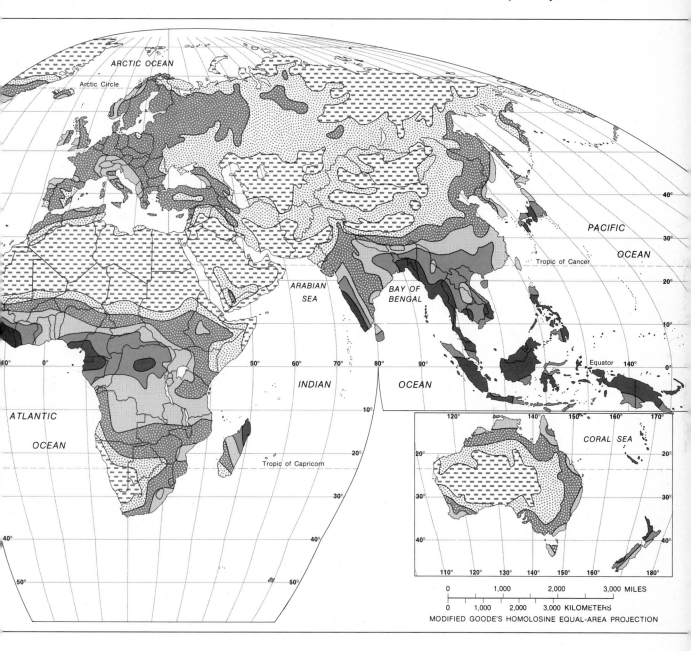

Orographic precipitation is caused by an air mass rising over a topographic barrier such as a mountain range or high plateau. Air is forced to rise on the windward side of the mountain, cooling until condensation, and perhaps precipitation, occurs. Conversely, air descends on the leeward side of the mountain and warms; clouds are evaporated and a dry *rainshadow* is created (Figure 5-20).

Orographic precipitation is especially common in parts of the world where moist air masses come in contact with high mountain barriers, such as along the southern edge of the Himalaya Mountains in northern India. For another example, the Sierra Nevada and Cascade Mountains of the west coast of North America block the passage of maritime air to the interior. This creates the immense, dry Great Basin desert on the downwind side of these mountains.

Convection is the third major cause of precipitation. Convectional precipitation is associated with direct surface

FIGURE 5-22
Climates of the world. (From James S. Fisher, Geography and Development: A World Regional Approach, *3rd ed., Columbus, OH: Merrill Publishing Co., 1989, Fig. 3-3, pp. 50-51. Reprinted with permission of Merrill Publishing Co.)*

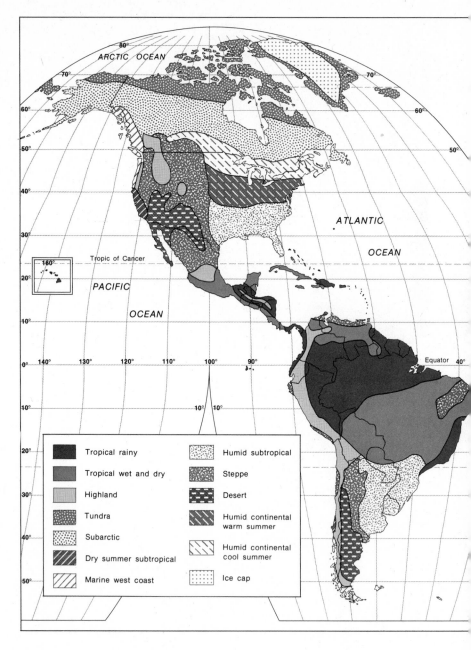

heating, a process common in warm regions of the world. As an air mass is heated at Earth's surface, it becomes lighter and begins to rise. This uplift results in cooling, condensation, and precipitation. Convectional precipitation is a common occurrence during the hot, humid summer days in the mid-latitudes where air becomes unstable due to the build-up of heat at the surface. Many tropical areas experience daily convectional showers.

A network of rain gauges measures the amount of precipitation at Earth's surface, including its quantity, duration, intensity, and geographic distribution. Of course, rainfall varies locally according to altitude, slope steepness, slope orientation, and the barrier effects of terrain (Figure 5-21). It may be seen that rainfall prediction can become a very complex and uncertain undertaking.

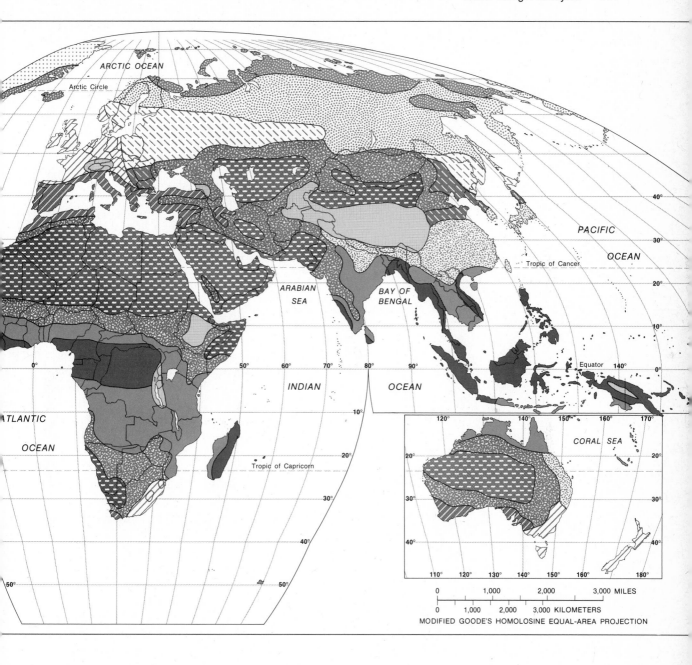

ARCTIC OCEAN

Arctic Circle

PACIFIC

OCEAN

Tropic of Cancer

ARABIAN
SEA

BAY OF
BENGAL

40°

30°

20°

10°

Equator

140°

0°

ATLANTIC

OCEAN

INDIAN   OCEAN

Tropic of Capricorn

10°

20°

30°

40°

50°

CORAL SEA

MODIFIED GOODE'S HOMOLOSINE EQUAL-AREA PROJECTION

0   1,000   2,000   3,000 MILES

0   1,000   2,000   3,000 KILOMETERS

WATER-BUDGET ANALYSIS

Regional Water Systems

It is possible to measure the water balance of a particular area on Earth or even to estimate a long-term water balance over the entire globe. This assessment, known as *water-budget analysis,* plays an important role in many practical stud-

ies, including the analysis of irrigation and the agricultural potential of a region. Groundwater studies, soil-moisture estimates, runoff, and stream-flow predictions also depend upon accurate water-budget analysis.

The main components of a **water budget** are precipitation, evapotranspiration, groundwater storage, and runoff. Precipitation and evapotranspiration are the two most important parts of the water-budget equation.

GEOGRAPHER AT WORK

While studying marine biology at Humboldt State University in California, Karl happened upon a course entitled *Geography and World Issues*. This entry into geography led to more classes, world travel, research, and a master's degree thesis. Karl had also taken some geographic information system courses, but he had not really considered GIS as a career until he had a chance to work as an intern for the Genasys Company. This led to his current position as technical support specialist and trainer. "I teach our GIS software to clients, ranging from city governments to the Defense Mapping Agency. When I am in Fort Collins [Colorado], I answer support questions regarding our various products, plotting, digitizing, and installation."

Karl sees GIS more than a job description or research tool. "Geography and its applications have taught me a way to think about the world. Nothing is static, education really begins when you walk out of the classroom with your diploma."

Karl Maser, Geographic Information Systems Practitioner

Water is stored in the system in groundwater, surface water, snow and ice, and biota, so it often is difficult to measure the exact water balance of a particular area. Other complicating factors include the variability and unpredictability of precipitation and the variation in evapotranspiration rates in different places. The water budget for a specific place is a daily or monthly bookkeeping system showing the supply and loss of water. Due to the variability of water input and output, this type of model is best used over longer time periods.

The Global Water Budget

Surface water and atmospheric moisture are unevenly distributed over Earth's surface. Globally, some areas have precipitation rates that exceed evapotranspiration, while others have evapotranspiration rates that exceed precipitation.

An excess of precipitation usually occurs in the westerly wind belts of the mid-latitudes. Also, due to the lower temperatures in the higher latitudes, both precipitation and evapotranspiration decrease, resulting in a small moisture excess there. The Intertropical Convergence Zone, with its converging winds and moist air, also has a precipitation surplus. In fact, the zone between 0° and 10° latitude on either side of the Equator has the greatest precipitation excess on Earth.

Conversely, between 10° and 30° north and south latitudes, evapotranspiration greatly exceeds precipitation. This zone of precipitation deficiency lies beneath the descending air and clear skies of the subtropical highs. Many of the surface climates are deserts in these regions.

Due to these global differences in precipitation and evapotranspiration rates, an exchange of energy and moisture occurs from moist regions to dry areas. The transport of water vapor acts as a vehicle for carrying energy from one area to another. Thus the global water system and water budget is a vital part of the Earth-atmosphere energy balance from place to place.

The patterns and processes of weather and climate on Earth do not exist in an isolated system. As shown on Table 5-1, other parts of Earth's physical environment, such as vegetation and soils, interrelate with weather and climate. Envisioning this close relationship among Earth's physical systems leads us into the following chapter, which discusses landforms, vegetation patterns, and the assemblage of all of these factors into world biomes. Refer to Table 5-1 often as you read through Chapter 6 to provide examples of the complex variety of interrelationships that occur in physical landscapes.

Figure 5-22 places each of the climates listed in Table 5-1 onto a world map.

TABLE 5-1
Interrelationships among climate, soils, and plant characteristics

Climate (Example)	Climate Characteristics	Climate Controls	Vegetation (Soil)	Plant Characteristics
Tropical equatorial (Panama)	Always tropical; wet year-round	ITCZ all year	Tropical rainforest (oxisols)	broadleaf evergreen
Tropical wet and dry (Southern Mexico)	Warm to tropical; high total precip.; summer maximum	ITCZ passage in summer; STH in winter	Savanna or tropical deciduous (ultisols, alfisols)	deciduous trees and grassland (role of fire important)
Subtropical deserts (Northern Mexico)	Warm to tropical; always dry	Stable side STH; BW (wastelands)	Desert xerophytes (aridisols)	ephemeral, succulents, spiney leaves, short roots
Mid-latitude arid (Nevada)	Dry; winter precip. maximum; cold to tropical	Westerlies, mP in winter; lee side of mountains	Shrublands (aridisols) BW (wastelands)	sagebrush (cold, wet, winter), creosote (warm, arid, winter), shadscale (cold, arid, winter)
Humid subtropical (Florida)	Mild to tropical; high total precip.; summer maximum	Unstable side of STH, mT air	Temperate forest (ultisols)	broadleaf evergreen or subtropical coniferous
Humid continental (Illinois)	Cold to warm; moderate precip.; summer maximum	cP, mP winter; mT summer; Westerlies	Deciduous forest (alfisols)	broadleaf deciduous
Mediterranean (California)	Cool, wet, winter; hot, dry, summer	*Westerlies; mP winter; STH in summer	Chaparral (alfisols, mollisols)	evergreen broadleaf; short growing season
Marine west coast (Washington)	Winter precip. maximum; mild temp.; year-round precip.	Westerlies present year-round, mP	Rainforest (spodosols)	needleleaf evergreen in North America
Semiarid (Nebraska)	Dry; summer precip. maximum (winter dry and cold)	mP dominates lee side of mountains	Prairie, selva steppe (mollisols)	deep- and shallow-rooted grasses; fire has dominant role
Continental subarctic (Minnesota)	Cold to mild; low total precip.; summer maximum	cP, some mP in summer	Coniferous forest (spodosols)	needleleaf evergreen (needleleaf deciduous in Eurasia)
Arctic (Northern Alaska)	Cold to cool; low precip.; summer maximum	cP all 12 months	Tundra vegetation (histosols, entisols)	low shrubs (deciduous), mosses, lichens

ITCZ Intertropical Convergence Zone   *cP Continental polar*
BW Dry all year   *STH Subtropical high*
mP Marine polar   *mT Marine tropical*

KEY CONCEPTS

Advection
Albedo
Anemometer
Anticyclone
Atmosphere
Atmospheric pressure
Barometer
Biosphere
Circle of Illumination
Climate
Climate modeling

Conduction
Convection
Cyclone
Declination
Dew point
Equinox
Evapotransporation
Front
Greenhouse effect
Hadley cell
Horse latitudes

Hydrologic cycle
Hydrosphere
Insolation
Jet stream
Latency
Linkages
Lithosphere
Monsoon
Normal lapse rate
Orographic
Photosynthesis

Revolution
Rotation
Solar constant
Solstice
System
Tropic of Cancer
Tropic of Capricorn
Water budget
Weather

FURTHER READING

The best new text in physical geography to be published in many years is Robert W. Christopherson's *Geosystems: An Introduction to Physical Geography* (New York: Macmillan, 1994). Also useful as an overview of the systems approach in the field is *Environmental Systems: An Introductory Text,* by I. D. White, D. N. Mottershead, and S. J. Harrison (London: Unwin Hyman, 1984). This book claims to be introductory, but it is really a very comprehensive text, using the systems approach to physical geography. These textbooks, along with Arthur N. Strahler and Alan H. Strahler's *Elements of Physical Geography* (New York: John Wiley, 1987), are all most teachers will need as physical geography references.

Other fine references, more specialized in coverage, are *Contemporary Climatology* by Ann Henderson-Sellers and Peter J. Robinson (New York: John Wiley, 1986); Paul Lydolph, *Weather and Climate* (Totowa, NJ: Rowman & Allanheld, 1985); Roger B. Barry and Richard J. Chorley's *Atmosphere, Weather, and Climate* (London: Methuen, 1987); *World Weather Extremes* by Pauline Riordan and Paul G. Bourget (Fort Belvoir, VA: U.S. Army Corps of Engineers, 1985); Willy Rudloff's *World Climates* (Stuttgart: Wissenschaftliche Verlagsgesellschaft, 1981); *The Weather Companion* by Gary Lockhart (New York: John Wiley, 1988); *A Field Guide to the Atmosphere* by Vincent J. Schaeffer and John A. Day (Boston: Houghton Mifflin, 1981); and John J. Hidore and John E. Oliver's *Climatology: An Atmospheric Science* (New York: Macmillan, 1993).

Sources that discuss climate as it relates to human activity are "Coping with Climate Change," in *Proceedings of the Second North American Conference on Preparing for Climate Change* (Washington, DC: Climate Institute, 1989); *Climate Change, 1992—The Supplementary Report to the IPCC Scientific Assessment,* published by the World Meteorological Organization/United Nations Environment Programme (New York: Cambridge University Press, 1992); Reid A. Bryson and Thomas J. Murray, *Climates of Hunger: Mankind and the World's Changing Weather* (Madison: University of Wisconsin Press, 1977); "Global Climate Change" in *Scientific American* 260 (April 1989): 36–44, by Richard A. Houghton and George M. Woodwell; Richard A. Kerr's "A New Greenhouse Report Puts Down Dissenters," in *Science* 249 (August 1990): 481.

A dated but delightful fictional work using weather as a major element of the story is George R. Stewart's *Storm* (New York: Random House, 1941).

Useful references on the hydrosphere include:

- Moustafa T. Chahine, "The Hydrologic Cycle and Its Influence on Climate," *Nature* 359 (October 1, 1992): 373–380.

- Aldo Leopold, *Water—A Primer* (San Francisco: W.H. Freeman, 1974).

- Laurence Pringle and the editors of Time-Life Books, *Rivers and Lakes,* Planet Earth Series (Alexandria, VA: Time-Life Books, 1985).

- D. Rind, C. Rosenweig, and R. Goldberg, "Modeling the Hydrologic Cycle in Assessments of Climate Change," *Nature* 253 (July 9, 1992): 119–122.

6
THE THEME OF PLACE: PHYSICAL SYSTEMS II

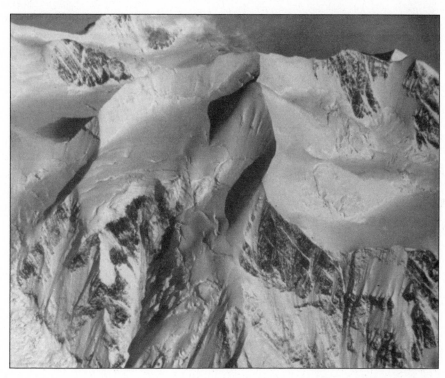

Mount Fairweather in southeastern Alaska (elevation 15,300 feet, or 4663 meters). (Photo courtesy of Stephen F. Cunha.)

Analysis of the planet's four geosystems—the atmosphere, the hydrosphere, the lithosphere, and the living biosphere—lays the foundation for our study of human activity in subsequent chapters of this text. We are using systems analysis—the study of interrelationships among variables—as an avenue for understanding the vast set of processes that have created Earth's physical landscape.

The preceding chapter introduced concepts in physical geography related to *National Geography Standard* 7,

pertaining to atmospheric and hydrologic systems. We have seen how the atmospheric and hydrospheric systems interact and interrelate in time and space. Air and water movements flow in an interconnected pattern in conjunction with Earth's surface and its life layer. In this chapter we investigate concepts important in understanding systems in the solid realm of the planet, the lithosphere—and in the living part of Earth, the biosphere.

EARTH'S STRUCTURE

As the atmosphere is layered, so is Earth made up of layers, divided into three distinct levels: the crust, the mantle, and the core (Figure 6-1). (These layers have been distinguished from each other on the basis of their seismic characteristics.) The *crust* of Earth, or **lithosphere,** is the main concern of this chapter because we live on this layer. It is very thin and makes up only about 1.55 percent of Earth's total volume. The lithosphere is thickest under land masses, especially beneath mountain ranges, and thinnest under oceans.

Below the crust lies a denser, heavier, plastic layer known as the *mantle.* We will pay special attention to the plastic zone of the mantle, the *asthenosphere,* which extends to a maximum depth of about 155 miles (250 km). Beneath the mantle lies the **core** of Earth, a hot, liquid layer constituting one-sixth of the planet's volume and one-third of its total mass (Figure 6-1).

The lithosphere may be studied as a system. Its outer layer forms an interface with the atmosphere, hydrosphere, and biosphere. It is this outer layer that contains Earth's continents, ocean basins, and lesser relief features. The lithosphere is an open system, with an exchange of matter and energy across both the inner boundary with the mantle and the outer boundary in contact with the atmosphere and hydrosphere.

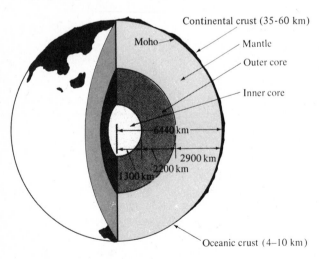

FIGURE 6-1
Cross section of Earth's layered interior shows the inner and outer core, mantle, and crust. Crust thickness is exaggerated several times (From Harold V. Thurman, Introductory Oceanography, *5th ed., Columbus, OH: Merrill Publishing Company, 1988, Fig. 1-6. Reprinted with permission of Merrill Publishing Co.)*

Rock Types

The lithospheric system includes various types of rocks. Minerals in Earth's crust combine in different ways to form a wide variety of rock types. Rocks in the lithosphere are divided into three major classes according to their origin:

- *Igneous* rocks are made of molten materials from within Earth that have solidified (magma and lava).

- *Sedimentary* rocks are made of layered accumulations of sediment, consisting of fragmented particles of rocks, minerals, and sometimes organic materials. The fragments composing most sedimentary rocks are deposited by water and cemented.

- *Metamorphic* rocks are preexisting rocks that have been metamorphosed ("changed in form") under conditions of high temperature and/or intense pressure. Metamorphism causes mechanical deformation and/or chemical change of the original rock material.

Rocks are changed and transformed within Earth in a system known as the **rock cycle** (Figure 6-2). Over millions of years rocks and minerals may pass through various stages in both the deep environment far below the surface of Earth and in the surface environment. Knowledge of rock types and an understanding of how they are formed are important to geographers who seek to understand the evolution of landforms.

PLATE TECTONICS

Tectonic Processes

The structure of the lithosphere varies and includes two subdivisions: oceanic crust and continental crust. *Oceanic crust* is dense, laterally continuous, highly mobile, geologically simple, and topographically subdued. Its most common landforms are mid-oceanic ridges, trenches, and deep basins.

Continental crust, on the other hand, is discontinuous, lower in density (so it tends to float atop oceanic crust), geologically complex, and topographically diverse. Continental landforms include the mountains, plateaus, plains, and basins familiar to all who appreciate landscape diversity.

Earth's crust is subdivided into a series of lithospheric plates. There are six major plates and several minor ones that fit together over Earth's mantle like a giant jigsaw puzzle (Figure 6-3). It may help to visualize the rigid lithospheric layer "floating" on the more fluid asthenosphere. Most of the major plates carry a continent (made of conti-

FIGURE 6-2
The rock cycle showing the relationships among igneous, sedimentary, and metamorphic processes. (From Robert W. Christopherson, Geosystems: An Introduction to Physical Geography, 2nd ed., New York: Macmillan, 1994, Fig. 11-7, p. 322. Used with permission of Macmillan Publishing Co.)

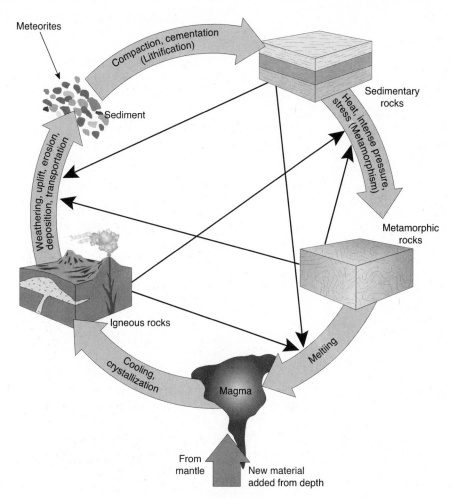

nental crust), and a single plate may carry both oceanic and continental crust. The major plates are the American, Pacific, Eurasian, Australian, African, and Antarctic.

Understanding the basic processes of **plate tectonics** has contributed to our understanding of disturbances such as volcanoes and earthquakes. Where two lithospheric plates meet, compression creates an unstable crust. Actual fracturing may occur along *fault* lines, which are stress fractures along which motion occurs. Two types of faults are common. *Normal faulting,* where the edge of one rock mass shifts upward and the other shifts downward, indicates vertical motion on the fracture. *Transform faulting* occurs when two plates in contact do not move vertically but one plate slides past the other. (An example is California's San Andreas fault.)

Plate tectonics is a ponderous process of titanic forces torturing millions of tons of rock over millions of years; the only visible action during a human lifetime is perhaps an earthquake or a volcano, or "tectonic creep" that very slow-

ly offsets rows in orchards and puts bends in streams. (See the box on page 119.)

The San Andreas fault in California is part of a long fault system separating two plates, the Pacific Plate to the west and the North American Plate on the east. Earthquakes usually occur along plate boundaries. This particular fault zone in California is infamous for its unstable seismic conditions and major earthquakes, most notably in 1906, 1971, 1989, and 1993. The dramatic and destructive 1993 earthquake in Northridge, California, is discussed in Chapter 10.

The well-known San Andreas fault is part of the Pacific Ring of Fire, a hazard zone of earthquakes and volcanoes that rims the Pacific Ocean. Along the perimeter of this zone lies Japan, Alaska, the west coast of Central and South America, and other places known for their frequent earthquakes and volcanic eruptions.

Plate collision gets a lot of attention, but plates are constantly pulling apart, too. An activity that occurs where

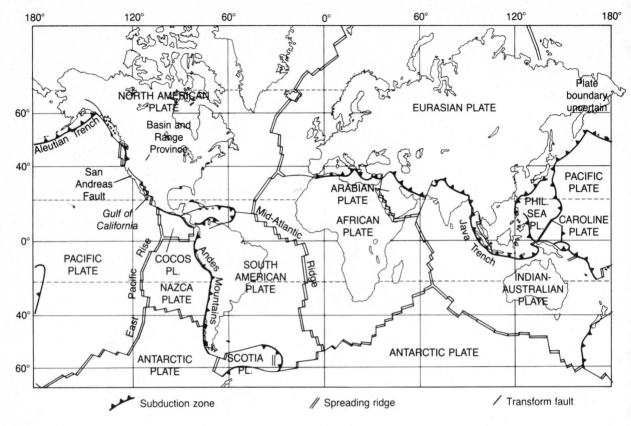

FIGURE 6-3
Earth's tectonic plates and fault line boundaries. (From Arthur N. Strahler and Alan H. Strahler,
Modern Physical Geography, *New York: John Wiley & Sons, 1987, Fig. 14-17, p. 252. Reprinted
with permission.)*

plates separate (diverge) is **seafloor spreading.** The Mid-Atlantic Ridge that slices through the Atlantic Ocean is an example of seafloor spreading at the plate boundary. Here the American Plate is parting company with the Eurasian and African Plates, and these continents are drifting apart at the rate of a few centimeters a year.

Material from the asthenosphere is rising into this rift zone in the Atlantic Ocean floor, causing the seafloor to spread apart. The extruded lavas are building a submarine ridge, generating new crustal material. When geologists dated material nearest the fault boundary they found that it was much younger than rocks farther away from the fault. The Mid-Atlantic Ridge, like other spreading boundaries, is a volcanically active rift zone. This was confirmed by the 1963 eruption that created Surtsey Island south of Iceland.

Where plate boundaries converge, crustal material may be forced upward either by simultaneous compression and uplift in a collision zone or by one plate overriding the other. The descent of one plate beneath another is called *subduction,* a process that returns crustal material to the as-

thenosphere. As one plate overrides another it is uplifted and folded to form mountain ranges. The range may be parallel to a related offshore trench. The lofty Himalaya Mountains in Asia formed over millions of years in this way as the Indian subcontinent collided with the continent of Asia. The crustal material has been forced upward and folded into complex structures that form these mountains.

Alfred Wegener's Impossible Hypothesis

The process of plate tectonics is so well accepted today that it is hard to imagine what ideas preceded its acceptance. While the concept of oceanic and continental plate movement has been accepted only since the late 1960s, a theory of drifting continents had been discussed for several centuries. Looking at a globe, you may have noticed the remarkable "fit" of South America and Africa if you were to push them together. This connection was observed by scientists as early as the 1600s. The formal hypothesis of a single continent breaking up and moving apart to form the

PLATE TECTONICS: THE SHAPE OF TOMORROW

According to an article in *Discover* magazine, the work of Christopher Scotese and Alfred Ziegler at the University of Chicago has yielded an incredible vision of Earth's future shape. Using computer models and plate tectonic theory to predict the position of the continents millions of years in the future, these scientists have concluded that our planet will feature:

- The city of Los Angeles as a neighbor (a suburb?) of the city of San Francisco
- A range of volcanic mountains connecting North and South America and linking Nova Scotia to Argentina
- Australia crashing into Southeast Asia, "folding the Philippines like a crumpled rug"

Using plate tectonics as a guide, Scotese and Ziegler maintain that Earth's continents should continue their travels for billions of years "until the Earth cools and the interior heat that drives them is finally exhausted."

Overall the future will see the North American continent pulled east and south, while the Eurasian and African regions will tilt clockwise, chilling the climate of Western Europe by exposing it to Arctic winds. As North America moves south its interior will be blocked from maritime air by coastal mountains and it will gradually become an arid desert. By 250 million years from now, Antarctica will have moved north to adjoin Australia.[*]

From Dennis Overbye, "Plate Tectonics: The Shape of Tomorrow," Discover *(Nov. 1982), pp. 20–25, © 1982 Discover Publications.*

---

pieces we see today was developed by German meteorologist and geographer Alfred Wegener, who became interested in the geologic connections between the continents in 1912. He was ridiculed by most geologists of his time.

Wegener suggested that a single, huge supercontinent had existed 200 million years ago. He named it Pangaea, which in Greek means "all Earth" (Figure 6-4). As Wegener viewed it, when Pangaea broke into pieces, continental drift began. The result is the arrangement of continents we have today.

The problem with Wegener's hypothesis was not a lack of geologic evidence to connect the continents in each hemisphere but a lack of explanation for the movements that separated these continents. The timeless concept of a rigid Earth crust that was "solid as rock" made his notion of continental drift seem impossible.

In the 1960s, half a century later, a series of geologic maps of the ocean floor and two studies of the ancient reversal of Earth's magnetic field revealed evidence that became vital to an explanation of continental drift. These studies proved the apparent movement not only of Earth's continents, but also of its ocean basins. Seafloor spreading was subsequently suggested as an explanation for the movement of entire lithospheric plates. Rigid plates moving over a soft, plastic asthenosphere would create patterns much like those first suggested by Wegener's theory.

Landform System Analysis

A systems analysis of plate tectonics affords us the opportunity to observe interrelationships that are crucial to our understanding of changes in the lithosphere. An internal energy flow, originating in the heat-producing radioactive decay of uranium deep inside Earth, powers a huge flow of crustal material at the surface. As subsiding plates are heated and melted, they are reabsorbed into the asthenosphere. Slow currents operating over millions of years deep within this lower layer eventually return (recycle) the mantle rock to the boundaries of spreading plates. This system balances the subduction of old material and the creation of new material (which is actually recycled old material) in the lithosphere.

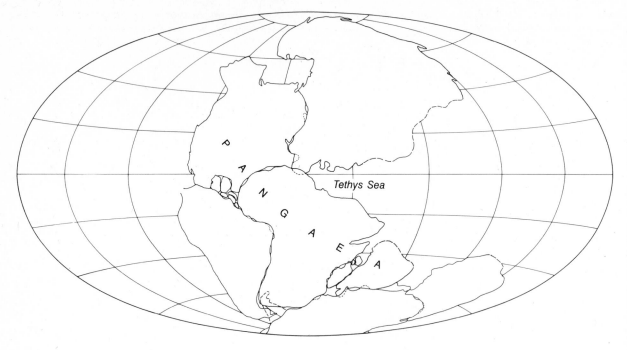

FIGURE 6-4
Wegener's hypothesis of continental drift postulated one huge supercontinent that he named Pangaea reconstructed here as it might have appeared 200 million years ago. (After R. S. Dietz and J. C. Holden, Journal of Geophysical Research *75: 4943. Copyright by the American Geophysical Union.)*

EARTHQUAKES

One consequence of the plate-tectonics system and its constant motion is instability created at plate boundaries. Seismic disturbances are byproducts of tectonic activity. As shown in Figure 6-5, the most earthquake-prone areas on Earth are located at plate divisions around the Pacific Ocean, across the Himalaya Mountains, around the Mediterranean Sea, and along mid-ocean ridges and rift zones.

Earthquakes begin deep within the crust as tension builds at plate boundaries. When stress exceeds the strength of the rock, an earthquake occurs as the rock slides to relieve pressure. These disturbances happen daily, but most are too small to be felt. However, major earthquakes may be devastating, especially if they occur in populated towns and cities built in the absence of rigorous building codes. The Santa Cruz–San Francisco earthquake of 1989 and the Northridge quake of 1994 would have caused considerably greater damage had it not been for strict building codes and other seismic-safety planning requirements.

Seismographs record earthquake magnitude according to the Richter scale. The Richter scale ranges from 0 to 9, each number representing an earthquake ten times the magnitude of the adjacent lower number. Often, people become so accustomed to small earthquakes in their homeland that they may fail to respond in a safe and sensible manner when a significant one occurs. This leads to disaster when a large-magnitude quake takes place.

VOLCANISM

The close relationship between plate tectonics and earthquakes can be extended to include volcanic activity. *Volcanism* occurs when molten rock flows from Earth's interior onto the surface through vents or fissures. Most volcanic activity is concentrated at the boundaries between plates, but hot spots such as the Hawaiian Islands are evidence of persistent volcanism in the middle of a plate.

The two major types of volcanoes are the composite-cone volcano and the shield volcano. In the spring of 1980 people living in the Pacific Northwest learned firsthand about the most dramatic type of volcanic event—the ex-

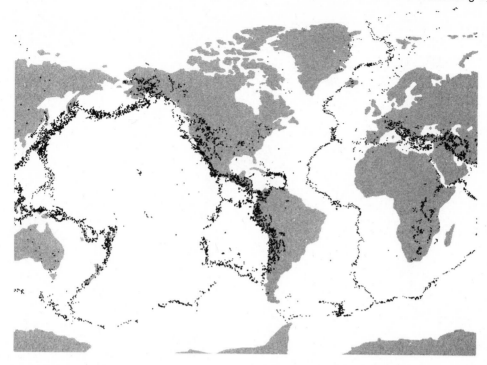

FIGURE 6-5

Map of world distribution of zones of earthquakes, volcanoes, and plate boundaries. (From Edward J. Tarbuck and Frederick K. Lutgens, The Earth: An Introduction to Physical Geology, *2nd ed., Columbus, OH: Merrill Publishing Co., 1987, p. 356. Reprinted with permission of Merrill Publishing Co.)*

plosion of a *composite-cone volcano.* Mount St. Helens in the Cascade Range of Washington State erupted with a powerful force that blew off the entire north side of its slope, illustrating the power of volcanic forces (Figure 6-6).

This type of volcano is the result of cumulative intrusion of magma beneath the cone. When the "breaking point" is reached, pressures are violently released by expulsion of steam, lava, gas, and pyroclastic debris from a central vent. The name composite-cone comes from the alternating layers of lava and ash in the cone. Other famous composite-cone volcanoes include Mount Fuji (highest peak in Japan; last eruption in 1707), Mount Kilimanjaro in Tanzania (dormant; highest point in Africa), and Mount Vesuvius in Italy (still active). Each of these volcanoes has remarkable conical symmetry, steep slopes, and is composed of layers of lava and ash.

The other type of volcano is the *shield volcano.* Relatively quiet with long, gentle, symmetrical slopes, shield volcanoes are usually characterized by less-intense eruptions than composite cones. Shield volcanoes also are known as "Hawaiian volcanoes." The scenic features of the Hawaiian Islands were formed by highly fluid basalt lava. Perhaps you have seen this type of volcano at Kilauea or

Mauna Loa in Hawaii. (See the box for more on volcanic hazards on page 123.)

THE WEATHERING SYSTEM

The preceding section analyzed the forces that change Earth from within. While these tectonic forces are working to change Earth from the inside, external forces also are working to change the surface of the planet. These systems erode the crust, transport the solid debris, and redeposit it as sediment. The external erosive forces include water, ice, wind, and the downslope pull of gravity.

Igneous and metamorphic rocks are formed below the surface under conditions of high temperature, high pressure, and in the absence of water. These rock types are unstable when they are exposed at the surface. Igneous rocks, for example, are formed under sealed, stressful conditions, but when they come to the surface, they are suddenly exposed to radically different conditions that cause weathering. The surface environment is characterized by variations in temperature, pressure, water, and atmospheric chemistry

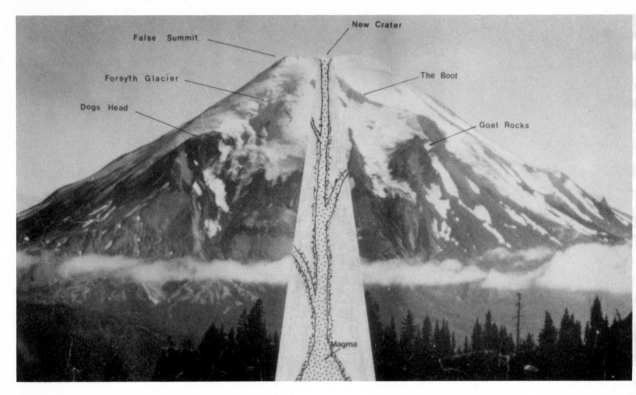

FIGURE 6-6
The eruption of Mount St. Helens in the Cascade Range of Washington State is the most recent explosive eruption of a composite-cone type volcano in the conterminous United States. (Photo courtesy of Geo-Graphics, Portland, OR.)

that act upon rocks and other surface materials. The term **weathering** refers to the alteration of rock material when exposed to the atmosphere, hydrosphere, and biosphere.

Weathering disintegrates "parent" rock material into smaller-sized particles and new minerals. These, combined with organic material to form soil, may again be changed into another rock type. Weathering proceeds progressively until a stable state is reached. There are two types of weathering: mechanical and chemical.

Mechanical Weathering

Mechanical weathering is the mechanical destruction of rocks by agents on Earth's surface. The destruction may arise from the release of internal stress within the rock, or it may be caused by an external agent, such as ice, crystal growth, heating and cooling, abrasion, or biota. When uplift or erosion occurs, removing overlying, confining rocks to expose the surface, a rock mass may expand vertically, cracking along joint fissures. This is mechanical weathering via release of stress. Granite is especially vulnerable to this type of weathering. The process of *exfoliation,* for ex-

ample, creates widely spaced joints parallel to the surface. These eventually slough off in large slabs. Hikers in California's Yosemite Valley and on other granitic slopes are killed every year when mechanically weathered slabs break loose unexpectedly where they are walking.

Another type of weathering that shatters rocks over time is caused by frost action. As frost crystals repeatedly grow and melt year after year inside the pore spaces of rocks, they cause expansion and contraction that ultimately results in breaking. Frost shattering is a major weathering agent in cold climates that results in tension along the joint. Plant roots also may grow within joints in rocks, achieving the same effect as frost action; we have all seen heaved and broken concrete slabs in sidewalks along tree-lined streets. Figure 6-7 illustrates the dramatic scenery that weathering can create in an arid landscape.

Chemical Weathering

The second type of weathering is chemical weathering. Mechanical weathering changes the physical size and shape of rocks, while chemical weathering actually changes the

A GRIM LOOK AT VOLCANIC HAZARDS

Volcanoes are the superstars of the physical world. They are dramatic and spectacular, often with profound effects on nearby human populations. They are more visible than earthquake faults, more temporal than glaciers, and more mysterious than floods or tornadoes.

Volcanic activity has continued through geologic time to the present, with some periods more active than others. Volcanologists tell us that we are probably in an era of relative dormancy.

Measuring the frequency of volcanic eruptions must be done with a combination of geological detective work and historical record. We must keep in mind that, in terms of geologic time, volcanoes such as Mount Fuji (last eruption in 1707) in Japan and Crater Lake (quiescent) in Oregon are still very active. Major historic volcanic events in the nineteenth and twentieth centuries have included:

1815   Tamboro Mountain, Indonesia

1883   Krakatau Island, Indonesia

1980   Mount St. Helens, Washington State

1983   El Chichon in Mexico

1985   Nevado del Ruiz, Colombia

In addition, volcanic activity in Hawaii has been almost continuous in this century.

As more is learned about earthquakes and volcanoes, confidence in predicting geologic events has increased. New signs of activity have been reported under more than 30 volcanoes in the United States, centering on the Cascade Range in the Pacific Northwest and the eastern Sierra Nevada of California. Many geologists believe that the molten rock or magma that lies below the Earth's crust is beginning to creep upward at plate boundaries. If, or when, the magma reaches the surface, dormant volcanoes could erupt with titanic violence. The specific effects of an eruption depend upon the amount of energy released and the type of rock ejected. Hawaiian-type shield volcanoes emit a runny, basaltic lava, but past eruptions from composite cone volcanoes have produced a silica-rich lava that is more explosive and chunky in texture.

Predictions of volcanic eruptions are still very general in human time, although bulging is recorded with strain gauges and rumbling is recorded with seismographs to note increases in activity. When subterranean activity reaches a critical level, there should be enough time to warn most nearby residents. Soon earthquake and volcano warnings may be as reliable as stormwarnings, thereby reducing property damage and loss of life.

chemical composition of rock minerals. These chemical reactions are accomplished by the interaction of water, atmospheric oxygen, and atmospheric carbon dioxide that are dissolved in rainwater and soil waters. Chemical weathering processes occur most rapidly in areas where temperature, rainfall, soil permeability, and biotic activity are high, for example, in the tropics.

The processes of chemical weathering include hydration, hydrolysis, oxidation/reduction, carbonation, and chelation.

Hydration occurs when water molecules attach themselves electrically to mineral molecules, thus weakening the mineral bonds. With the addition of water, some clays undergo a volume increase, cyclically swelling and contracting.

FIGURE 6-7
Evidence of weathering in Bryce Canyon National Park in southern Utah. Weathering has removed symmetrical parts of the rocks to create the dramatic scenery typical of many arid landscapes. (From Edward J. Tarbuck and Frederick K. Lutgens, The Earth: An Introduction to Physical Geology, *2nd ed., Columbus, OH: Merrill Publishing Co., 1987, Fig. 4-27, p. 123. Reprinted with permission of Merrill Publishing Co.)*

These clays form unstable foundation materials and often are associated with slumps and mudflows. Hydration often precedes hydrolysis.

Hydrolysis can be simply described as the deep rotting of rocks in the presence of water. Chemical dissolution and recombination take place between minerals and the acid soil water. This destroys parent materials, breaking them into soft, fine residuals (clay minerals) and into solubles (salts). The solubles move through the streams and rivers, finally collecting in the various salts of the oceans. The acid soils water becomes neutralized: the acid hydrogen ion (H^+) is combined into clay minerals, while the alkaline hydroxide ion (OH^-) moves to the oceans, making these bodies alkaline.

Oxidation and reduction reactions—the addition or removal of oxygen—are a third type of chemical weathering. *Oxidation* takes place when minerals containing iron are exposed to oxygen in air or water. Iron is prone to rapid oxidation. The red and yellow oxides of iron are the dominant inorganic coloring agents in soils. *Reduction* is exactly the opposite of oxidation and produces grey and green colors in soils. Reduction occurs only in the waterlogged soils of swamps or rice paddies.

Carbonation takes place when atmospheric carbon

FIGURE 6-8
Chemical weathering in limestone often creates the exciting and mysterious landforms known as karst topography—underground caverns, sinkholes, and passageways (Photo courtesy Florida Geological Survey.)

dioxide and oxygen combine to produce weak carbonic acid (pH 5.5; for comparison, a pH of 7.0 is neutral). This acid, though weak, vigorously attacks terrain, totally dissolving the rock given a sufficient number of years. *Karst topography*—regions of underground drainage, limestone caverns, and sinkholes—are the result of this type of weathering in limestone (Figure 6-8).

Chelation is an organic chemical reaction requiring biotic activity. In this process, organic acid molecules produced by decaying plant material link to mineral molecules and disperse them through the soil.

Weathering is ultimately controlled by climate, lithology, vegetation, time, and slope position. The end products of all forms of weathering are the formation of new minerals and new soils, and the deposition of dissolved solids into the oceans. Spatial and temporal variations create weathering zones that vary with depth beneath the surface of Earth. As a general rule, the depth of soil is directly related to the amount of water leaching through it. Deep soils are found in areas of high rainfall.

Slope Systems

Slopes allow gravity to move weathered rock particles down-hill, where they accumulate in a new environment having different weathering conditions. Closely related to weathering processes are "slope systems." Inputs into the slope system, or factors in slope behavior, are the particles produced by weathering, solar energy, the potential energy of gravity, atmospheric motion, and the biosphere. Outputs from the system—what washes down the hill—may go into the drainage system of streams and rivers, may be incorporated into plant biomass, may be stored on the slope, or may be removed by the atmosphere. The storage and balance of these inputs and outputs depends on the internal strength of slope materials and the magnitude of forces attempting to move them. Slopes reflect past processes and influence present processes. They are a part of all larger landform systems.

There is a considerable variation of initial slope form. Water is instrumental in slope processes, acting in both surface wash and within soil pore space. Water sets in motion the movement of soil particles through mechanisms such as creep, sliding, flow, and heaving (Figure 6-9).

The Evolution of Slopes

Slopes evolve through two principle gradational agents: denudation by water or wind, and mass movement of material on slopes. *Soil creep* is a grain-by-grain downhill displacement of soil, with rates of movement greatest near the surface. This form of mass movement is affected by several factors, including slope angle, frost heaving, and water content. Soil creep is triggered by moisture and temperature

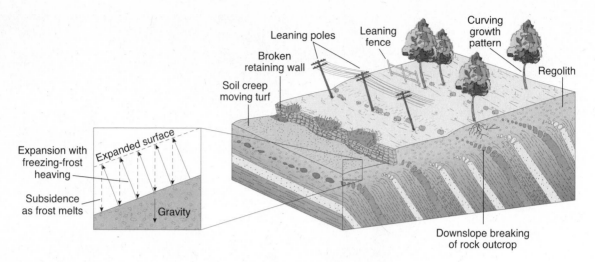

FIGURE 6-9

Mechanisms such as pure slide (creep), pure flow, and pure heave (slide) move soil particles down a slope. (From Robert W. Christopherson, Geosystems: An Introduction to Physical Geography, *2nd ed., New York: Macmillan, 1994, Fig. 13-24, p. 409. Used with permission of Macmillan Publishing Co.)*

changes; its slow action is responsible for tilting fenceposts and trees, and for trees that display curved trunk sections due to vertical growth after initial tilting.

Mass wasting is the movement of weathered material downslope due to the pull of gravity. Forms of mass wasting include rockfalls, slides, earth and mudflows, slumping, and snow avalanches. Water commonly assists gravity in mass movements, both by increasing the weight of the soil and by lubricating the particles. Gravity can most easily overcome resistance to inertia of the mass when slopes are steep. Movement may be rapid or quite slow (depending on the water content of the materials, the temperature, and the cohesion of sediments).

Mass movements often result from human activity on the surface of Earth. Building a subdivision on a slope, mining and quarrying hillsides, and burying areas in heavy landfills have frequently led to downslope movement. Oversteeping of slopes in roadcuts is the most common example of human-caused instability leading to mass movement. Human-induced mass movements often result from poor planning and unawareness of basic geographic principles.

FLUVIAL SYSTEMS

The word *fluvial* comes from the Latin word for river. When we speak of fluvial systems, we use the term in a broad sense to mean all liquid surface water (streams, rivers, lakes, and ponds), and frozen surface water that behaves in a riverlike manner (glaciers).

The action of running water on the lithosphere is the most important of the external geomorphic agents. Consider the dramatic fluvial expression visible in the Grand Canyon, as shown in Figure 6-10. This chasm reflects the immense power of running water to carve the land.

The amount of water flowing in channels is variable because of its distribution. Large-magnitude flows (floods) occur infrequently, while low-magnitude flows are common. Streamflow regimes may thus be *ephemeral* (flow only during and immediately after a precipitation event), *intermittent* (flow seasonally), or *perennial* (flow year-round because of groundwater input to supplement precipitation). Energy in the system necessary to move sediments and carve landforms comes from the conversion of the potential energy of gravity into kinetic energy.

A drainage basin is an open fluvial system with inputs of water and energy and outputs of water and sediment load. A model of this fluvial system is diagrammed in Figure 6-11.

Discharge—the volume of water flowing in a stream—is the primary variable controlling stream-channel processes. Discharge is derived from precipitation falling directly into the channel, overland flow into the channel, and the migration of groundwater into the channel. Most water en-

FIGURE 6-10
The Grand Canyon in Arizona is a dramatic example of the power of fluvial activity. (Photo by author.)

ters streams by subsurface flow (otherwise, streams would be empty on days when rain was not falling).

Water Erosion, Transportation, and Deposition

The process of stream erosion begins with overland flow and ends when material that has been eroded is deposited in a new area. The transportation mode for each particle size has a name: *suspension* (the carriage of fine particles—the silt which forms clay—that will not settle out until water movement stops); *saltation* (the intermittent, leaping movement of sand or gravel); or *traction* (the dragging of large particles, such as rocks).

Movement of rock and sediment at the bottom of the stream channel causes abrasion and scouring of the channel. Small particles travel faster than large ones because less energy is required to move them. The amount of sediment carried in the water varies with the system discharge: the greater the water volume, the greater the amount of sediment carried.

Material also is carried in chemical solution. Generally, the dissolved load equals or exceeds the solid load over a

year, and the greater the amount of groundwater contributing to a stream system, the greater the amount of dissolved load.

The amount of erosion accomplished by a stream depends on the volume of water, plant cover in its basin, soil and rock types, steepness of slope, and the velocity of the water. Different sediment sizes have different critical threshold velocities that are necessary to erode them. If the velocity of the stream is increased, either by increased water volume or by increased channel steepness, the size of particles that can be moved also increases. Streams cut their channels by a combination of three means: *widening* their channels, *deepening* their channels, and by *extension* (for example, the process whereby the head of the stream progressively erodes its way up a ravine).

Urbanization, flood control projects, freeway construction, lumbering, mining operations, and other human modifications of vegetation result in bare ground and lower infiltration of precipitation, thus augmenting the erosional power of fluvial systems. Considerable flood damage in developed areas may be traced to improper land use (Figure 6-12). As sediment moves downstream, the channel responds to accommodate the input of new material.

The patterns formed by streams and their tributaries may be *dendritic, radial,* or *trellis* (Figure 6-13), depending on the underlying structure and geology.

Materials being carried by a stream drop out as its velocity decreases. The heaviest particles (pebbles, gravel) are deposited first; then the intermediate particles (sand) drop out next; and then lightest particles (silt clay) drop out last. Rivers carrying large amounts of alluvium often exhibit well-developed floodplains and natural levee systems as a result of sediment deposits. The coarse particles are always next to the stream or at the beginning of a delta, where water velocity is greatest.

Stream meanders are further evidence of a stream adjusting its slope and velocity to match the load. Regular periods of flooding are common in this type of stream pattern as water spills out of the main channel into surrounding floodplains. As water is spread over a larger area, stream efficiency decreases, allowing new sediment to drop out and be deposited.

Flood-prone streams have long attracted human settlement throughout the world. Chinese villages prospered along the Yangtze and Hwang Ho rivers. The well-known agricultural hearth of seed plants evolved at the confluence of the Tigris and Euphrates rivers. More recent urbanization in the Central Valley of California often has been alongside rivers that flood seasonally. The importance of

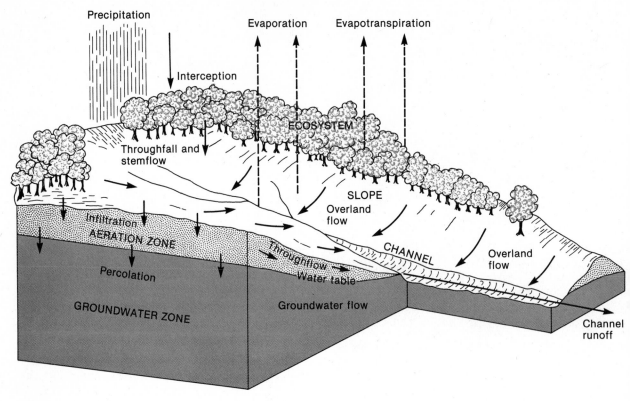

FIGURE 6-11
This diagram of a drainage system illustrates the inflows and outflows typical of an open system.
(From I. D. White, D. N. Mottershead, and S. J. Harrison, Environmental Systems: An Introductory Text, *London: Unwin Hyman Ltd., 1984, Fig. 10-9, p. 220. Reprinted with permission of Unwin Hyman Ltd.)*

rivers to human settlement is presented as a special case study in Chapter 13.

GLACIAL SYSTEMS

Another major agent of landscape change is ice in the form of glaciers. Glacial systems develop when winter snowfall exceeds snow melting in the summer months. A glacier is a flowing mass of ice; the snow is converted into ice as a result of the weight and pressure of accumulated, never-melting snow.

Glacial systems require high snowfall and low solar radiation. A summary of glacial system inputs would include snow, minerals and rocks, energy, and topographical relief or slope. Storage is in the glacier itself; glaciers and ice caps store approximately 75 percent of the Earth's freshwater. Outputs of the glacial system include water vapor, meltwater, blocks of ice, sediments, and energy. It is obvious from

this list that a glacial system is an open system, because both energy and materials are exchanged across the system's boundaries.

Glacial Budgets

The yearly gain or loss of snow and ice in a glacier is the annual *glacial budget*. Some glaciers have large budget changes; they gain and lose a great deal of material annually. A *positive* budget describes snow accumulation, a growing glacier, and the advance of ice. In the opposite case, when net wasting, or *ablation* (loss of ice mass), occurs due to low snowfall and/or high output, the glacial budget is said to be *negative*. The glacial budget defines the state of a glacier and determines whether it is in equilibrium or is increasing or decreasing in size.

Input into a glacial system consist primarily of precipitation and radiation, although accumulation processes also include sublimation (dry evaporation) and avalanches. Ablation results from melting, evaporation, sublimation, wind erosion, and *terminal calving* (when large blocks of ice

FIGURE 6-12
Flooding illustrates the erosive power of river activity during a high-water stage.

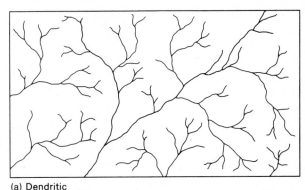

(a) Dendritic

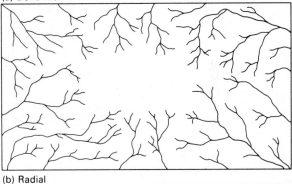

(b) Radial

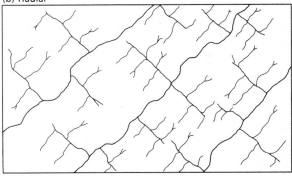

(c) Trellis

FIGURE 6-13
Stream patterns on the land may be dendritic, trellis, or radial. (From E. Willard Miller, Physical Geography: Earth Systems and Human Interactions, *Columbus, OH: Merrill Publishing Co., 1985, Fig. 13-5, p. 261. Reprinted with permission of Merrill Publishing Co.)*

break off as icebergs). Accumulation and ablation may differ in time and space; their comparative rates determine the rate of movement at any point on the glacier. Net accumulation is high as ice advances. Glaciers melt downward (appearing to move backward, or to retreat) when the budget is negative.

Glacial Movement

To be classified as a glacier, ice must be in motion. The mechanisms of this movement are *internal shearing* (creep and fracturing) and *basal sliding* (the sliding of the glacier over its bed). The velocity is highest at the surface of the glacial ice and in the center. It is most rapid at the equilibrium line, where accumulation is balanced with ablation.

Types of Glaciers

The logical locations of glaciers are high elevations and high latitudes. Today, extensive areas of glacial ice are found only in Greenland, Iceland, and Antarctica. On the island of Iceland, a large glacier covers over 7700 square miles (20,000 km²). Mountain glaciers are found at high elevations even near the Equator in Kenya. Mount Kilimanjaro and Mount Kenya both support glaciers, where the high precipitation and elevation combine to create a positive ice budget.

Glaciers that form on high mountains like Mount Kilimanjaro are known as *alpine glaciers* or *valley glaciers*. Mountain climbers and tourists appreciate the dramatic landscapes of these glaciers. Figure 6-14 illustrates a typical alpine glacier in Alaska on Mount McKinley.

The massive ice caps on Greenland and Antarctica are remnants of past ice ages and are known as *continental glaciers*. A succession of alternating glaciations and their disappearance during interglacial periods has occurred at least 20 times during the past 3.5 million years. In fact, we may be living in the final years of the current interglacial period. The last ice age, during the Pleistocene geologic period, ended 10,000 years ago.

FIGURE 6-14
Mount McKinley in Alaska displays the amazing mass of ice contained in some alpine-type glaciers. (Photo courtesy Stephen F. Cunha.)

Many scientists today credit our changing weather and climatic patterns with predicted glaciation. They have found that our "normal" weather on Earth during much of its history has actually been much colder and wetter than that of the twentieth century. The trend toward increased global warming in recent decades (discussed in Chapter 10) offers an alternate detailed glimpse into our planet's changing climatic conditions.

Figure 6-15 shows the extent to which ice covered Europe and North America at the maximum spread of the last ice sheet. Continental glacial ice advanced all the way south to northern Pennsylvania, covering all of Canada, New York, New England, and a good portion of the Midwest. The Scandinavian countries in Europe were covered, along with much of central Germany, the former USSR, and even Great Britain. South America also had an ice sheet at this time, which flowed out of the Andes Mountains poleward of 40° S latitude (southern Argentina and Chile). Distinctive landforms that are the direct result of this massive continental glaciation are shown in Figure 6-16.

Glacial Erosion, Transportation, and Deposition

Glaciers are one of the most dramatic agents of lithospheric erosion. At the point of contact between the ice and its bed a great force is exerted and erosion of surface materials proceeds at a dramatic pace. Consider, for example, the beds of the Great Lakes, vivid reminders of the power of glacial erosion during past ice ages.

Glacial erosion mainly takes place through the processes of abrasion and quarrying or plucking. *Abrasion* occurs when the base of the glacier scrapes the surface like a giant sanding machine. *Quarrying* or *plucking* invades fractures and removes large bedrock blocks. As the ice moves, blocks of enormous size are moved down the glacial slope.

Transportation of eroded material may occur at the surface of the ice or deep within the glacier; debris is found at all levels of the glacier's cross section. Material that falls on top of the ice by rockfall directly above the glacier may be carried at the surface or fall into cracks (crevasses) and be incorporated into the bulk of the ice mass. Material quarried from the ground may be carried at the base of the glacier.

The deposition of material carried by the glacier occurs at the glacial margin (the sides) and along its base. At the terminus, the debris is released as ice begins to melt. The debris then is moved and sorted by running water and wind. *Glacial till* is a term used to describe material deposited directly beneath the ice mass. A *moraine* is made up of material deposited near the edges or beneath the ice and consists of all sizes of unsorted material.

Another kind of glacial deposit is the *glacial outwash*, which is material deposited by meltwater from the glacier. These glaciofluvial deposits are sorted by size and stratified (deposited in layers). Their appearance and genesis follow the same patterns as other fluvial deposits discussed earlier.

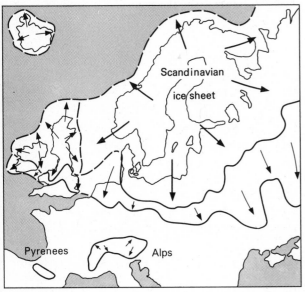

FIGURE 6-15
*Extent of Pleistocene glaciers in North America and Europe.
(From E. Willard Miller,* Physical Geography: Earth Systems
and Human Interactions, *Columbus, OH: Merrill Publishing
Co., 1985, Fig. 14-3, p. 280. Reprinted with permission of
Merrill Publishing Co.)*

Glacial Landforms

Processes of glacial erosion, transportation, and deposition
result in distinct landforms at the Earth's surface. Most of
us cannot help but be impressed by the dramatic views offered by glacial activity. Images of Yosemite Valley, the
European Alps, and the Canadian Rockies offer exciting examples of glacial erosion and deposition.

OTHER GEOMORPHIC SYSTEMS

Wind Denudation

Denudation of Earth's surface is also caused by wind. The
action of wind on the landscape causes transportation and
deposit of silt and sandy materials. Clay is generally too cohesive (sticky) to be moved by wind. Wind systems use the
kinetic energy of air motion to entrain and move sediment.
The kinetic energy is expended first by moving sediments
and second by conversion to heat energy as a result of friction.

Evidence of the wind at work is usually found in locations having a combination of little protective vegetation
and high wind velocities. Arid and semiarid landscapes,
barren fields, beaches, floodplains, and glacial outwash
plains are particularly susceptible to this type of denudation.

Coastal Denudation

Along the edges of continents where land and water meet,
geomorphic change occurs both seasonally and daily.
Perhaps you have returned to a favorite camping spot at the
beach, only to discover it has changed drastically since you
last visited it. Marine geomorphic agents such as waves,
tides, and currents may erode, transport, and deposit materials, rapidly forming new landforms and destroying old
ones.

Erosional landforms such as cliffs, platforms, terraces,
sea stacks, arches, and beaches are some of the natural won-
ders of a visit to the seashore. Depositional features also
may be striking additions to the changing landscape.
Beaches, spits, bars, and sand barriers are caused by the deposition of transported coastal material. Human interfer-
ence often creates unwanted areas of deposition behind
breakwaters and piers (Figure 6-17). The city of Santa
Barbara, California, for example, must dredge its harbor
regularly to keep sand from building up. Many coastal har-

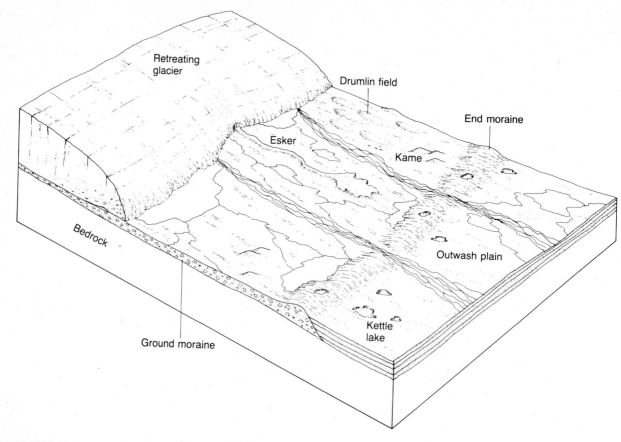

FIGURE 6-16
Landforms of glacial deposition are often so large that an observer would be amazed to learn of their glacial origin. (From Edward J. Tarbuck and Frederick K. Lutgens, Earth Science, *5th ed., Columbus, OH: Merrill Publishing Co., 1988, Fig. 4-15, p. 108. Reprinted with permission of Merrill Publishing Co.)*

bors such as San Francisco Bay and the port of Los Angeles experience this same problem.

SOIL SYSTEMS

The study of soil systems *(pedology)* provides physical geographers with a link between processes in the lithosphere and processes in the biosphere. Soils form the interface between ecosystems and weathering systems. Soil is vital for the production of food for human life on Earth.

Soils are frequently defined as natural substances consisting of minerals, a soil atmosphere, a soil solution, and organic material. Horizons—layers of distinct composition—evolve in vertical sequences as soils form. Soils may differ from their *parent materials* (usually rock) in morpho-

logical, physical, chemical, and mineralogical properties, and in their biological characteristics.

Soil scientists study five major factors in the evolution of any soil system: parent material, climate, organic material or biota, relief or topography, and time. Soil evolves and changes through time according to influences of these variables.

A soil system is three dimensional. It has vertical boundaries with the atmospheric-surface interface and the lower limit of root penetration. Horizontal boundaries are established at the interface with water bodies, parent rock material, and human-made surfaces. Soil is connected to the atmosphere, lithosphere, hydrosphere, and biosphere, so it is made up of solids, liquids, and gases.

Water flowing down through the soil from the surface tends to remove material from the upper part (*eluviation*) and redeposit it in the lower part of the soil (*illuviation*).

FIGURE 6-17
*Landforms that we often enjoy at
the beach include estuaries, penin-
sulas, spits, hooks, and bay bars
(From Robert W. Christopherson,
Geosystems: An Introduction to
Physical Geography, 2nd ed., New
York: Macmillan, 1994, Fig. 16-9, p.
489. Used with permission of
Macmillan Publishing Co.)*

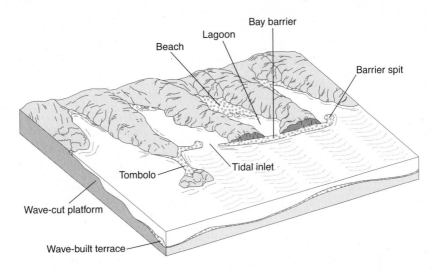

This action leads to the development of distinct layers or
horizons. A *horizon* is a distinguishable layer in the soil that
has certain chemicals, and a distinct color, texture, pH, and
structure. A **soil profile,** which is the vertical arrangement
of horizons down to the parent material, is shown in Figure
6-18. A soil profile is most easily observed in an exposed
roadcut in arid lands or at a construction site.

Soil Structure

Soils are classified according to structure or the arrangement of soil particles. Structure is based on the chemical
characteristics of the parent rock material, the amount of
weathering, the availability of organic material, and the size
of soil particles. Structure is the most important of soil's
physical properties. Structure also depends on the shape of
soil particles, or *peds.* As shown in Figure 6-19, peds may
be spherical, blocky, platelike, or prism-shaped.

Soil Texture

Soils are grouped into three general texture classifications—
coarse, medium, and fine. Texture may best be defined as
the "feel" of soil. A coarse soil with a high percentage of
sand particles will feel loose and grainy when dry and will
fall apart when wet. Medium-sized soil particles (silt) feel
like talcum powder. Fine soil texture is dominated by clay
particles; clay soils crack when dry and stick together when
wet.

The proportions of sand, silt, and clay determine the
texture characteristics of soils and are commonly depicted
on the soil triangle shown in Figure 6-20. **Loam** soils provide the best substrate for plant growth.

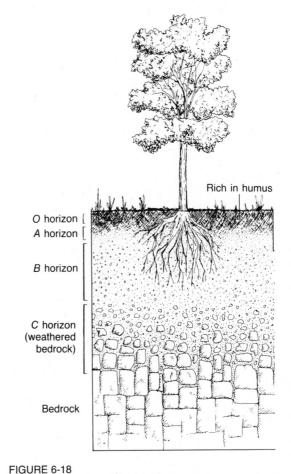

FIGURE 6-18
*A typical soil profile with at least three distinct levels of soils
on top of bedrock. (From Edward J. Tarbuck and Frederick K.
Lutgens,* Earth Science, *5th ed., Columbus, OH: Merrill
Publishing Co., 1988, Fig. 2-10, p. 50. Reprinted with permis-
sion of Merrill Publishing Co.)*

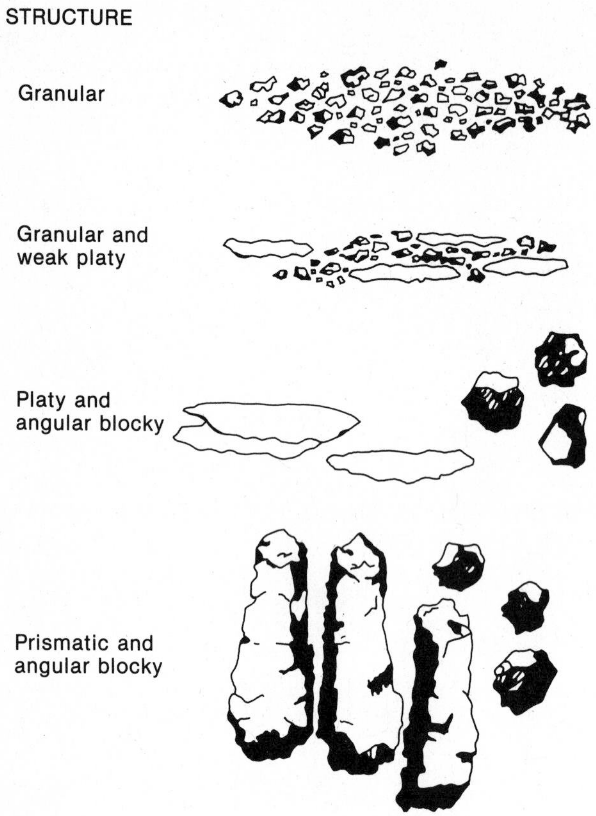

STRUCTURE

Granular

Granular and
weak platy

Platy and
angular blocky

Prismatic and
angular blocky

FIGURE 6-19
The structure of the soil depends on the shape of the soil particles or "peds." These individual pieces come in many shapes. (From Edward J. Tarbuck and Frederick K. Lutgens, The Earth: An Introduction to Physical Geology, 3rd ed., Columbus, OH: Merrill Publishing Co., 1988, Fig. 5-19, p. 127. Reprinted with permission of Merrill Publishing Co.)

Soil Color

The different colors of soil reflect processes of formation and evolution. Most of us readily recognize a dark brown or black soil as fertile due to the large amount of organic material it contains. Grasslands where organic material from plant roots is incorporated into the soil are usually the darkest. *Chernozem* soils (from the Russian, meaning "black earth") in the western part of the North American wheat belt and in Ukrainian wheat fields are among the world's darkest and most fertile. The amount of **humus** that has decomposed in the soil determines its organic content and creates the darker color.

Red and yellow soils are a product of iron oxidation. When organic material is weathered away by physical processes, mineral residues, especially iron oxides, are left

behind. These red and yellow soils are especially common in subtropical and tropical areas where frequent and intense precipitation removes humus through *leaching*.

Soil Classification Systems

It is very laborious to classify soils; most classification schemes are confusing. The soil in your front yard may be different from the soil in your neighbor's yard. Soils may be classified according to structure, color, or texture, or according to the climatic or vegetation zone in which they are located. Through the centuries soil classification systems have been defined according to different characteristics and criteria.

The most recent system was introduced by the U.S. Department of Agriculture in 1960. Officially known as the United States Comprehensive Soil Classification System, it is commonly called the "Seventh Approximation" because it is the seventh revision of the first system presented to American pedologists. The Seventh Approximation is based upon current properties of soils, not upon their origin or environment. Table 6-1 compares soil types with vegetation and climate.

Soil surveys have been conducted in most counties of the United States by the U.S. Department of Agriculture's Soil and Conservation Service (SCS). If you consult your local SCS office to discover the characteristics and potential uses and limitations of your local soils, you will probably be surprised at their variability.

THE LIFE LAYER

The Biosphere

The **biosphere** includes portions of the atmosphere, the lithosphere, and the hydrosphere. It is the "veneer of life" on the planet. It consists of all living organisms and their remains. As with the other "spheres," the biosphere is best considered in the systems context. The complex of factors that operate in and affect the biosphere are described as open systems, with energy as the main input and output. Within each system, flows of energy and matter are transferred to and from living material. For example, energy is most often transferred through photosynthesis by plants into organic molecules. These compounds are then processed when they are consumed by animals.

The systems at work in the biosphere support associations of plants and animals called biotic communities or

FIGURE 6-20

This soil triangle often is used by geographers and soil scientists (pedologists) to determine the texture of soil. (From E. Willard Miller, Physical Geography: Earth Systems and Human Interactions, Columbus, OH: Merrill Publishing Co., 1985, Fig. 10-4, p. 172; after Bridges, 1978. Reprinted with permission of Merrill Publishing Co.)

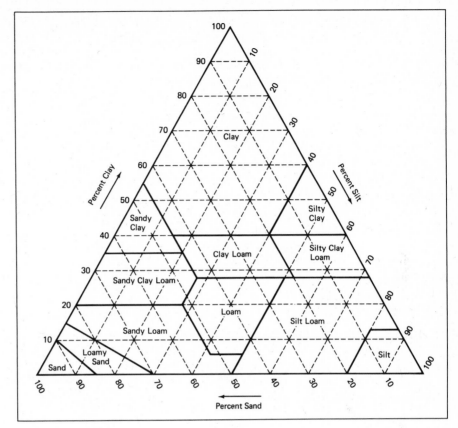

TABLE 6-1

Summary of soil types showing how Earth systems of soils, vegetation, and climate interrelate

Climate	Vegetation	Typical Area	Soil Type	Topsoil	Subsoil	Remarks
Temperate humid (> 25 in. or 63 cm rainfall)	Forest	Eastern U.S.	Pedalfer	Sandy, light colored, acid	Enriched in iron, aluminum, clay; brown	Extreme development in conifer forests because abundant humus makes groundwater very acid. Produces light gray soil due to iron removal.
Temperate dry (< 25 in. or 63 cm rainfall)	Grass and brush	Western U.S.	Pedocal	Commonly enriched in calcite; whitish	Enriched in calcite; whitish	*Caliche* is name applied to the accumulation of calcite.
Tropical (heavy rainfall)	Grass and trees	Tropics	Laterite	Zones not developed. Enriched in iron and aluminum; brick red; all other elements leached out.		Apparently bacteria destroy humus, so no acid is available to remove iron.
Extreme arctic or desert	Almost none, so no humus develops	Arctic; desert	No real soil forms, due to back of organic material. Chemical weathering is very slow.			

Source: Edward J. Tarbuck and Frederick K. Lutgens, Earth Science, *5th ed., Columbus, OH: Merrill Publishing Co., 1988, Table 2.2, p. 52. Reprinted with permission of Merrill Publishing Co.*

biomes. (A particular species of plant or animal is rarely found all by itself.) All of the factors affecting a **biotic community** and that community's effects on its surroundings are termed its **ecosystem.**

Thus, the structure of the biosphere may be thought of as a hierarchy of levels describing relationships between individual organisms and their larger contexts. The levels of complexity and size, in increasing order, are organism, population, community, ecosystem, biome, and biosphere. A number of individual organisms make up a population. Several populations of different organisms make up a community. One or more communities in conjunction with influences from the abiotic realm make up an ecosystem. Several proximate ecosystems constitute a biome that makes up the biosphere. This section discusses how ecosystems operate and how they are differentiated into biotic communities and into global scale biomes.

Plant Species

All similar organisms are identified as **species.** Most studies of function, evolution, and ecologic relationships among organisms are done at this level. Biologists and biogeographers may study the evolution of species of plants or animals, or they may study their adaptive mechanisms to various environments. The geographic component of their research almost always lies in viewing distributions of species either globally or within a region.

There are about 400,000 species of plants on Earth today. There are more than a million species of invertebrate animals, 850,000 of which are insects. There are approximately 40,000 species of vertebrates. The location and distribution of each species is dependent on a great number of genetic, environmental, and geographic factors, including the organism's place within the larger biotic community; the temperature, moisture, isolation, and chemical nature of the environment; and, of course, the human influences upon the species.

Often the location of a population of a species will reflect a long and complex history of evolution, dispersal, or retreat as Earth environments have changed through time. Changes in climate and topography, the extension and retreat of continental ice sheets, and changes in sea level have all influenced the distribution of species. Current patterns of distribution reflect the history of the environment as well as the species' relationships to present environmental conditions.

In general there are more species of plants and animals, both aquatic and terrestrial, in the world's tropical zones. Species with restricted distributions are described as *endemic,* whereas species present throughout most of the globe are known as *cosmopolitan.*

Figure 6-21 shows similarities among the species found in North America and in Eurasia. Note that these species are significantly different from those found in South America and Africa. Comparison of this map with Figure 6-22 shows a relationship between length of isolation due to continental drift and the degree of botanical and zoological differences in speciation. Note that Australia, India, and Madagascar are biological "islands" as a result of their geographic isolation.

Various species that occupy the same place and interact in some way make up a **biotic community.** The variety and diversity of communities is almost as great as their component species. The community concept can be applied at different scales, ranging from a small freshwater pond to an Amazon rain forest. There are usually several distinct communities within the larger mapped areas, the biomes. For example, the Mediterranean scrub biome, found on all continents except Antarctica, may contain communities of coniferous (needle-leaved) forest, deciduous (broad-leaved) forest, riparian (stream-bank) forest, and grassland, as well as its unique Chaparral scrub community.

Trophic Levels

A community may be modeled as an ecosystem with energy and matter exchanges within it, and also between the system and its surroundings. In the ecosystem model, organisms are placed within the community structure depending upon their place in the food chain or upon their source of energy or their **trophic level.** The following classifications are according to the following three trophic levels: autotrophic, heterotrophic, and saprotrophic (see Figure 6-23).

The *autotrophic* level (composed of living things that create their own food from inorganic materials) contains the primary production system of the global biomass, the green plants. Here light is absorbed in the process of photosynthesis. Heat, water, minerals, and carbon dioxide are utilized by the organism. Heat, oxygen, and water vapor are given off to the system environment. Biomass is accumulated until the plant dies or drops leaves or fruits. The organic matter decomposes in the *saprotrophic* (decomposition) level to become the storehouse of chemical nutrients for new plant growth.

The *heterotrophic* level (made up of living things that eat other living things), sometimes referred to as the grazing-predation pathway, is composed primarily of animal

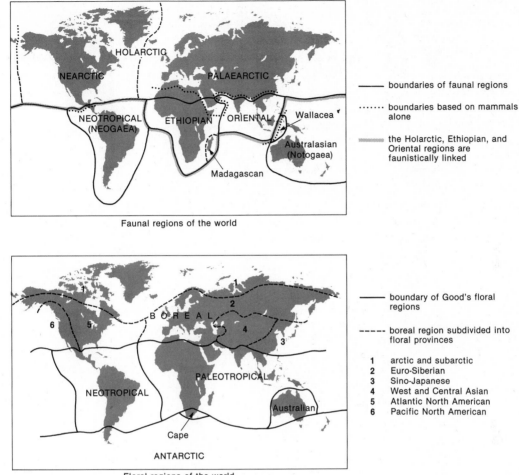

boundaries of faunal regions

.... boundaries based on mammals alone

▓▓▓▓▓ the Holarctic, Ethiopian, and Oriental regions are faunistically linked

Faunal regions of the world

——— boundary of Good's floral regions

----- boreal region subdivided into floral provinces

1 arctic and subarctic
2 Euro-Siberian
3 Sino-Japanese
4 West and Central Asian
5 Atlantic North American
6 Pacific North American

Floral regions of the world

FIGURE 6-21
Species similarity among the continents. Note the similarity of species between North America and Eurasia as compared with South America and Africa. This is due in part to changes in the relative positions of the continents through geologic time. (From I. D. White, D. N. Mottershead, and S. J. Harrison, Environmental Systems: An Introductory Text, *London: Unwin Hyman Ltd., 1984, Fig. 7-4, p. 148. Reprinted with permission of Unwin Hyman Ltd.)*

consumers of plant materials, and the secondary consumers such as the carnivorous animals. In these cases, food in the form of plant and animal tissue provides the energy and nutrients for the animal. Wastes and the entire body find their way to the decomposers.

The saprotrophic level (made up of living things that eat decaying matter) is also called the **detrital system.** At this level dead organic matter is stored, and decomposer microorganisms (such as bacteria) and macroorganisms (such as worms and insects) reduce the material to forms reusable by the primary producers (autotrophs). The detrital system is the source of soil organic matter (humus).

Chemical elements in the humus become available as plant nutrients in the primary production system.

A simplified model of the relationships among vegetation, biomass, and soil is shown in Figure 6-24.

Systems in the Biosphere

Matter is transferred cyclically through the various levels of the ecosystem. The water or hydrologic cycle already has been introduced. In a similar manner, one may trace through cycles in the ecosystem the movement of carbon and oxygen (photosynthesis) or nitrogen and phosphorus

FIGURE 6-22

The greater the distance from the northern latitudes of Eurasia and North America, the greater are the differences in plant and animal forms. (A note on taxonomy: very similar organisms are labeled "varieties" within the same species, or may be termed "subspecies." The terms species, genus, family, *and* order *indicate increasing differences in genetic makeup and appearance of organisms. Neogaea consists of entirely separate biotic realms.) (From I. D. White, D. N. Mottershead, and S. J. Harrison,* Environmental Systems: An Introductory Text, *London: Unwin Hyman Ltd., 1984, Fig. 7-4, p. 153. Reprinted with permission of Unwin Hyman Ltd.)*

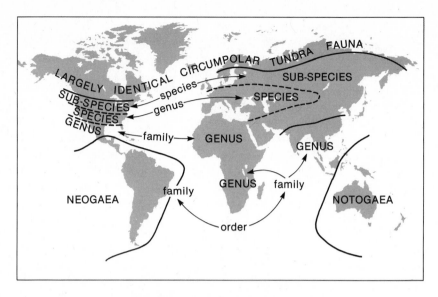

FIGURE 6-23

Model of the world biosphere system showing trophic levels and energy transfers. (After I. D. White, D. N. Mottershead, and S. J. Harrison, Environmental Systems: An Introductory Text, *London: Unwin Hyman Ltd., 1984, Fig. 7-7, p. 154. Reprinted with permission of Unwin Hyman Ltd.)*

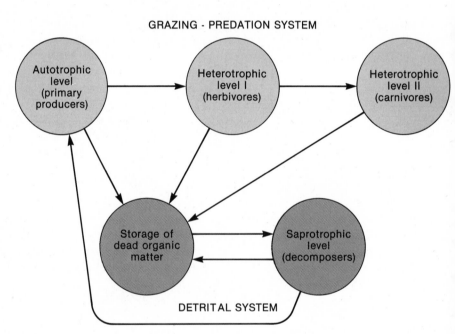

(which are primary plant nutrients used as fertilizers in agriculture).

GLOBAL ECOSYSTEMS

A composite of many biotic communities, each with its own ecosystem, becomes one of the major global biomes.

Figure 6-25 shows a world map of the major terrestrial biomes. The two most important factors in the determination of vegetation type in all biomes are temperature and precipitation.

Annual rainfall totals are especially significant. Rainfall availability determines whether an area is desert, grassland, or forest. Generally an area with 10 inches (25 cm) or less of annual rainfall will support only the most specialized drought-resistant plants. This is true regardless of temperature, soil characteristics, or topography.

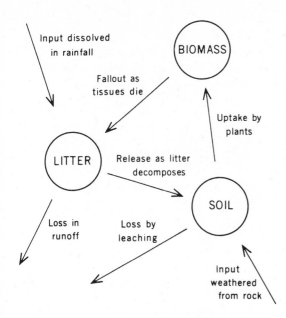

FIGURE 6-24
The mineral cycle showing interrelationships among living plants, dead organic matter, and soils. (From Philip J. Gersmehl, "An Alternative Biogeography," Annals of the Association of American Geographers *66 (June 1976): 227. Reprinted with permission of the Association of American Geographers.)*

Grasslands are found in regions that experience 10–30 inches (25–75 cm) average annual precipitation. In these areas rainfall is irregular, fires may be frequent, and grazing animals may inhibit tree seedlings.

Forests occur in undisturbed areas where rainfall patterns are regular and average annual over 30 inches (75 cm) per year. You may see from examining the map of biomes that deserts, grasslands, and forests occur naturally in tropical, temperate, and polar regions.

Each of these biomes is briefly described below by noting its location on the globe, its typical vegetation and animal communities, its climatic and lithologic influences, and its relationships to neighboring biomes. Finally, a few implications for human settlement and land use are given.

Deserts

More than a third of Earth's land surface is desert by virtue of low precipitation and high evaporation. Tropical deserts such as the Sahara in northern Africa or the Great Sandy Desert in Australia are generally the hottest and support the least vegetation cover. Mid-latitude temperate deserts, such as are found in North America and Asia, are generally hot

in the summer but cold in winter. They support a variety of succulents, shrubs, and seasonal wildflowers. Cold deserts, such as the Gobi Desert (China) and Antarctica, are located in high latitudes and away from the moderating influence of large water bodies. They are cool in the summer and cold in winter.

Most deserts are some distance from water bodies. Exceptions include the Atacama Desert in Chile and the Kalahari Desert of Botswana in southern Africa, which are located on the west coast of continents and are not subject to the passage of low-pressure systems. Desert soils almost always are nutrient deficient because growth rates of plants are slow and there is little biomass to decompose into humus.

Desert animals are almost always nocturnal and may exhibit unusual adaptations for coping with drought and extreme temperatures (the camel is a fine example). Human habitation of the deserts has traditionally been limited to nomadic hunter-gatherers, such as the Kung of southwest Africa and the Aborigines of Australia. Other migratory desert peoples are pastoralists who graze animals in the deserts of Central Asia and North Africa. A major world problem associated with human activity in arid areas is the process of **desertification,** the unwelcome conversion of grassland to desert by overgrazing and inefficient farming. This alarming cause of much starvation and resettlement is discussed in Chapter 10.

Grasslands

Bordering most desert areas in middle and low latitudes are regions of grassland. Mid-latitude grasslands may be called plains, prairies, pampas, selva, steppes, or veldt. Tropical grasslands are called savanna and may support scattered trees. Grasslands also grow in other areas of moderate rainfall.

Temperate grasslands tend to have limited and irregular rainfall and a large seasonal temperature range (warm summers and cold winters). Most are in the centers of land masses away from the moderating influence of oceans. They are usually very windy places; the American Midwest is typical. Many temperate grasslands historically supported large herds of grazing animals, such as the bison in North America. Soils in the temperate grasslands are the best on Earth for field agriculture and produce a major portion of the world's wheat, maize, livestock, and vegetables. The deep topsoils are very fertile due to their large humus content. Nutrients are stored in the soil rather than in the living biomass (as in forest ecosystems).

FIGURE 6-25
Major Earth biomes distribution map. (From James S. Fisher, Geography and Development: A World Regional Approach, *3rd ed., Columbus, OH: Merrill Publishing Co., 1989, Fig. 3-4, pp. 54–55. Reprinted with permission of Merrill Publishing Co.)*

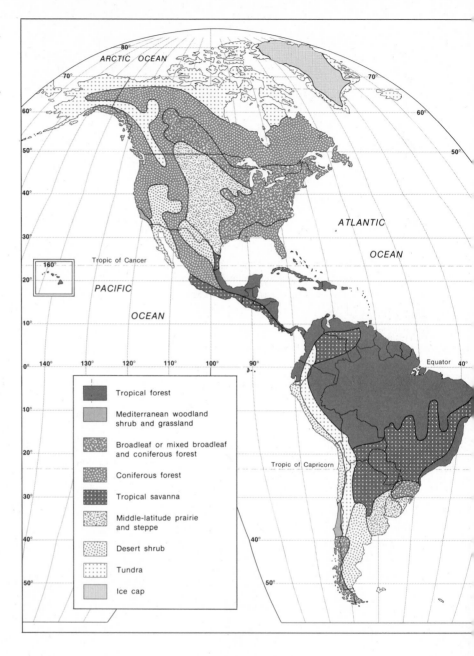

Prior to the Age of Exploration in the fifteenth century, the world's mid-latitude grasslands were in Central Asia, northern and southern Africa, southern South America, and the central part of North America. Today, less than 1 percent of the original mid-continent grasslands in North America remains in its original state, the greatest part of it having been heavily used because of its high soil fertility for grazing livestock and farming, made possibile with improved irrigation technology. There is a trend worldwide toward increased use of other such grasslands. Figure 6-26 shows the grasslands of central China.

Tropical savanna grasslands are found in tropical climates that have distinct rainy and drought seasons. Total annual rainfall is high. The savanna area lies in wide belts north and south of the Equator. Some of these areas are treeless plains, while others support scattered, drought-resistant trees. In Africa the savanna biome was home to a wide variety of large mammals such as zebra, giraffes, ante-

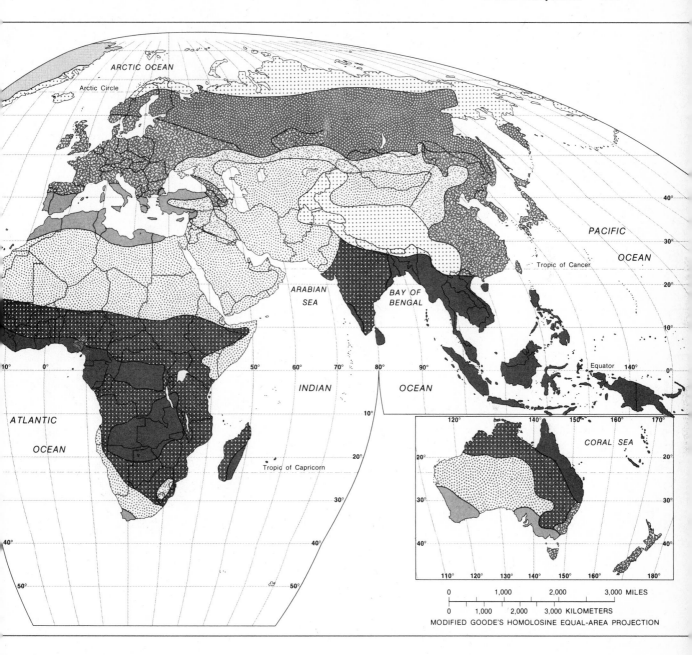

lope, and their predators. However, the ecosystems within the savanna biome are rapidly changing under the increased pressures of livestock grazing, farming, and poaching.

Tundra

Tundra is found in polar regions having low average annual temperatures, long and cold winters, low average annual precipitation, and wet or frozen soil. Growth rates of plants are very slow, and growth is limited to the short summer growing season. Plant life is limited to short sedges, herbs, mosses, lichens, and annual flowering varieties. The typical tundra mammal is a small herbivore, a rodent.

The slow rate of organic decomposition, the shallow soils, and the slow growth rate of plants make the tundra very fragile and vulnerable to human impact. Thus far, traditional peoples have survived here by hunting and animal herding. Mineral exploitation (especially for petroleum) has

FIGURE 6-26
Kirghiz herders setting up their yurts in the grasslands of China. (Photo courtesy Stephen F. Cunha.)

FIGURE 6-27
A Huli Wigman (tribal name) bow hunter in the rain forest of Papua New Guinea. (Photo courtesy of Stephen F. Cunha.)

had, and will continue to have, considerable impact on the people who depend on this biome for their survival.

Forests

Approximately 33 percent of Earth is covered with forest-land. These forests, along with sea plankton, replenish the global oxygen supply. The primary requisite for tree cover is adequate year-round rainfall. As in the cases of other biomes, the number, range, and specific characteristics of species are influenced by soil, topography, temperature, and human influences.

Tropical rain forests are located in equatorial zones having unvaryingly warm temperatures, high humidity, and almost daily rainfall (Figure 6-27). These forests comprise only 7 percent of the land surface but contain about 43 percent of Earth's vegetation (biomass). They hold an incredible diversity of plant and animal species, many of which remain undiscovered even today. The soils under this multistoried and diverse ecological system are thin and lack the humus layer found in most other biomes. This occurs because dead matter decomposes very rapidly in the tropical rainforest and the nutrients return to the living biomass very quickly. Tropical soils are infertile and ill-suited to field agriculture.

One of the most serious problems associated with the development of some Third World nations is the clearing of rain forests for lumber, grazing, or farming. A portion of Chapter 10 summarizes this dilemma and what is being done about it.

Boreal forests are cold, high-latitude, needle-leaved (coniferous) forests, often called **taiga.** They stretch across vast expanses of North America and Asia in zones of low precipitation, poor soils, and short summers. Species di-

versity is limited in these areas due to the long, harsh winters. Thus far the economic development of the high-latitude forests is limited to resource extraction, such as found in forestry, fur trapping, farming, and mineral exploitation.

Temperate forests are mid-latitude *deciduous* forests located on most continents where there is a full cycle of seasons and ample rainfall throughout the year. Winters are not as severe as in the boreal forest, nor are summer temperatures extremely high. While often called deciduous, most of these forests support conifers as well. In some cases, conifers such as pines may become the dominant species. Most of these mid-continental forests in the Americas, Europe, and Asia have long been occupied and

FIGURE 6-28
A mid-latitude deciduous forest, thick with trees and low-level shrubs along the Allegheny River in central Pennsylvania. (Photo by author.)

modified by humans. Historically, many were cleared to make way for farms, the trees being used as building materials and for firewood (Figure 6-28).

Mediterranean Scrub

Neither desert nor forest, the Mediterranean scrub biome is a greater reflection of climatic influence on vegetation than the other biomes. Its deep-rooted, small-leaved, perennial shrubs, mixed with scattered, savannalike woodland, is particularly adapted to summer drought and mild winters. This biome is present in the Mediterranean basin, southern Australia, South Africa, and the west coasts of North and South America. With application of water this environment becomes highly productive for agriculture because many of its soils are deep and fertile. Warm, rain-free summers have resulted in these regions becoming the focus of tourism and resort industries.

Aquatic Biomes

Until very recently the aquatic world (the hydrosphere) has received less attention from scientists and planners than the terrestrial world simply because it is so large (two-thirds of Earth's surface) and inaccessible. Today we know more about this zone, but we also realize how much remains to be learned. Unlike terrestrial biomes, the aquatic realm is best divided into freshwater and saltwater components.

Freshwater ecosystems make up only 1 percent of the hydrosphere but are critically important for the survival of many terrestrial ecosystems. They may take the form of lakes, rivers, ponds, marshes, or seasonal wetlands. The major environmental factors that determine what life forms will appear in a specific water body are temperature, salinity, oxygen supply, food supply, and human impacts (especially pollution). The number and diversity of freshwater habitats has decreased as many wetlands have been filled or drained and as rivers have been controlled to reduce damaging floods along their floodplains.

Saltwater ecosystems usually are connected to the oceans, although exceptions exist such as the Great Salt Lake, the Caspian Sea, and California's Salton Sea. Of all the water on Earth, 97 percent is in the oceans, and 250,000 species of plants and animals are known to exist therein. Most of this life is present in the coastal zones, estuaries, reefs, and zones of offshore upwelling. (Offshore upwelling occurs where cold water from the depths is moved to the surface by subsurface currents and certain geological features.) The deep oceans are "deserts" in terms of plant and animal life. As pressures for limited land resources continue to increase, the resources of the sea will become more important in the global economy. Offshore oil, metallic minerals, and fisheries are drawing particular attention from planners and investors.

CONCEPTS IN BIOGEOGRAPHY

Productivity

Biological **productivity** is the amount of mass (biomass) or its equivalent that is produced by living things. Of great importance is the rate of productivity at the primary trophic level (autotrophic); this includes agricultural products. Figure 6-29 is a world map showing primary productivity as best it can be estimated at this time. Note that the tropics, shallow coastal waters, and some mid-latitude temperate forests have the highest rates of productivity.

Isolation

Biotic islands may be *isolated* by water barriers (for example, the Galapagos Islands), by intervening topography, by contrasting biomes, or by human-caused habitat change. Many natural parks and preserves have become biotic islands as surrounding private lands have been developed. In general, the larger the area, the greater the number of plant and animal species that may be expected. Also, the longer such an island remains isolated, the greater is the likelihood that the local species will become genetically different from species in similar environments elsewhere.

Succession

No biotic community is static. Change may occur due to natural physical processes such as fire, flooding, landslides, or sedimentation, or changes may be human-caused. Disturbances may occur in large areas due to climatic change. In either case, a succession of vegetation and associated animal types usually occupies a site over time. A filled-in lake may initially support marsh vegetation, giving way to meadows, shrubs, and finally trees, depending upon the local soil and light conditions.

Ecosystem characteristics such as species diversity, speciation, mineral cycling, and soil development may change during the process of succession. Historical records can sometimes tell us what the natural vegetation of a place might have been like before intensive settlement or other modification.

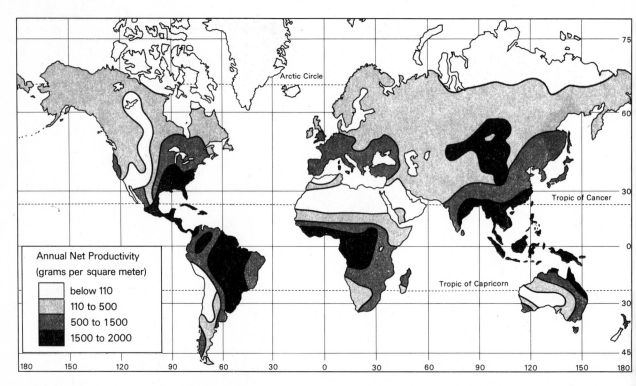

FIGURE 6-29

World biomass productivity. (From E. Willard Miller, Physical Geography: Earth Systems and Human Interactions, *Columbus, OH: Merrill Publishing Co., 1985, Fig. 11-3, p. 191. Reprinted with permission of Merrill Publishing Co.)*

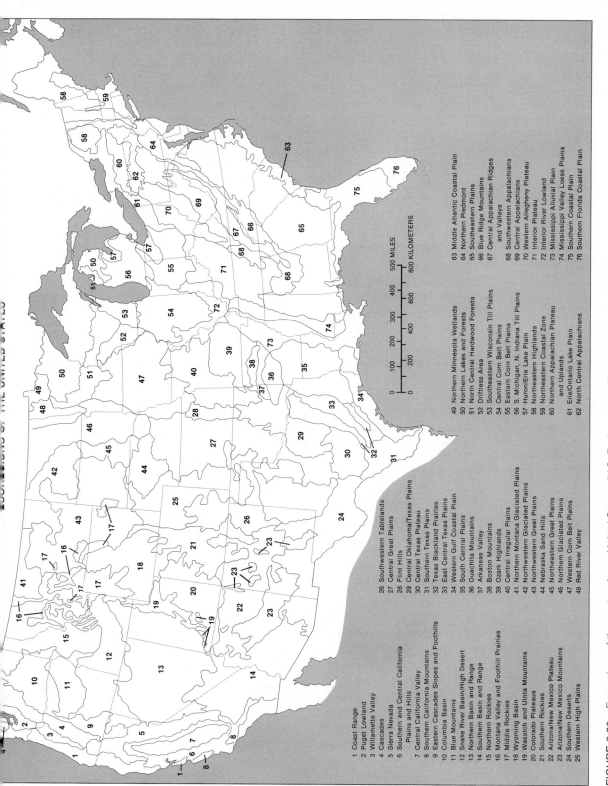

FIGURE 6-30 Ecoregions of the conterminous United States. (U.S. Environmental Protection Agency, 1988.)

FIGURE 6-31
World biogeographical provinces. (From Miklos D. F. Udvardy, Whole Earth Review, *1978. Reprinted with permission of* Whole Earth Review, *27 Gates Road, Sausalito, CA 94965.)*

WORLD BIOGEOGRAPHICAL PROVINCES

NEARCTIC REALM

1. Sitkan
2. Oregonian
3. Yukon Taiga
4. Canadian Taiga
5. Eastern Forest
6. Austroriparian
7. Californian
8. Sonoran
9. Chihuahuan
10. Tamaulipan
11. Great Basin
12. Aleutian Islands
13. Alaskan Tundra
14. Canadian Tundra
15. Arctic Archipelago
16. Greenland Tundra
17. Arctic Desert and Ice
18. Grasslands
19. Rocky Mountains
20. Sierra-Cascade
21. Madrean-Cordilleran
22. Great Lakes

OCEANIAN REALM

1. Papuan
2. Micronesian
3. Hawaiian
4. Southeastern Polynesian
5. Central Polynesian
6. New Caledonian
7. East Melanesian

24. Pacific Desert
25. Monte
26. Patagonian
27. Llanos
28. Campos Limpos
29. Babacu
30. Campos Cerrados
31. Argentinian Pampas
32. Uruguayan Pampas
33. Northern Andean
34. Colombian Montane
35. Yungas
36. Puna
37. Southern Andean
38. Bahamas-Bermudan
39. Cuban
40. Greater Antillean
41. Lesser Antillean
42. Revilla Gigedo Island
43. Cocos Island
44. Galapagos Islands
45. Fernando de Noronja Island
46. South Trinidade Island
47. Lake Titicaca

NEOTROPICAL REALM

1. Campechean
2. Panamanian
3. Colombian Coastal
4. Guyanan
5. Amazonian
6. Madeiran
7. Serra do Mar
8. Brazilian Rainforest
9. Brazilian Planalto
10. Valdivian Forest
11. Chilean Nothofagus
12. Everglades
13. Sinaloan
14. Guerreran
15. Yucatecan
16. Central American
17. Venezuelan Dry Forest
18. Venezuelan Deciduous Forest
19. Equadorian Dry Forest
20. Caatinga
21. Gran Chaco
22. Chilean Araucaria Forest
23. Chilean Sclerophyll

0 1000 2000 MILES
0 1000 2000 3000 KILOMETERS

Tropical Humid Forests

Subtropical and Temperate Rainforests or Woodlands

Temperate Broad-leaf Forests or Woodlands and Subpolar Deciduous Thicket

Temperate Needle-leaf Forests or Woodlands

Evergreen Sclerophyllous Forests, Scrub, or Woodlands

Tropical Dry or Deciduous Forests (incl. Monsoon Forests) or Woodland

Tropical Grasslands and Savannas

Temperate Grasslands

Ecoregions

One difficulty with mapping biogeographic information lies in selecting an appropriate scale that will be compatible with the mapping of other topics of analysis, such as cultural or economic regions. An intermediate level between the very generalized descriptions of global biomes and the mapping of local, site-specific biotic communities is the *ecoregion*. An example of the use of ecoregions is the map in Figure 6-30. Unlike earlier maps of natural vegeta-

tion, this map also considers land-surface form, land use, and soils to delimit 76 physical landscape regions in the United States, which can be compared with maps of human occupancy and activity. Compare this map with the interesting view of regions in Figure 6-31, which shows New Age "bioregions."

Estimates of the degree of correspondence or overlap of the two sets of regions often gives insight into what regional factors may have influenced human decisions regarding settlement patterns, land use, or cultural imprints.

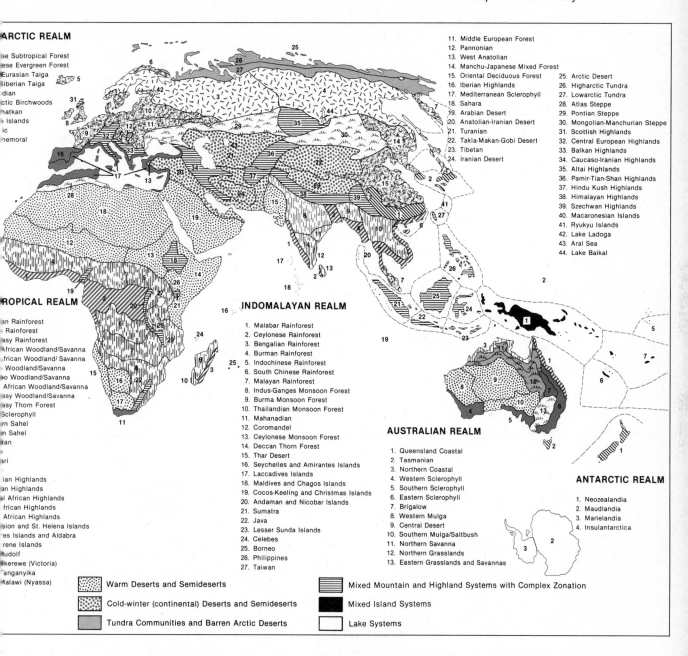

ARCTIC REALM

...ese Subtropical Forest
...ese Evergreen Forest
...Eurasian Taiga
...Siberian Taiga
...dian
...ctic Birchwoods
...hatkan
... Islands
...ic
...hemoral

11. Middle European Forest
12. Pannonian
13. West Anatolian
14. Manchu-Japanese Mixed Forest
15. Oriental Deciduous Forest
16. Iberian Highlands
17. Mediterranean Sclerophyll
18. Sahara
19. Arabian Desert
20. Anatolian-Iranian Desert
21. Turanian
22. Takla-Makan-Gobi Desert
23. Tibetan
24. Iranian Desert

25. Arctic Desert
26. Higharctic Tundra
27. Lowarctic Tundra
28. Atlas Steppe
29. Pontian Steppe
30. Mongolian-Manchurian Steppe
31. Scottish Highlands
32. Central European Highlands
33. Balkan Highlands
34. Caucaso-Iranian Highlands
35. Altai Highlands
36. Pamir-Tian-Shan Highlands
37. Hindu Kush Highlands
38. Himalayan Highlands
39. Szechwan Highlands
40. Macaronesian Islands
41. Ryukyu Islands
42. Lake Ladoga
43. Aral Sea
44. Lake Baikal

INDOMALAYAN REALM

1. Malabar Rainforest
2. Ceylonese Rainforest
3. Bengalian Rainforest
4. Burman Rainforest
5. Indochinese Rainforest
6. South Chinese Rainforest
7. Malayan Rainforest
8. Indus-Ganges Monsoon Forest
9. Burma Monsoon Forest
10. Thailandian Monsoon Forest
11. Mahanadian
12. Coromandel
13. Ceylonese Monsoon Forest
14. Deccan Thorn Forest
15. Thar Desert
16. Seychelles and Amirantes Islands
17. Laccadives Islands
18. Maldives and Chagos Islands
19. Cocos-Keeling and Christmas Islands
20. Andaman and Nicobar Islands
21. Sumatra
22. Java
23. Lesser Sunda Islands
24. Celebes
25. Borneo
26. Philippines
27. Taiwan

...ROPICAL REALM

...an Rainforest
... Rainforest
...asy Rainforest
...African Woodland/Savanna
...frican Woodland/Savanna
... Woodland/Savanna
...o Woodland/Savanna
...African Woodland/Savanna
...asy Woodland/Savanna
...asy Thorn Forest
...Sclerophyll
...n Sahel
...n Sahel
...ian
...ari
...ian Highlands
...an Highlands
...l African Highlands
...frican Highlands
...African Highlands
...sion and St. Helena Islands
...es Islands and Aldabra
...rene Islands
...udolf
...kerewe (Victoria)
...anganyika
...Malawi (Nyassa)

AUSTRALIAN REALM

1. Queensland Coastal
2. Tasmanian
3. Northern Coastal
4. Western Sclerophyll
5. Southern Sclerophyll
6. Eastern Sclerophyll
7. Brigalow
8. Western Mulga
9. Central Desert
10. Southern Mulga/Saltbush
11. Northern Savanna
12. Northern Grasslands
13. Eastern Grasslands and Savannas

ANTARCTIC REALM

1. Neozealandia
2. Maudlandia
3. Marielandia
4. Insulantarctica

Warm Deserts and Semideserts

Cold-winter (continental) Deserts and Semideserts

Tundra Communities and Barren Arctic Deserts

Mixed Mountain and Highland Systems with Complex Zonation

Mixed Island Systems

Lake Systems

Chapter 12 offers more ideas about the use of map comparisons in regional study.

SUMMARY

This chapter and the previous one have introduced the physical makeup and discussed ongoing processes that are shaping our natural world. Much of our physical landscape cannot really be separated from the human landscape because each has been affected by the other since human existence began.

The global system, with ourselves as part of it, is made up of innumerable subsystems, most of which are very complex and sensitive to change. Repeating the first law of ecology, "everything is connected to everything else," reminds us that our occupation of this planet is also complex and subject to change. Implications of the many relation-

GEOGRAPHER AT WORK
The Earth in a Glass of Wine

Biogeographer Deborah Elliott-Fisk found a way to keep her love of the outdoors in her university education by attending the geography department's Institute of Arctic and Alpine Research at the University of Colorado at Boulder. Upon attaining her doctorate, she began teaching at a major university and doing extensive biogeographical research in remote mountain and Arctic regions.

Becoming increasingly fascinated by the relations among vegetation, soils, landforms, and climate, Debbie also focused her research upon the study of viticulture (grape-growing and winemaking). She began to contribute to the ongoing debate about the potential affects of the physical environment on grape and wine composition. Particularly at issue was the matter of defining and designating viticultural areas on wine bottle labels for consumers in the United States, as is the practice in France and Italy.

U.S. regulations allow approval of a registered viticultural area if the area can be defined by a boundary that is historically and geographically unique in wine heritage, popular identification, and physical characteristics. Debbie, an energetic biogeographer, is particularly accomplished in supplying information that settles disputes about the locations of physical-geographic boundaries between grape-growing areas.

She bases her expert testimony on a geographic system consisting of "layers" of spatial data that can be viewed as a whole and used to establish system boundaries using such factors as microclimates, vegetation, soils, and geomorphology. With knowledge of the plant ecology and physiology of the European grape vine *(Vitus vinifera)* and its many varieties (such as Cabernet Sauvignon and Chardonnay), it is possible to establish that a particular grape-growing district is unique.

Each ecosystem receives a somewhat specific amount of solar energy that affects the processes of plant growth, plant transpiration, soil evaporation, soil development, soil temperature, and the local microclimatic elements of diurnal temperatures, humidity, winds, and precipitation.

In a recent case, Debbie was able to map the environmental factors, to show the evolutionary development of the natural landscape patterns, and to explain how these factors are the direct result of environmental dynamics. The logical conclusion was that these same factors will continue to influence grape vine growth and fruit quality in the same ways, thus establishing a distinct viticultural area. In her expert testimony she concluded:

> In sum, the soils, geomorphology, and climate which are reflected in natural vegetation patterns of the area north of the proposed district boundary suggest different affinities than the areas to the immediate north of the district. This specific district has a distinct geography as a function of the alluvial fan and other environmental factors.

Six months later, the U.S. Bureau of Alcohol, Tobacco, and Firearms confirmed the viticultural area with the boundaries she had suggested.

ships between human activity and our environmental systems are found later in this book, especially in Chapter 10 where the theme of Human-Environment Interaction is discussed in more detail. (See the box above to learn more about the fascinating career of a physical geographer who studies wine and biogeography.)

=== KEY CONCEPTS ===

Asthenosphere	Ecosystem	Mantle	Seafloor spreading
Biome	Fault	Mass wasting	Soil profile
Biotic community	Fluvial	Moraine	Subduction
Chernozem	Glacial budget	Pedology	Succession
Coniferous	Glacier	Plate tectonics	Taiga
Deciduous	Humus	Richter scale	Tundra
Detrital system	Karst topography	Rock cycle	Volcanism
Earthquake	Loam	Savanna	Weathering

=== FURTHER READING ===

In addition to sources offered in Chapters 5 and 10, here are selected references on the geography of the lithosphere and biosphere: Don J. Easterbrook, *Processes and Landforms* (New York: Macmillan, 1993); Dale F. Ritter, *Process Geomorphology* (Dubuque, IA: William C. Brown, 1986); *The Earth's Dynamic Systems* by W. Kenneth Hamblin (New York: Macmillan, 1992); Stuart C. Rothwell, *A Geography of Earth Form: Preface to Physical Geography* (Dubuque, IA: William C. Brown, 1973); and George R. Rumney, *The Geosystem: Dynamic Integration of Land, Sea, and Air* (Dubuque, IA: William C. Brown, 1970).

A few introductory geology textbooks are useful as references for the study of geomorphology. Particularly useful for its excellent graphics and maps is Edward J. Tarbuck and Frederick K. Lutgens, *The Earth: An Introduction to Physical Geology,* 4th ed. (Columbus: Merrill, 1993). A newer and more comprehensive source is *Environmental Geology* by Edward A. Keller (New York: Macmillan, 1992).

Earthquakes and volcanoes are the subjects of numerous books and articles. Particularly useful for teaching geography are *Planning for Earthquakes* by Philip R. Berke and Timothy Beatley (Baltimore, MD: The Johns Hopkins University Press, 1992); Peter L. Ward and Robert A. Page's *The Loma Prieta Earthquake of October 17, 1989* (Washington, DC: U.S. Government Printing Office, 1990); "Volcanoes: Crucibles of Creation," in *National Geographic* 182 (Dec. 1992): 3–41, by Noel Grove and R. H. Ressmeyer; and Fred M. Buller's *Volcanoes of the Earth* (Austin: University of Texas Press, 1976).

A useful reference for learning more about landform regions in North America is *Geomorphic Systems of North America* by William L. Graf, ed. (Boulder, CO: Geological Society of America, 1987).

The relation between geology and ecology as it pertains to soil development and classification is explained in Peter W. Birkeland, *Soils and Geomorphology* (New York: Oxford University Press, 1984). A comprehensive textbook on the relations between soils and human activity is Henry D. Foth, *Fundamentals of Soil Science* (New York: John Wiley, 1984). The systematic approach to soils as a part of the larger ecosystem is offered in two fine works: Roy L. Donahue and others, *Soils: An Introduction to Soils and Plant Growth* (Englewood Cliffs, NJ: Prentice-Hall, 1983), and Hans Jenny, *The Soil Resource: Origin and Behavior* (New York: Springer-Verlag, 1980).

A survey text concerning ecology and biogeography is Peter A. Furley and Walter W. Newey, *Geography of the Biosphere: An Introduction to the Nature, Distribution, and Evolution of the World's Life Zones* (London: Butterworth, 1982). Other useful references include Michael G. Barbour and William D. Billings, eds., *North American Terrestrial Vegetation* (Cambridge: Cambridge University Press, 1988); *Reforesting the Earth,* Worldwatch Paper 83, by Sandra Postal and Lori Heise (Washington, DC: Worldwatch Institute, 1988); *Natural Ecosystems* by W. B. Clapham, Jr. (New York: Macmillan, 1983); *Global Warming and Biological Diversity,* by R. I. Peters and T. E. Lovejoy, eds. (New Haven, CT: Yale University Press, 1992); *Communities and Ecosystems,* by Robert H. Whittaker (New York: Macmillan, 1975); Heinrich Walter, *Vegetation of the Earth and Ecological Systems of the Geo-Biosphere,* translated from the German revised 5th edition by Owen Muise (New York: Springer-Verlag, 1985); Joseph Wood Krutch, *The Great Chain of Life* (Boston: Houghton Mifflin, 1978); and Thomas R. Vale, *Plants and People: Vegetation Change in North America* (Washington, DC: AAG, 1982).

Another collection of readings with excellent illustrations from *Scientific American* is *The Biosphere* (San Francisco: William H. Freeman, 1970). Other newer sources on the geography of the biosphere are discussed in Chapter 10.

7

THE THEME OF PLACE: CULTURAL SYSTEMS

Warriors during a sing-sing near Mendi, Papua New Guinea. (Photo courtesy of Stephen F. Cunha.)

A student walking to school in Chicago passes by colorful storefronts and facades, fenced front yards, large thorough-fares and narrow winding streets, and sometimes even an old cemetery. The many people she encounters have diverse skin colors and dress. Without being aware of it, this student on her daily school route is observing the **cultural landscapes** of her city.

This chapter focuses on the important role of culture in shaping patterns on Earth. Concepts and themes woven through this discussion of Cultural Geography grow out of the following three *National Geography Standards:*

Standard 9: *A geographically informed person knows and*

understands the characteristics, distribution, and migration of human populations on Earth's surface.

Standard 10: *A geographically informed person knows and understands the characteristics, spatial organization, and complexity of Earth's cultural mosaics.*

Standard 13: *A geographically informed person knows and understands how forces of cooperation and conflict among people shape human control of Earth's surface.*

In the last two chapters we concentrated upon the physical characteristics of place. Learning about landforms, vegetation patterns, soils, and climate creates an important foundation for understanding our planet, but human activities are as much a part of regional differentiation as topography and climate. Both visible and invisible evidences of culture are fundamental to understanding human diversity and to observing and analyzing the human imprint on Earth. Culture is also a vitally important aspect of understanding place. Physical features are not rigid structures that compel all people to conform to a particular pattern; in fact, many Earth features have been shaped by humans.

Each group sees and thinks about its "world" differently, and therefore behaves differently from other groups. Because of the cultural "baggage,"the group carries its own cultural characteristics. Cultural geographers are interested in cultural similarities and differences and in why and how such diverse cultural elements evolved through time.

Cultural geographers are also interested in the study of human ecology, the diffusion of cultural traits and innovations, settlement, land use, and agricultural landscapes. These topics are investigated in subsequent chapters.

CULTURE

A well-known cultural geographer, Wilbur Zelinsky, wrote that "the cultural process is one of the few great first causes that shape those place-to-place differences of phenomena on or near the Earth's surface that we geographers study."[1] This viewpoint first became popular in geography in the late nineteenth century after the field had become strongly dominated by physical geographers. Early conservationist George Perkins Marsh (1801–1882) first raised the point of human influence in shaping the environment, although his ideas were not widely accepted.

Near the turn of the century, French geographer Paul Vidal de la Blache (1845–1918) clearly defined the new goals for the field of cultural geography. His ideas subsequently were redefined, but his belief that Earth does not dictate human behavior—that instead it offers possibilities for human choice—was a radical departure from the more traditional "environmental determinism" that had dominated the field for decades. Vidal de la Blache's viewpoint has been labeled "possibilism."[2]

In the United States, Carl Sauer (1889–1975) had the most influence on the development of the field of cultural geography. During his more than three decades at the University of California at Berkeley, Sauer encouraged American geographers to expand their early interest in the physical environment to include an interest in human-environment relationships. He stressed the importance of understanding the elements of culture in an area and the are-

al differentiation of cultures over the surface of Earth. Sauer suggested that geographers first collect as many facts as possible about a specific region to interpret each characteristic landscape. This regional approach favored the study of the cultural regions of the world as shown in Figure 7-1. Sauer and his students also had a special interest in the origin and dispersal of cultural traits on Earth.

Culture: A Design for Living

The **culture** concept is central to the work of various social scientists and humanists.[3] Culture is the total of all human experience and all learned behavior patterns. It is, according to Zelinsky, the totality of a complex system of learned behavioral patterns, assumptions, ideas, and attitudes, along with their associated artifacts and institutions.[4] Elements or traits of culture can be both material (housing, clothing) and nonmaterial (stories, music, attitudes).

Culture is transmitted from one generation to another by imitation, instruction, and example; it has nothing to do with genetics or with instinct. Culture is simply the lifeway of a group of people, a *people's design for living*. (It is important to note that individuals may not learn all aspects of their own culture. Age, sex, social status, or ethnicity may dictate which facets of the cultural whole are shared.)

Cultures are dynamic. They change constantly, either from new inventions within or by introduction and diffusion of culture traits from without. Usually the group must accept these innovations for change to occur, although change sometimes can be forced onto a culture group, as in the case of conquest.

Sometimes such change is most evident in the landscape in rural-to-urban development. Topographic maps of Anaheim, California (Figure 7-2) illustrate the rapid change evident in twentieth-century cultural and urban landscapes. This rapidly changing urban landscape is almost unrecognizable to visitors who have not seen it since the 1950s.

These changing elements of culture may be grouped for study and analysis. This chapter examines the building blocks of culture and their resultant cultural landscape. We look first at population distribution on Earth and then turn to a spatial study of three important cultural elements: languages, religions, and political patterns. The chapter concludes with a discussion of the evidences of culture on the landscape.

Looking ahead, the most important cultural trait influencing culture change is technology, so that is the subject of Chapters 8, 9, and 10. Varying technological capability has been the most influential factor dividing the developed world from the developing world.

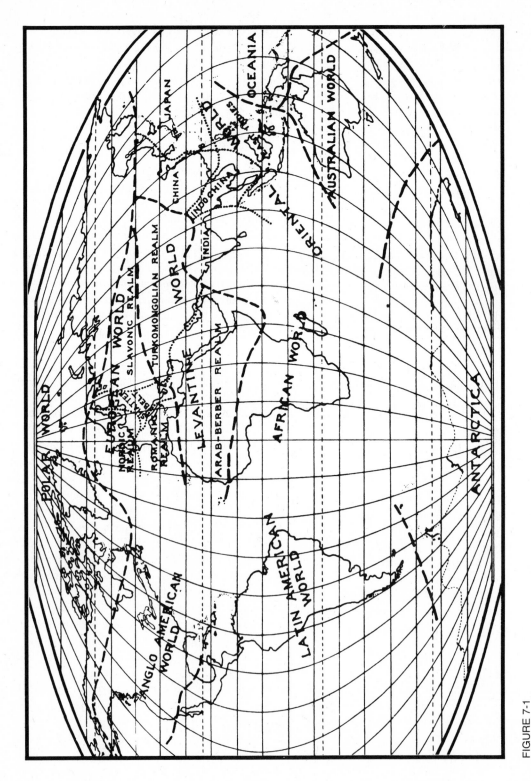

FIGURE 7-1
Culture regions of the world (1930s) based on divisions used by Carl Sauer in his geography class at University of California–Berkeley. (Map by Warren Thornthwaite, from Leslie Hewes, Journal of Geography 82 (July–Aug. 1983): p. 141. Used with permission of Journal of Geography.)

PEOPLE ON EARTH

Population Distribution

Geographers ask specific questions about people on Earth:

- Where do people live and why do they live there?
- Where are the most densely populated areas of Earth located?
- Why do people live in certain places on Earth in such great numbers while avoiding other places for settlement?

The irregular spatial arrangement and distribution of the world's population is largely based upon physical resources and varying cultural traditions. Almost half of the world's people live on only 5 percent of the land.

World Population Growth

Humans appeared on Earth more than 1 million years ago, but only in postglacial times did we emerge as a dominant species with worldwide distribution. Our numbers increased very slowly at first. At the beginning of the Christian era, our population was about half a million. It took 16 centuries to double this, and the first billion was not reached until 1820. Only 110 years later, by 1930, the second billion had been reached. Then came the real acceleration: in 1960 there were 3 billion people on Earth, another billion was added in the 1970s, and still another in the 1980s! In 1995 the world population exceeded 5.7 billion.

Table 7-1 shows world population distribution in 1994 and the projected population of each area for the years 2010 and 2025. Every year, more than 85 million new people are added to the carrying capacity of this already crowded planet. The recent period of tremendous population growth is unique in Earth's history and probably cannot be sustained indefinitely. Figure 7-3 shows world population distribution; Figure 7-4 maps world birth rates and death rates.

Changes in the physical environment seem to have been the most important factor influencing early population distribution. The earliest people were sparsely distributed across the tropical, subtropical, and mid-latitude portions of the eastern hemisphere at least 150,000 years ago. The cold latitudes were avoided, awaiting such innovations as advanced hunting skills, warm clothing, and houses. While cold temperatures were a severe limiting factor in population densities, early humans coped well with high temperatures, especially where water was accessible. Early, dense populations in Egypt, in the Middle East, and in river valleys in India illustrate our tropical heritage.

TABLE 7-1
World distribution of population (millions)

Region	1993	2010	2025
World	5506	7041	8425
Africa	677	1081	1552
Asia	3257	4175	4946
North America	287	331	371
Latin America	460	589	682
Europe	513	523	516
Former USSR	285	307	320
Oceania	28	34	39

Source: Population Reference Bureau, World Population Data Sheet, 1993. Used with permission.

Other limiting factors for population growth have been oxygen and water. Mountains and other lands at high elevations have remained much less important for human settlement than lowlands, evidently due to harsher circumstances of thinner soils, extreme temperatures, and sparse vegetation. Most important, prehistoric people always favored sites with an ample water supply.

Thus the most densely settled areas on Earth have been river valleys and coastal plains (today two-thirds of all people live near a coast). Earth's sparsest populations have been in deserts, cold areas, and mountains.

Why this sudden surge in population? What caused the number of people to increase so dramatically in the past century? Before answering these questions, it is important to define several basic terms in the study of population geography. Birthrate, death rate, and the rate of natural increase are some of the simplest measures used to define population distribution.

Birthrate is the annual number of live births per 1000 population. Over 30 per 1000 is considered a high birthrate, and two-thirds of the world's countries have birthrates that high or higher. In the United States, Canada, Japan, and some European countries, the birthrate averages less than 20 live births per 1000 people.

Death rate is the annual number of deaths per 1000 people. Rates of more than 25 per 1000 are considered high. The most important measure of death rate is *infant mortality* (deaths per 1000 of infants aged one year or younger) because this is where the greatest decline in deaths has occurred for several decades. This has been largely due

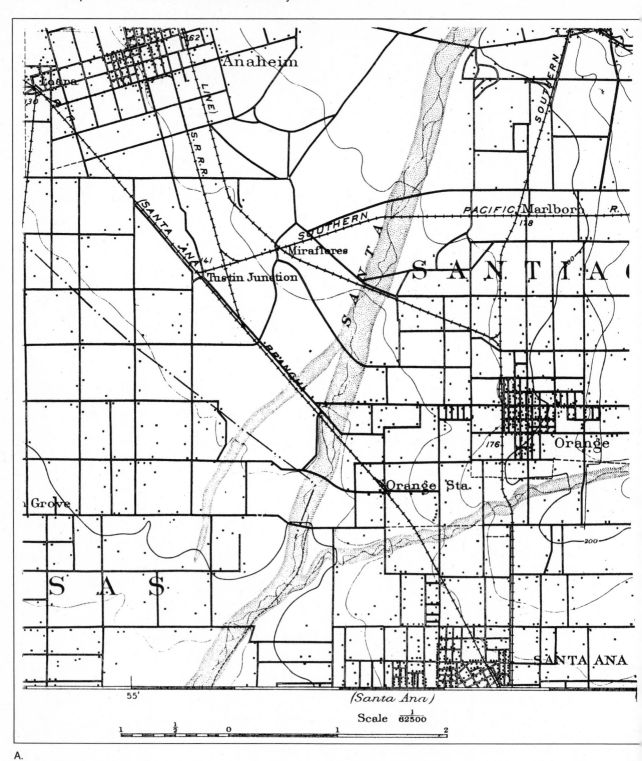

A.

FIGURE 7-2
A. Topographic map of Anaheim, California showing development in the year 1900. B. Topographic map of Anaheim, California showing development in the year 1981. (U.S. Geological Survey topographic map.)

11111111111111111

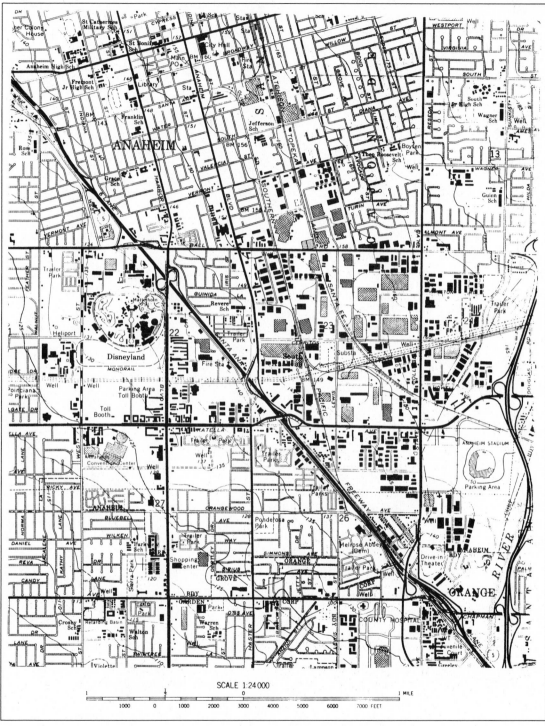

SCALE 1:24 000

B.

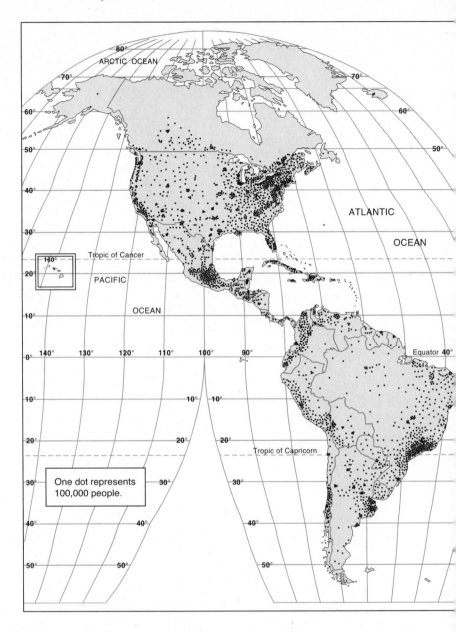

FIGURE 7-3
Distribution of the world's population. (From James M. Rubenstein, The Cultural Landscape: An Introduction to Human Geography, *4th ed., New York: Macmillan, 1994, Fig. 2-1, pp. 52–53. Reprinted with permission of Macmillan Publishing Co.)*

One dot represents 100,000 people.

to the development of modern medicine and sanitation since the 1940s in even the remotest parts of Earth.

Rate of natural increase is determined by subtracting the death rate from the birthrate. (This term does not include migration figures.) When we discussed the rate of population doubling, the rate of natural increase was used as a basis for the figures. This is an important statistic because it measures overall population growth or decline. Population doubling time is easily estimated. For example, at a growth rate of 1 percent, population doubles in 140 years; at 2 percent, 70 years; 3 percent, 35 years; and so forth.

Malthusian Growth and the Demographic Transition

Most of our thinking about the rate of natural increase of population is based upon two theories. The first was developed by a theologian-economist named Thomas Malthus (1766–1834). The **Malthusian doctrine,** as it is called, was first presented in a remarkable book entitled *An Essay on the Principle of Population,* first published in 1798. In this foresighted publication, Malthus first presented his pessimistic viewpoint on population growth in two simple postulates and one assumption:

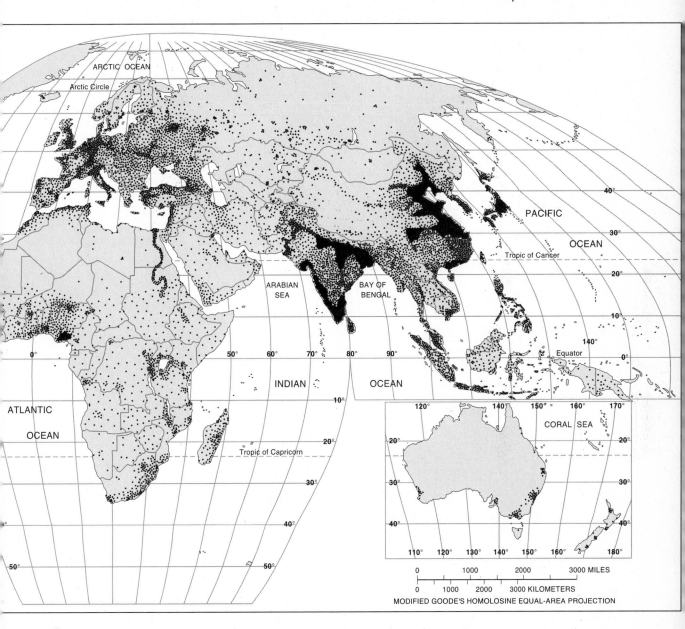

MODIFIED GOODE'S HOMOLOSINE EQUAL-AREA PROJECTION

POSTULATE: Food is necessary to the existence of man.

POSTULATE: The passion between the sexes is necessary and will remain in its present state.

ASSUMPTION: Population, when unchecked, increases at a geometric rate (2, 4, 8, 16, 32), whereas food production increases at an arithmetic rate (1, 2, 3, 4, 5, 6).

What Malthus said was that populations grow faster than food supply. However, his most extreme consequences of population growth can never be realized, for reasons con-

tained in the second theory of natural increase. This theory explains the braking mechanism that must eventually limit population growth. It is called the Theory of the **Demographic Transition.** The gap between birthrates and death rates in Europe actually has been closed by a decreasing birthrate, not by the increasing death rate gloomily predicted by Malthus.

The transition to a stabilizing population seems to be a byproduct of industrialization as societies successfully have evolved from agrarian economies to commercial economies. The Demographic Transition therefore has

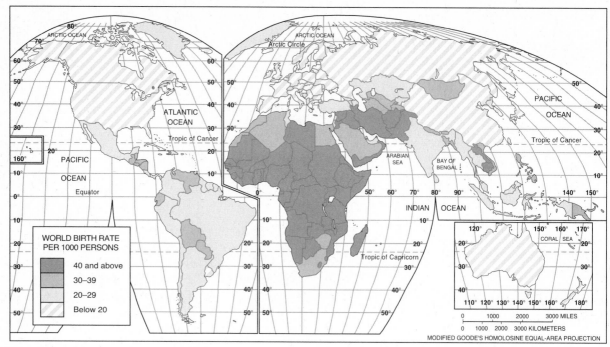

A.

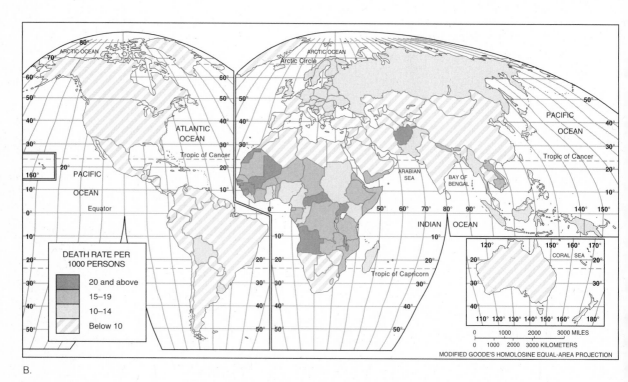

B.

FIGURE 7-4

A. World birthrates.   B. World death rates. (From James M. Rubenstein, The Cultural Landscape:
An Introduction to Human Geography, *4th ed., New York: Macmillan, 1994, Fig. 2-5, p. 61, and
Fig. 2-8, p. 64. Reprinted with permission of Macmillan Publishing Co.)*

happened not only in Europe but also in such countries as the United States, Australia, New Zealand, Canada, Japan, Singapore, and Hong Kong. As shown in Figure 7-5, this demographic model can be summarized in four stages of population change:

I. Until about A.D. 1750 both birthrates and death rates were high and population grew slowly. Growth was slow and unsteady as disease, famines, wars, and other disasters took heavy tolls.

II. The Population Explosion: after the Industrial Revolution, birthrates remained high, but death rates began to decline. This resulted in rapidly increasing population. Decreasing death rates were directly encouraged by improvements in sanitation, medicines, and food storage/distribution facilities; a rising per capita income; and increasing urbanization.

III. Birthrates declined but total population continued to increase due to lowered death rates. Children became economic liabilities rather than assets as urbanization and industrialization continued.

IV. Populations leveled off and the rate of increase stabilized with low birthrates and death rates.

The demographic transition probably still represents the best basis for understanding world population change today. Although it was originally applied only to Europe, many demographers feel it is a useful tool for analyzing other parts of the world. However, others doubt its applicability to places outside of the industrialized world because of rapid changes in medical technology in remote parts of the world (and thus the very sudden lowering of death rates in those areas in modern times).

A fifth stage could be added to the model to show recent decades in which death rates have exceeded birthrates in some nations. West Germany, for example, experienced this stage in the mid-twentieth century.

Data on population are often gathered by census takers. In recent decades, statistics from census records have become highly sophisticated. Information on such variables as age of population, ethnicity, income, housing, employment, and hundreds of other bits of information are currently gathered by the United States Census Bureau. (For an interview with a geographer employed in this type of population work see the box on page 160.)

Case Studies in Population Geography

Now that we have begun to understand the background for demographic analysis, let us turn our attention to real-world population issues. A toddler living in an overcrowded slum in Rio de Janeiro does not think of birthrates or infant mortality, but only of where the next meal is coming from. Other people and places in the world face similar problems.

Family Planning in China. China is the most populous country in the world today with over 1 billion people. Despite this staggering figure, results of the Chinese government's efforts to decrease the birthrate over the past 40 years have been dramatic. From 1949 to 1978, China made a demographic transition from high mortality and high fertility to low mortality and low fertility (Figure 7-6).

The decrease in deaths after 1950 was due to several factors, including more equitable distribution of land among workers, the cessation of warfare, and disease control programs. Following one of the greatest fertility peri-

FIGURE 7-5
Stages of the demographic transition.

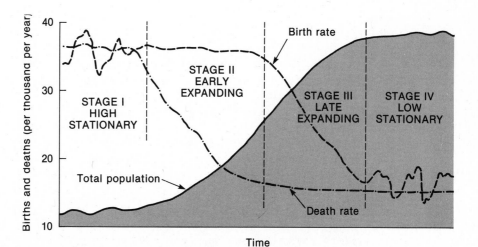

MEET THE GEOGRAPHER
Spatial Population Data at the U.S. Census Bureau

Geographer Joel Sobel was interested in geography as a child. Travels with his family throughout New York State and a fascination with maps imprinted the geographer's perspective in his observations and viewpoints before he knew there was such a subject in the university curriculum. Although he intended to study journalism, Joel discovered degree programs in geography before matriculating at Pennsylvania State University.

After completing an undergraduate degree program that emphasized population and cultural geography, and a rigorous doctoral program in urban and historical geography at the University of Minnesota, Joel began work at the U.S. Bureau of the Census in Washington. Within the Geography Division, he participated in developing automated geographic files for metropolitan areas, a major Census Bureau advance for the 1980 census. As Chief of the Geographic Base Development Branch, he directed automated update of these digital files for the 1990 census.

Currently, as chief of the Geographic Areas Branch, Joel directs all programs designed to identify and delineate geographic areas for the year 2000 census, and to ensure that this information appears accurately in the Census Bureau's digital cartographic database, the TIGER file. Joel and his staff work with local government officials and representatives to obtain the legal boundaries of all counties, towns, and districts in the United States and to devise boundaries for statistical areas such as census tracts, census designated places, and urbanized areas.

Their boundary decisions profoundly affect people's lives: many federal and local government programs rely on accurate geographic boundaries and demographic data from the Census Bureau to distribute funds and services equitably. Legislative reapportionment at all levels depends upon the delivery of census data in meaningful geographic units. Census data must be presented in meaningful geographic units to be used effectively.

In addition to his duties at the Census Bureau, Joel is a member of the Federal Executive Committee on Metropolitan Areas (MAs), an interagency

---

ods in the history of the country in the 1950s and 1960s, the Chinese government expanded urban-based family planning programs into rural areas, set up birth control committees, and expanded the role of women in the economy. In 1979, the post-Maoist government announced the "one-child program." Authorities saw that a large number of people were rapidly approaching childbearing age and realized that a one-child program was necessary to reach 0.5 percent growth by 1985 and zero growth by 2000.

Incentives were given to families who signed a pledge to limit their families to one child. Penalties resulted from the birth of a third child and abortions were required. If a couple insisted on having more than one child, their plot size for farming was reduced, thereby decreasing their in-

come. Local cadres (family planning activists) were given cash incentives to ensure that the couples in their village married late and had the fewest children. These incentives are vividly documented in the Public Broadcasting System film *China's Only Child.*

In 1983, China began a program of required sterilization for all families having two children. Abortions were allowed in all three trimesters. Requirements for later marriages were established. The ultimate goal was and is to keep the annual population increase to 0.5 percent per year.

Figure 7-6 illustrates the results of these strict population control measures. It is obvious that birthrates and death rates in China have been drastically lowered by government control of family size.

group that recommends changes in MA criteria to the Federal Office of Management and Budget. He also is on the Steering Committee of the Applied Geography Conference.

In a 1994 discussion, Joel described how the Census Bureau's Geography Division developed the TIGER file system to automate support operations for the 1990 decennial census. A cornerstone of this project was a historic agreement with the U.S. Geological Survey. The USGS accelerated production of its new 1:100,000–scale maps covering the lower 48 states, and then produced digital files of selected data from these maps. The Census Bureau then assigned geographic attributes, such as feature names, classification codes, and geographic area codes to the points, lines, and areas in these files.

The Geography Division has begun updating the TIGER data base for the 2000 census by adding address ranges that were collected during the 1990 census. As a result, the TIGER database now contains address ranges for approximately 85 percent of all residences in the United States, up from 60 percent for the 1990 census. The Geography Division continues to work with the U.S. Geological Survey on updating and improving the TIGER System, and the Census Bureau also is working with the U.S. Postal Service to develop methods of identifying new residential streets and addresses.

In working with the public and his colleagues at the Census Bureau, Joel sees increased awareness of what a geographic approach to problem solving can do. This, he says, is especially true of location research and market research where census data and products are used. Joel hopes that the training of future geographers will include even more attention to techniques and to the exchange of ideas:

> We are measured by what we do as much as by what we think. We must stop talking about geographic illiteracy, and instead demonstrate by our actions that geographic understanding is essential to our survival as well as our progress.

Joel suggests that many more "undeclared" geographers than we (or they) realize are in the workplace, performing geographic work. Perhaps you are such a geographer by association. If so, welcome!

Zero Population Growth in Sweden. Sweden was a pioneer in the international family planning movement and was the first developed nation to provide assistance to the less-developed world, beginning with a pilot project in Ceylon (now Sri Lanka) in 1958. Yet with all its progressive reforms in the area of birth control and population planning, Sweden now faces another problem—that of a declining population at home.

With more immigrants arriving in Sweden each year, and a declining native-born population, the nation is considering providing incentives for childbearing, including more favorable tax regulations, longer-term subsidies for children, shorter workdays for parents, and more day-care services. All this is occurring in an age when most coun-

tries are desperately trying to lower their birthrates. What events in Sweden's history have led to this "problem"?

In the latter nineteenth century, Sweden was overpopulated in rural areas. The land could not support the people living there. The result was a tremendous outmigration of Swedes in the 1880s and 1890s when "America Fever" spread through rural Sweden. Birth control discussions also began in the 1880s as a weapon against poverty. Family planning clinics were organized along with a rapid and radical transformation in the Swedish economy itself. Urbanization and industrialization decreased the need for large families, and a record low birthrate was experienced during the Depression in the 1930s.

In the mid-1930s, government policies were estab-

FIGURE 7-6
*Estimated birthrates and death
rates in China, 1949–1984.
(Adapted from L. C. Bannister,* The
China Quarterly *100 (Dec. 1984):
739. Reprinted with permission of
The China Quarterly.)*

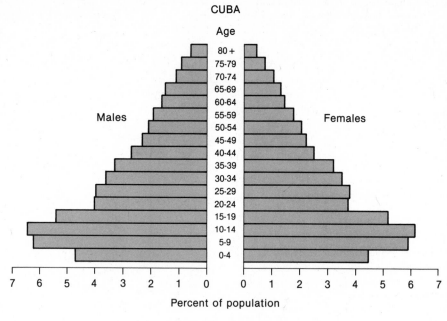

POPULATION AGE PYRAMID: 1979

lished to maintain Sweden's low rate of natural increase. The resulting programs were set into a broad social and economic context, including education, job incentives, and tax incentives, which still stand as a model for governments concerned about overpopulation and the quality of life.

In the 1960s, fertility rates once again declined in Sweden as more women entered the work force. The rapid retreat from formal marriage, the increasing separation of sexuality from childbearing, and the easy availability of contraceptives all contributed to this second wave of population control. Coupled with increasing educational opportunities for women and the economic disadvantages of having children, population totals continued to decline.

A comparison of the population growth and decline in China and in Sweden offers many opportunities for furthering our understanding of the important relationships between people and their environment.

Global Population and Politics

It is not the *number* of people in the world that poses problems for shrinking resources, but the **distribution** of population. There are simply too many people in the wrong places. Total Earth population rises by about 220,000 new people every day. Of these, about 90 percent are in countries that find it hard to offer their people rising standards of education and living.

The gulf is widening between regions with slow-grow-

ing populations and those with fast-growing ones. In recent years, slow-growing Europe and North America have been joined by slow-growing China, Japan, Taiwan, and South Korea. By contrast, Africa has the highest growth rate in the world. For example, if present conditions continue, by the year 2040 Nigeria will have as many people as did the whole continent of Africa in 1988. If that is not dramatic enough, we can note that the Nigerian growth rate is exceeded by several other African countries including Kenya, Zimbabwe, and Mozambique!

This case study in population distribution may be applied to the demographic transformation model presented earlier in the chapter (Figure 7-5). Countries in Europe, North America, and Asia moved somewhat rapidly into the third stage of the transformation, but this transition was recent. Africa and Latin America have, in fact, been in the second stage for far less time than Great Britain was. In many African and Latin American nations, however, at least 60 more years will pass before populations stabilize.

Population pyramids for four nations are shown in Figure 7-7. Compare population totals in each age group in Kenya, China, the United States, and Russia. Why are the patterns in each of these nations so different?

The Worldwatch Institute monitors the world's population growth and environment. Lester Brown, the Institute's president, suggests that if countries with fast-growing populations do not move into the third stage soon, they may revert to the first stage. Deteriorating living con-

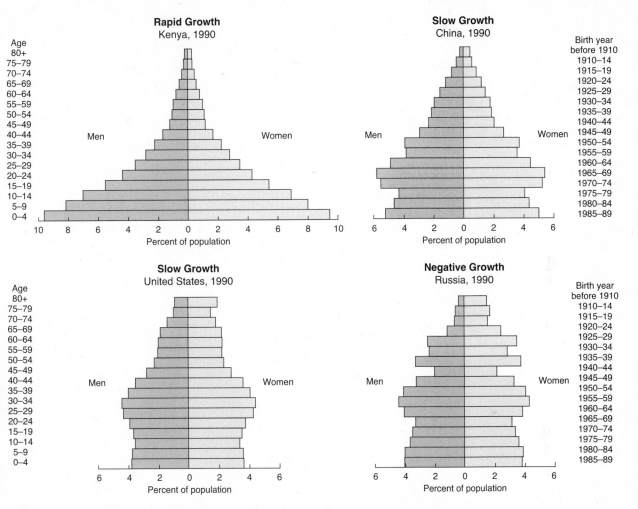

FIGURE 7-7
*Patterns of population change: population pyramids of Kenya, China, United States, and Russia
(From the Population Reference Bureau,* World Population: Toward the Next Century, *1994, p. 5.
Used with permission of the Population Reference Bureau.)*

ditions, the fearful spread of AIDS, famine, and political conflict will cause death rates to rise.

THE GEOGRAPHY OF LANGUAGES

How many languages do you speak? If you are a typical American student, you probably know only one language fluently. At the present time only 15 percent of all American high school students are studying a foreign language. Yet, if you lived in another country, such as Belgium or India, the knowledge of more than one language would be absolutely necessary. Imagine staying in a beautiful man-

sion and entering only one of its rooms during your entire life—such is the limitation you face in a shrinking world if you know only English!

Of course you could never learn all the world's languages, no matter how many courses you studied. Today there are believed to be 3000 or 4000 languages in the world—even in this age of supercommunication, we are not sure exactly. To establish an exact count is impossible because so many are scarcely known. We do know that there are almost a thousand languages in Africa alone, and that the single island of New Guinea contains at least 700 more. Only 200 are used widely enough to be classified as languages of international importance. Fewer than 100 are spoken by 95 percent of Earth's population (Figure 7-8).

FIGURE 7-8
*The world's principle language families. (From James M. Rubenstein,*The Cultural Landscape: An Introduction to Human Geography, *2nd ed., Columbus, OH: Merrill Publishing Co., 1989, Fig. 4-9, p. 138–139. Reprinted with permission of Merrill Publishing Co.)*

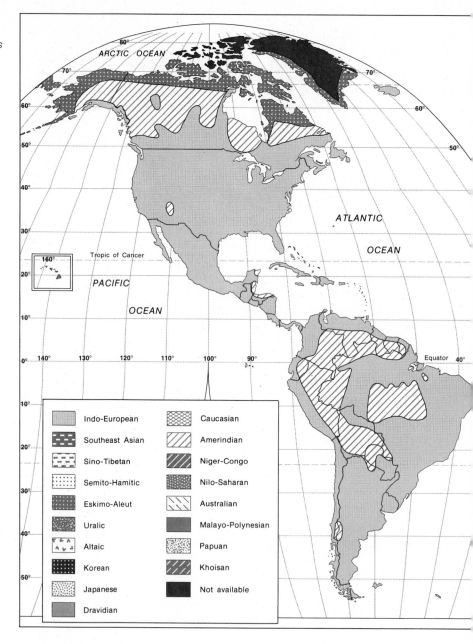

The languages spoken by the most people on Earth include Mandarin Chinese (800 million), English (425 million), Hindi (315 million), Spanish (310 million), and Russian (225 million).

Language is a vital part of culture. It is a vehicle for communication between generations and between cultures. Language may be defined as a meaningful learned system of communication by which humans express ideas and concepts through the use of written symbols and the human voice.[5] To be classified as a language, there must be a central core of sounds and written symbols that are mutually intelligible to all persons in the group.

Geographers are interested in languages because they form one of the universally important building blocks of culture. The spatial distribution of languages is especially important to geographers. Spoken languages first developed when people were scattered across Earth in small groups, and they have evolved differently in different areas. Written

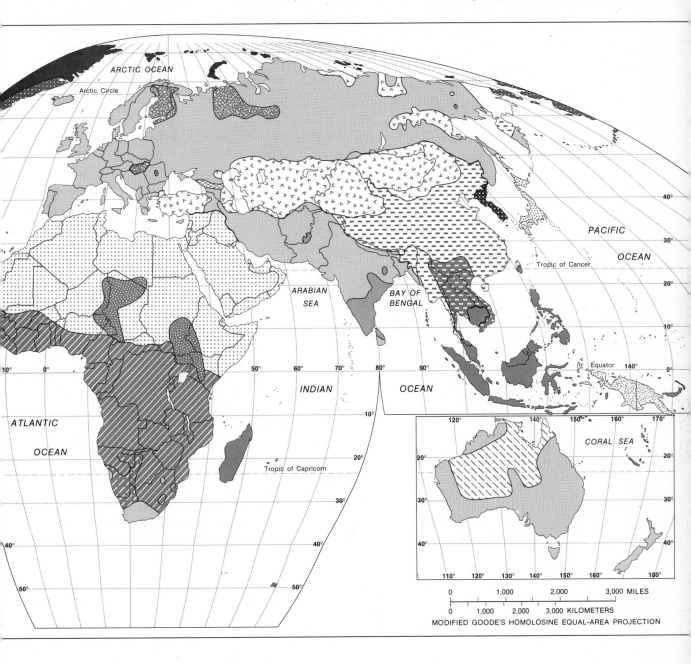

languages occurred much later in history and actually are a rather recent phenomenon, being known only in the last 1 percent of human time on Earth (less than 10,000 years before the present).

There are four basic types of true writing:

1. *Picture writing* (for example, hieroglyphics in ancient Egypt)

2. *Ideographs* (for example, Chinese)

3. *Scripts transitional* (for example, cuneiform writing of the ancient Sumerians)

4. *Alphabetic* writing (for example, Roman, Greek, Hebrew)

The most recent style of writing developed in the Middle East prior to 2000 B.C. and spread outward by diffusion. Alphabetic writing is the easiest to learn and thus facilitates education.

Language is one of the strongest forces that bind cultures together. When groups speak different languages the pursuit of common goals is difficult. In the United States, the existence of a single dominant language has helped blend and bind its diverse ethnic groups. Other nations, such as Belgium, are divided. In Belgium, Flemish speakers in the north and Walloon speakers in the south. India has fifteen official languages, although over half of her people speak Hindi.

Feelings of ethnic identity seem to be a major motive in resisting the imposition of another language. The Welsh and Irish resist the advance of English, French speakers in Quebec resist English, and the Basques in Spain resist Spanish. Imposition of an official language may cause the decline of local language through time.

Grouping Languages into Geographic Patterns

Many of the world's languages can be grouped into language families. These linguistic groupings resemble each other through sounds, grammar, or vocabulary structure and may have originated from the same ancestral tongue. Latin, for example, was the language spoken by the upper classes of the Roman Empire. In each of the provinces of this far-flung empire, slightly different variations of the original Latin evolved; hence we have today the Romance languages of French, Spanish, Portuguese, Rumanian, and Italian. Languages in a particular family resemble each other in some way. Some languages, like Basque in northern Spain, cannot be assigned to a family. Figure 7-8 shows the distribution of the principle language families on Earth.

The largest and most widespread language family on the globe is Indo-European. This family dominates Europe, Russia, North and South America, Australia, and parts of southwestern Asia and India. About one-third of all people on Earth speak a form of Indo-European.

Studying word clues and grouping vocabulary into patterns led geographer-detectives to believe Indo-European languages originated somewhere in the middle of Asia or in Eastern Europe at least 5000 years ago. Empire-building, colonization, and migration have spread these languages worldwide. English is the most widely spoken Indo-European language.

English: A Case Study in Linguistic Diffusion

A British television documentary based on the book *The Story of English* by Robert McCrum, William Cran, and Robert MacNeil explored in vivid detail the remarkable success of this language. English is spoken today by at least 750 million people, and barely half of these speak it as a mother tongue. As shown in Figure 7-9, in the late twentieth century, English is more spatially dispersed than any other language on the planet.

English evolved as a direct result of three invasions and a cultural revolution. The language first was brought to Britain by Germanic tribes (the Angles, Saxons, and Jutes); in fact, the name "England" is derived from the name "Angles" (meaning "Angle-land"), and the English language is a direct result of the fusion of the languages spoken by these three tribes. English was further influenced by Latin and Greek when England was converted to Christianity in the seventh century A.D. It was considerably enriched by the Norman conquest of England in 1066. The Normans spoke French, which became the dominant language of the upper class in England for 150 years, although the masses continued to speak English. What we call "English" was ultimately established for all people in England after the year 1200.

English eventually spread around the world by the establishment of British colonies in North America, Australia, Africa, Asia, and South America. Unique dialects evolved in many of these new branches of the language region. The best known of all English dialects, of course, was the one used by educated speakers in London. Most residents of the city, along with its educational institutions and government publications, used a single form of the language, which became known as official English or *British Received Pronunciation* (BRP). In modern times, however, strong differences in the language continue to exist despite standardization through textbooks and dictionaries printed in BRP.

Why is the English language so different in North America? We all know that the British colonists brought the original language to the Atlantic seaboard in the seventeenth century and that it became the official language of the new colonies. Because the United States developed in isolation from its mother country, however, American English developed in a completely different way than in England. Pronunciation, vocabulary, and even spelling differ. These differences between American English and "English English" are immediately recognizable and often are portrayed humorously in both countries.

During the Colonial period, the settlements along the Atlantic Coast remained isolated and restricted for many years. Dialects evolved there that still are discernible, but contact with other languages and other places over the years brought about a large variety of sounds in local areas. It is interesting to note that recent additions to popular culture in the United States also reflect dialect differences. One example of this tendency is the naming of popular American

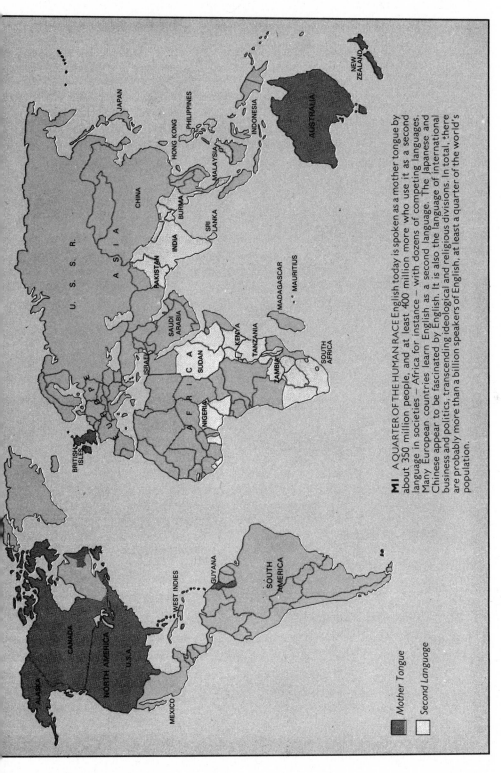

FIGURE 7-9

Distribution of the English language in the world today. English is the mother tongue for about 350 million people, and at least 400 million more use it as a second language in societies that have dozens of competing languages—for example, in Africa. Many Continental Europeans learn English as a second language. The Japanese and Chinese appear fascinated by English. It is also the language of international business and politics, transcending ideological and religious divisions. In all, over 1 billion people probably speak English, at least a quarter of the world's population. (From Robert McCrum, William Cran, and Robert MacNeil, The Story of English, New York: Viking Penguin, Inc., 1986, p. 23. Used with permission of Viking Penguin, Inc.)

FIGURE 7-10
*Regional variations in the names of
"big bun" sandwiches. Names re-
tain regional variations despite the
proliferation of franchise chains.
The maps are a product of a survey
of American regional English conducted at the University of
Wisconsin. Interviews were conducted at over 1200 locations.
Computer plotting intentionally distorted the sizes of states to correspond with the number of respon-
dents in each state, while retaining
their relative positions. (From John
G. Rooney, Jr., Wilbur Zelinsky, and
Dean R. Louder, eds.,* This
Remarkable Continent: An Atlas of
United States and Canadian
Society and Culture, *College
Station: Texas A & M University
Press, 1982, p. 127. Used with per-
mission.)*

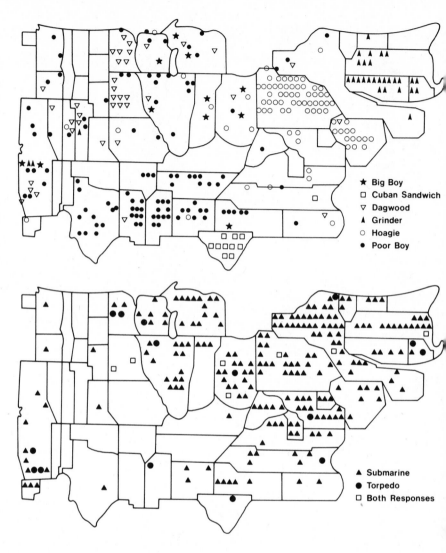

foods. Figure 7-10 shows regionalization of the names for "big bun" sandwiches.

Linguistic geographers are most interested in the study of place names. These place names, known as *toponyms,* are conspicuous and mappable elements of the cultural land-scape. Toponyms tend to persist for many years even as other aspects of the landscape change. Cultural geographer Henry Mencken classified American placenames into one of eight general categories:

1. Names embodying personal names, mainly the sur-names of pioneers and heroes

2. Names transferred from the area of settlement, either Europe or the eastern states

3. Native American names

4. Dutch, Spanish, French, German, and Scandinavian names

5. Biblical and mythological names

6. Names descriptive of localities

7. Names suggested by local flora, fauna, or geology

8. Purely fanciful names

A map from an excellent atlas of cultural geography entitled *This Remarkable Continent: An Atlas of United States and Canadian Society and Cultures* (Figure 7-11) illustrates some of these themes.

The geography of language is an element of cultural geography that links the present to the past. It is a fascinating topic for study by students, teachers, and the general public. Languages may actually be observed on the land-

FIGURE 7-11

Origins of placenames on the Louisiana landscape. Hammocks, located along the coast and in hilly areas, are elevated surfaces surrounded by lower land and covered with pine, oak, or cypress. Lagoon, introduced by both Spanish- and French-speaking settlers, generally applies to freshwater lakes, not to saltwater inlets. Marais is seldom found in placenames, although it has widespread vernacular use for unnamed, irregular-shaped prairie ponds. It denotes both coastal marsh and inland swamp. Chênière in Louisiana designates oak-covered natural levee remnants in southeastern Louisiana and oak-covered stranded beach ridges in the southwestern part of the state. In English-speaking Louisiana (as distinct from French Creole-speaking), the term applies to a creek and a cypress break. Prong, *an Anglo-Louisiana term, signifies both small tributary streams and moderate-velocity nontributary streams, except in southwestern Louisiana where prongs are tidal. (From John G. Rooney, Jr., Wilbur Zelinsky, and Dean R. Louder, eds.,* This Remarkable Continent: An Atlas of United States and Canadian Society and Culture, *College Station: Texas A & M University Press, 1982, p. 138. Used with permission of Randall Detro.)*

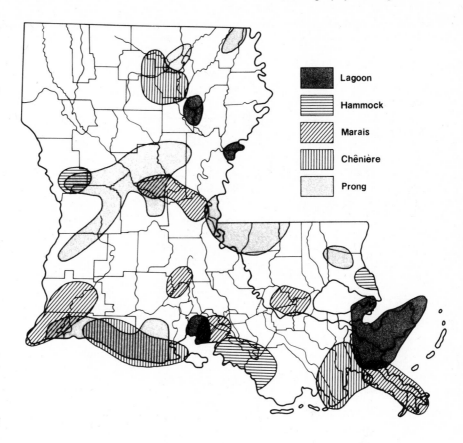

scape as placenames, reminding us of the incredible variety of people and cultures on Earth. Languages serve as a vital part of the cultural characteristics of place. (Some interesting placenames in the Appalachian state of West Virginia are highlighted in the box on page 170.)

THE GEOGRAPHY OF RELIGION

Religion also ultimately affects almost all aspects of a culture. Like language, religion is a cultural universal found in every culture on every continent. Religion not only has influenced specific culture areas but also has shaped landscapes, political systems, and most other aspects of the cultural scene. Not many days pass in the news without some mention of religious conflicts in the former Yugloslavia be-

tween Muslems, Eastern Orthodox Christians, and Roman Catholics. Religious differences between Muslims and Hindus forced the division of the Indian subcontinent into three distinct countries: India, Pakistan, and Bangladesh.

A set of beliefs, values, and sentiments commonly called an *ideology* is basic to the existence of organized society. Ideology may be defined as a common set of beliefs and values that unify a society, stabilize it, and keep it separate from other groups. Included in this definition are both secular and religious systems, traditions, and the overall themes of a particular society.

Religion is one of the most important ideologies. Because it is a personal experience, however, it is difficult to define. Essentially, religion refers to human belief in the supernatural, sacred, or spiritual. Thus any form of faith may be classified as religion, from monotheism (belief in one god), to animism (worship of the spirits of certain inanimate objects), to magic, and even to ancestor worship.

APPALACHIAN APPELLATIONS
An Editor's Choice of West Virginia's Zestiest Toponyms

If you are fascinated by placenames, West Virginia offers a fanciful feast. For starters, there are the communities of Skygusty, Cinderella, Whirlwind, Peapatch, Frost, Paw, Twiggs, Burning Springs, Johnnycake, Quick, and Red Jacket. These colorful names are listed in the West *Virginia Gazetteer of Physical and Cultural Place Names* (West Virginia Geological and Economic Survey, 1986), which lists 30,806 named places in the Mountain State.

West Virginia's ethnic heritage is written in its placenames. American Indian names applied to rivers, counties, and springs include Ohio, Kanawha, Pocahontas, Mingo, Logan, Wyoming, Monongalia, Allegheny, Pocatalico, Powhatan, Seneca, and Minnehaha. English names abound, or course. Scottish names number nearly 300, from McAdoo Ridge to McWhorter community. There is O'Brien community, and Irish Corner Tax District. French names are sprinkled about—Lesage, Letart, and Roncevert (which means green briar, a thorny vine of remarkable greenness and persistence). There are some German names—Scherr, Switzer. The community of Helvetia bears the Latin name of Switzerland, bestowed by Swiss immigrants.

The state's woodland tradition is recorded in hundreds of animal names, including Bearwallow Knob, Beaver Creek, Cowskin Fork, Dog Patch, Elk River, Foxgrape Run, Hogpen Hollow, Horse Creek, Polecat Hollow, Possum Hollow, Raccoon Creek, Wildcat, and Wolf Creek. Tree names also are common—Birchroot Run, Cherry Run, Chestnut Grove, Hickory Ridge, Oak Forest School, Pine Mountain, Poplar Tree Church, and Sassafras Ridge.

West Virginia's economic enterprises have given names many places: Coal City, Coketon, Colliers, Carbon, Cannelton (for cannel coal), Petroleum, Limestone, Sandstone, Mineralwells, Nitro, Oil Creek, Saltpeter, Salt Rock, and Natrium (Latin for sodium). Smaller local enterprises also have named some places: Beanpatch Hollow, Whiskey Hollow, Hemp Patch Run, Speakeasy Hollow.

Some names are direct tributes to companies that named them—Besoco, Clearco, Orgas, Parcoal—or to company officials whose names are now immortalized in the towns of Erbacon (E. R. Bacon), Huntington (a railroad executive), and Jenkinjones (a mine operator).

Religious beliefs shape cultures in many ways. Sacred structures that may be observed in the cultural landscape are evidence of local beliefs. A traveler observes religious geography upon seeing cemeteries, temples, churches, monasteries, religious names on the land, or sacred nature-worship shrines (Figure 7-12). Settlement patterns also are linked to religion: Jews in U.S. urban areas, Baptists in the southern United States, Mormons in Utah, Anglicans in New England.

Religion also affects other aspects of culture and society. Attitudes about the role of women in society may be strongly affected by the dominant religious system. The underrepresentation of women in politics and other leadership roles in many of the world's nations often stems from the maintenance of their powerless position in society based on traditional religious teachings. Even the foods we eat often are determined by our culture's religious beliefs. Religion becomes intimately interwoven into the tradition and life of a culture.

Universalizing and Ethnic Religions

Religions may be divided into two distinct types of belief systems, universalizing and ethnic.

Some names reflect natural features: Horseshoe Bend, White Sulfur Springs, Potato Knob, and Devils Darning Needle, a sharp-pointed mountain. Others are tributes to prominent West Virginians: the communities of Gassaway and Davis (for Senator Henry Gassaway Davis), Stonewall (for Civil War hero Stonewall Jackson), and Morgantown (for an early settler named Morgan).

There are names of beauty and happiness: Beauty, Romance, Valentine, Venus, Gem, Rosebud, Posey, Caress, Happy Hollow, Quiet Dell, and Friendly.

Some West Virginia placenames appear to be puns: Dollar Gap, Little Orphan Island, Notomine, Rabbit Run, Weekly Run. And then there are the communities of Wewanta, Uneeda, and Needmore.

Many names are simply amusing: Wahoo, Mud, Odd, Droop, PeeWee, Dingy, Omps, Waggy, Boomer, Bozoo, Gooney Otter Creek, Slippery Gut Branch, Tickle Britches Fork, Buck Shuck Run, Possumtrot Branch, Buggy Branch, Fuss Creek, Taylor Drain, Looney Cemetery, Big Bottom, and Superior Bottom.

A few placenames are somewhat mysterious: Shabbyroom Branch, Tightsqueeze Hollow, Polemic Run, Wizardism Run, Neggletetwist Run, Counterfeit Branch, Seldom Seen Hollow, Chiselfinger Ridge, and Sallys Backbone Ridge.

On the darker side, there is War, Shades of Death Creek, Bears Hell Ridge, Big Dark Hollow, Big Ugly Creek, Hateful Run, Danger Run, Big Deadening Creek, Hell for Certain Branch, Blister Swamp, and Booga-Boo Hollow.

For children, there are some interesting schools: Pigtail School, New Interest School, Retreat School, Oozley School, Muzzle School, Spider Den School, and Big Scary School.

And finally, here are some places of final repose in the Mountain State: Lively Cemetery, Accident Cemetery, Bone Cemetery, Buzzard Cemetery, Snuffer Cemetery, Mordue Cemetery, Grim Cemetery, Goodnight Cemetery, and At the End of the Trail Cemetery.

Adapted from Fred Schroyer, "The Mountain State from Aaron Creek to Zuspan Cemetery," Mountain State Geology *(Morgantown, WV: West Virginia Geological and Economic Survey, 1988), p. 5. Used with permission from West Virginia Geological and Economic Survey.*

Universalizing religions actively seek new members. Christianity, Buddhism, and Islam are powerful universalizing religions active in the world today. Indeed, many battles have been fought among these faiths to gain converts; witness the terrible loss of life during the Crusades when Muslims and Christians battled for "truth." Universalizing religions are usually thought of as "religions of revelation."

Members of *ethnic religions,* on the other hand, are not interested in seeking new members. Hindus, for example, do not actively proselytize (Hare Krishnas, no matter what they preach at street corners and airports, are not Hindus in the truest sense). Other ethnic religions include Judaism and most forms of animism or nature worship. Universalizing religions often grow out of ethnic religions; Christianity, for example, emerged from Judaism. Buddhism grew out of Hinduism.

The World's Religions

It is beyond the scope of this chapter to detail all the beliefs and teachings of the world's major religions. Figure 7-13 illustrates the great diversity of belief systems in the world today. Overall, there seems to be more mixing of

FIGURE 7-12
Russian Orthodox cemetery in Sacramento, California, is landscape evidence of the large number of Russian emigres living in California's capital city. (Photo by author.)

faiths in East and South Asia than in the Middle East and Europe. This is probably due to the greater religious tolerance of Hinduism in Asia. Likewise, the Chinese traditions of Confucianism and Taoism blend together with little conflict. When the Chinese migrated to Gold Rush California in the 1850s, their "joss houses" had separate rooms for Confucianist, Taoist, and Buddhist worship.

It is significant to note that the large monotheistic religions appear to have been the most intolerant and warlike throughout history. Wars on religious grounds between polytheistic societies are unknown, but since the rise of monotheistic Islam, Judaism, and Christianity, religious wars have proliferated in Europe, the Middle East, Africa, and Asia.

Europe and the Middle East have been dominated by many more exclusive religions than the rest of the world. Christianity in Europe and Islam in the Middle East not only are universalizing in the truest sense, but also are exclusive. The longstanding ill will between the United States and some countries of the Middle East is based in part on these religious differences.

Religious Names on the Land

Religious placenames lead a traveler through Quebec in Canada. One would find it hard to travel from St.-Pamphile to Ste.-Lucie-de-Beauregard without thinking about Quebec's French Catholic heritage. Places here take

their names from isolated parish churches built by early French settlers. This has created a most noticeable religious landscape.

In other parts of the world, religious geography may be seen on the land as well. The names of five of California's seven largest cities have religious connotations (*St., San,* and *Sacramento* all have obvious Roman Catholic connections). Religion geographer David Sopher suggested that the religious naming of places in Europe did not occur at the same rate as in North America.[6] He writes that where the Spanish, French, or Portuguese settled in the New World, they christened the landscape with the names of numerous saints, but in Europe, existing Roman toponyms minimized the number of Christian names on the land. Figure 7-14 shows the popularity of using Roman Catholic saints' names in Quebec.

Landscapes of the Dead

Burial sites are examples of small-scale religious landscapes (Figure 7-12). All over the world, cemeteries preserve very ancient cultural traits. Christians, Muslims, and Jews prefer setting aside special plots of land for their dead, with monuments and grave markers. Islamic countries often create parks from burial sites to save space, especially in urban areas.

Pressures on cemetery land are increasing in the twentieth century as urban populations increase. In San

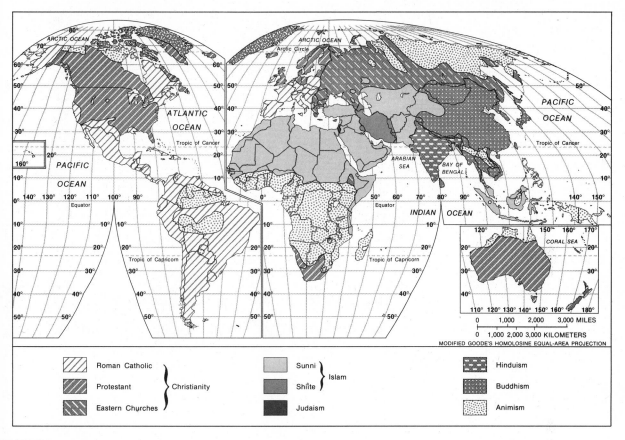

FIGURE 7-13
Religions of the world. (From James M. Rubenstein, The Cultural Landscape: An Introduction to
Human Geography, *2nd ed., Columbus, OH: Merrill Publishing Co., 1989, Fig. 5-1, pp. 162–163.
Reprinted with permission of Merrill Publishing Co.)*

Francisco land became so valuable that all bodies in city cemeteries were moved a few miles south to the city of Colma. Visitors are invariably struck by the diverse array of cemeteries: three Chinese, three Jewish, and one each for Greeks, Russians, Serbians, Italians, and Japanese. Colma literally has become a "city of the dead" in the 1990s.

The dead are cremated by Hindu, Buddhist, and Shintoist believers, so landscapes of the dead are uncommon in most Asian countries. An exception is China, where the dead are buried and commemorated as an important part of ancestor worship.

Religious Settlement Patterns: The Mormons

The impact of religion on the land can easily be observed during a trip to the state of Utah. While most urban plans are designed with economics and efficiency in mind, the Mormons came to Utah to design a religious landscape.

Their first city at the Great Salt Lake is based upon Mormon leader Joseph Smith's plan for "the city of Zion."

Expressions of this religious group's beliefs are widely dispersed in the American West. Geographer Donald Meinig has studied the Mormon culture landscape, identifying a list of traits observable in the Mormon region. These include Mormon hay derricks, houses on the corners of lots, red barns, poplar trees along streets and roadways, wide streets, houses large enough for multiple wives, irrigation ditches, and the ever-present church.

Other evidence of religious expression in the landscape may be seen in unique areas of the United States or may be observable on historic maps. In California and the Southwest, for example, early settlement was conducted by Spanish missionaries and soldiers. Their legacy includes ruined and sometimes reconstructed missions, irrigation systems, roads, and cemeteries. In Louisiana, the importance of the church is reflected in government divisions called *parishes* instead of counties.

FIGURE 7-14
*A map of toponyms—of place-names—near Quebec's boundaries with Ontario and New York State shows the impact of religion on the landscape of Quebec. (From James M. Rubenstein,*The Cultural Landscape: An Introduction to Human Geography, *2nd ed., Columbus, OH: Merrill Publishing Co., 1989, Fig. 5-12, p. 185. Reprinted with permission of Merrill Publishing Co.)*

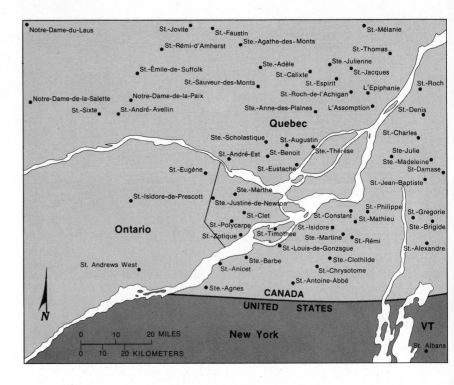

Religion and Food Habits

Religious teachings have a great deal of control over food habits in certain parts of the world. For example, Judaism's dietary codes are directed primarily toward meat. In the Old Testament books of Leviticus, Numbers, and Deuteronomy, the rules regarding meat take two forms: all meat must be kept separate from dairy products, and all meat must be free of blood when cured, cooked, and eaten. To be "kosher," all food must be prepared using the proper procedures.

The New Testament book of Acts prohibits Christians from eating three foods: carrion (meat not intentionally slaughtered for consumption), meat from strangled animals, and meat previously offered to idols. Dietary codes printed in Islam's holy book, the Koran, are simple and direct. Muslims are restricted from consuming pork, blood, carrion, meat offered to idols, and alcoholic beverages.

The zealousness of its followers makes the Asian religion of Buddhism an excellent case study of the interplay among food, diet, and religion. Their dietary code was first established by George Osawa and states that every disease can be cured by proper therapy based on natural food and activity level. Medicine and surgery are unnecessary. This "Zen Macrobiotic Diet," as it is called, involves ten levels of eating and drinking, each of which constitutes a different percentage of the various types of foods. As the desired

state of well-being develops, one progresses up the levels of the ladder until the body is purged of toxins at the ultimate level. This final aspiration level is limited to a diet of 100 percent cereal products.

In this section we have seen how religion influences many aspects of culture and place. One hypothesis about the interaction of religion and societal development, the "Protestant Ethic," is detailed in the box on page 175.

GEOPOLITICS AND THE THEME OF PLACE

You probably played with a puzzle of the world when you were in elementary school. Each piece of the puzzle was painted a different color and each represented a different country. These questions must have come to your mind: Why are there so many divisions in the world today? What caused it to be like this? Why have people partitioned Earth into territorial units and merged tribal lands into huge states?

The study of political patterns on Earth, "geopolitics," is one of the most exciting subfields in human geography. Understanding why the world has been divided into hundreds of fragmented political units, why boundaries serve as such strong dividers of cultures and peoples, and what ge-

RELIGION AND ECONOMICS: THE PROTESTANT ETHIC

Attitudes derived from religious teachings affect an individual's view toward life and influence the evolution of the society involved. For example, the "Protestant Ethic," with its emphasis on hard work and frugality, appears to have contributed to the rise of capitalism. For the past half century, the question has interested scholars: "Did the Protestant Reformation, especially its Calvinist branch, influence the development of modern capitalism?" This controversy began in 1904–1905 with the publication of two articles by German sociologist Max Weber. His articles on the sociology of religion began a long debate on the influence of the Protestant religion on politics and government.

According to Weber, religion produces a distinct attitude toward life, and this orientation affects the further development of a society. His thesis was a reaction against Marxist teachings, which emphasized that the methods of production ultimately affected the development of other institutions (including religion).

The reality is that neither of these viewpoints tells the whole story. Most cultural situations result from interrelationships between religion and the social, political, and economic factors in a society.

graphic elements contribute to national powers are fascinating issues that contribute to our understanding of the human characteristics of place.

Political patterns are a basic element in understanding cultural variations. They are one important aspect of political geography, along with political phenomena and their areal expression. To better understand our world, geographers are especially interested in relating political patterns to other aspects of culture, including language, religion, and ethnicity.

Political Systems

The idea that people owe allegiance to a country or "space" on Earth rather than to a king or leader is known as **nationalism.** This term first took on political significance in Europe during the Middle Ages in kingdoms that were unified within defined borders. People who spoke the same language, who believed in the same religious teachings, and who communicated regularly developed a feeling of group identity. The idea of a nation-state was first developed in the eighteenth century by political philosophers during the French Revolution. The concept soon spread from France to Germany, England, Spain, and ultimately around the world as colonial powers imposed upon new lands their belief in the nation-state.

Definitions of "state" and "nation-state" often are confusing. A **state** is an area organized into a political unit and ruled by an established government. (The term refers to *any* political unit and should not be confused with the "states" of the United States.) A *nation* is a cohesive group of people. A **nation-state** is an organized political unit that *also contains a sense of national cohesion.* Many states lack this cohesiveness and so cannot be categorized as nation-states.

Earth currently has 190 individual states or nation-states (see Figure 7-15 and the box on page 182). The political landscape of our twentieth-century world is largely a result of the development of nation-states in Europe in the Middle Ages. Most of the new states that have been organized in Asia, Africa, North America, and South America were formed after the European model.

Today every person on Earth lives under some type of political system. We are all affected daily by the organization of political space. We are residents of our city, our county or parish, our state, our country, and our world.

Figure 7-16 illustrates the myriad boundaries among ethnic groups in Africa. As shown on this map, it is not always possible to define nationality by the imposition of artificial boundaries. Many African states possess artificial boundaries imposed at the convenience of their former landlords, the European colonial powers; these do not correspond at all to traditional tribal boundaries. The emerging African states, however, have kept their imposed boundaries after independence despite tribal differences in order to simplify the process of creating a new state. The

FIGURE 7-15
*Political map of the world.
(From James M. Rubenstein,
The Cultural Landscape: An
Introduction to Human
Geography, 4th ed., New
York: Macmillan, inside pa-
per cover. Reprinted with
permission of Macmillan
Publishing Co.)*

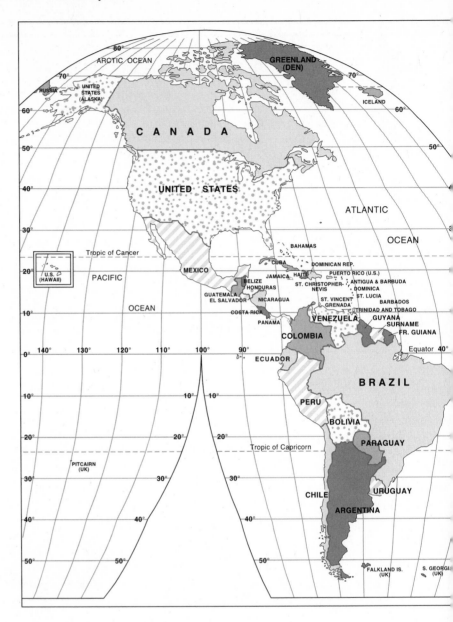

task of uniting diverse tribes into one politically cohesive unit is both urgent and complex.

Other tribes or ethnic groups in other parts of the world similarly have been cut off from each other because of political boundaries. The island of Ireland, for example, is divided into two parts due to religious and political differences between the Irish and the English. The former nation of Yugoslavia is a virtual jigsaw puzzle of ethnic groups and will serve as a case study in geopolitics later in this chapter.

National Cohesiveness

What holds a nation-state together politically? What makes a country stable or unstable? Forces that hold a nation together are called **centripetal forces,** whereas forces that divide a nation are called **centrifugal forces.**

Languages may act as either strong centripetal or centrifugal forces. English has been a centripetal (cohesive) force in the United States and Australia. In Canada English and French work against each other. In Ireland, re-

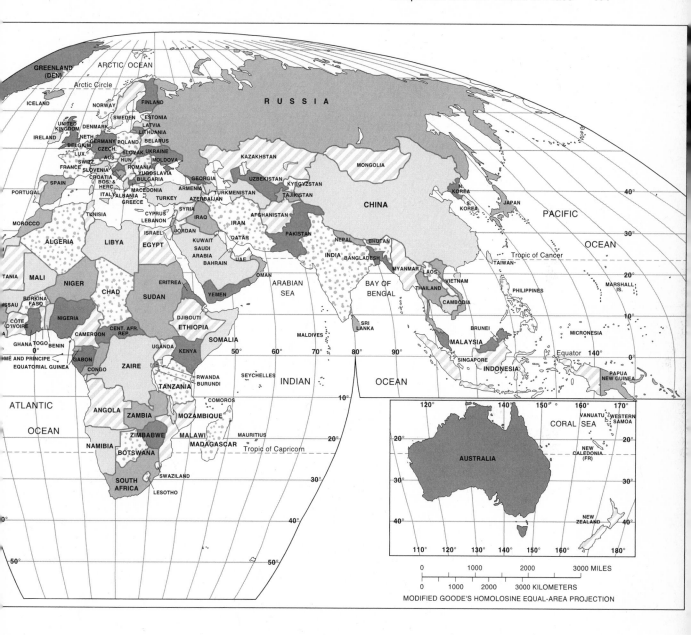

MODIFIED GOODE'S HOMOLOSINE EQUAL-AREA PROJECTION

gion acts as a centrifugal (divisive) force. It also is a centrifugal force throughout the Middle East.

Another consideration is the shape of the bounded area (Figure 7-17). Elongated nations like Chile and Vietnam often have a difficult time holding together politically. Circular or hexagonal nations like France facilitate interaction and more easily remain united. Small, compact political units are the easiest to maintain. This geographical consideration is just one of the many spatial advantages or disadvantages that nations experience.

A third important issue for national cohesiveness is the presence of enclaves or exclaves. An **enclave** is an area of land surrounded by a different state but not ruled by it. The Vatican is an example of an enclave. It is not a part of Italy but is surrounded by Rome on all sides. **Exclaves** are a part of a national territory that is located some distance away from its center. Alaska and Hawaii are both exclaves of the United States and often feel isolated from national decisions made in faraway Washington, DC.

Shared **iconography** also helps hold nations together.

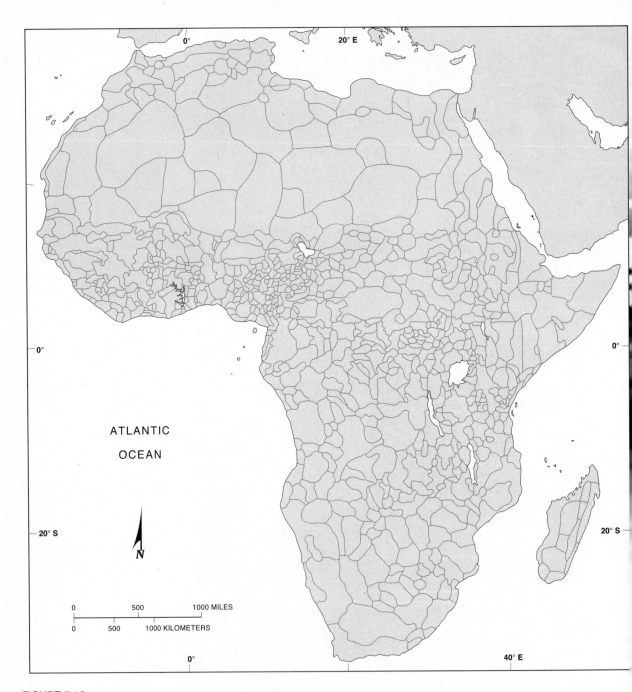

FIGURE 7-16
Tribal divisions of Africa showing the diverse political divisions of land on this huge and complex continent. (After G. P. Murdock, Africa, Its Peoples and Their Cultural History, *New York: McGraw-Hill Book Co., 1959. Used with permission of McGraw-Hill, Inc.)*

FIGURE 7-17
The shapes of states. (From James M. Rubenstein, The Cultural Landscape: An Introduction to Human Geography, *2nd ed., Columbus, OH: Merrill Publishing Co., 1989, Fig. 7-15, p. 254. Reprinted with permission of Merrill Publishing Co.)*

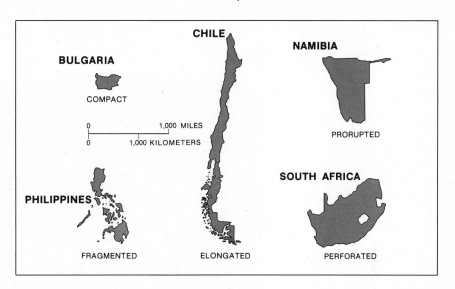

National symbols like flags and patriotic songs unite people under one government. So do shared institutions (schools, churches) and the organization of resources. An efficient nationwide transportation and communication system ensures a continuing exchange of common goals and symbols. Most nations today also have some governmental control of the press and news.

Richard Hartshorne, one of the best-known political geographers, defines national cohesiveness as[7]

> the conviction of integration in the minds of all groups in all areas. The feeling, that is, of identification of themselves with the region as a whole and with its organization as a political state.

Geopolitics

The division of Earth into political units—**geopolitics**—has been accomplished by drawing a series of **boundary** lines on the globe. These boundaries involve a practical application of geographic knowledge in resolving problems. Sometimes boundaries are formed by definite physical barriers (mountain ranges, rivers, or deserts), but boundaries most often come from human decisions. Boundaries drawn for the greater good can create awkward local situations; a good example is the U.S.–Canadian boundary between British Columbia and the state of Washington, which isolates a tiny part of Washington—Point Roberts. It is approachable by land only through Canada (Figure 7-18).

Boundary drawing is never easy. Boundaries are the only locations between countries where contact must take place, so they are a potential area for conflict. Many geographers have worked to settle boundary disputes during the past 100 years. For example, at the Paris Peace Conference after World War I, President Woodrow Wilson appointed several geographers to provide advice on settling boundary disputes in Eastern Europe.

Historical boundary disputes between China and India, North and South Korea, Argentina and Chile, and Israel and its neighbors have been in the news for decades. Another example, in the United States, is the dispute between California and Nevada, which have long been involved in a "boundary war" that slices Lake Tahoe down the middle.

One of the most challenging long-term boundary disputes involves a 2500-mile (4000-km) boundary between India and China in the wilderness of the Himalaya Mountains. As much of this land has never even been mapped, numerous delegations of Indian and Chinese diplomats have as yet failed to settle the boundary issue. In 1962 Chinese armies moved southward over the Himalaya Mountains and seized control of large areas that had been under Indian control. After slicing through Indian territory, the Chinese withdrew from the eastern sector but seized a large part of the desolate land of Ladakh, which they still hold.

As we have seen, boundaries are invisible lines on Earth, yet they are probably the most recognizable of all features on the cultural landscape. Most boundaries are unmarked, although some have fences, guardhouses, or police patrols. It is the more subtle cultural differences expressed on each side of a boundary that are most observable.

Geographer Julian Minghi has found eight categories of study most common to political geographers interested in boundary issues:[8]

FIGURE 7-18
Point Roberts, Washington, and surrounding territory. (From Association of American Geographers, High School Geography Project, "Political Geography," unit IV of Geography in an Urban Age, *New York: Macmillan, 1972, p. 50. Used with permission of the Association of American Geographers.)*

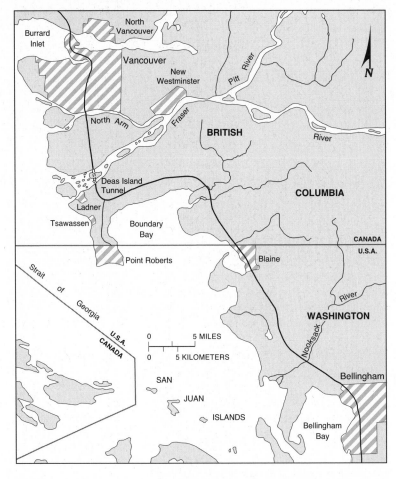

1. Disputed areas
2. Effect of boundary change
3. Evolution of boundaries
4. Boundary delimitation and demarcation
5. Exclaves and tiny states
6. Offshore boundaries
7. Disputes over natural resources
8. Internal boundaries

The Earth's Final Political Frontiers

Antarctica. Why would any nation be interested in "capturing" the windiest, coldest, driest, highest, and most undeveloped continent for their own? Humans seem incapable of leaving any part of Earth unclaimed, and the continent of Antarctica is one of the globe's final land frontiers. The discovery of valuable resources—scattered deposits of coal, iron ore, and manganese; petroleum and natural gas; marine fauna and flora; and open space on the continent—have encouraged nations to grab as much o this remote continent as possible.

After decades of confusing land grabs by numerou countries, Antarctica was divided into sectors by a treaty signed in 1959, following the International Geophysica Year (IGY). Commemorating a period of unusual sunspo activity, the IGY was a model effort in world diplomacy that is worth examining. This coordinated study of Eart and sun involved peaceful collaboration of over 10,000 sci entists and technicians from 67 countries. The Antarcti Treaty was signed in Washington by 12 nation-states tha participated in IGY research on the continent. Since the seven more nation-states have been added, includin Poland. Figure 7-19 illustrates the sectoral political bound aries, some overlapping, on the Antarctic continent.

Oceans. Humans have used the sea freely for many cen turies. In recent years the need to "own" this last frontie on Earth has gripped the world's nations, but drawing wa ter boundaries is a complicated issue. The historical devel opment of the territorial sea concept is fascinating an

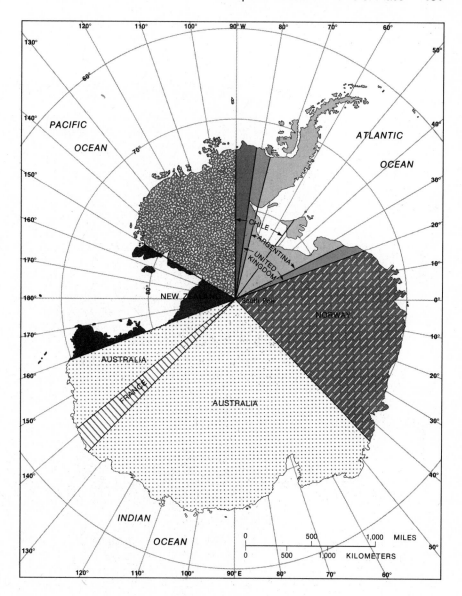

FIGURE 7-19
Political divisions in Antarctica. The area between the Chilean and New Zealand sectors is unclaimed. (From James M. Rubenstein, The Cultural Landscape: An Introduction to Human Geography, *2nd ed., Columbus, OH: Merrill Publishing Co., 1989, Fig. 7-2, p. 233. Reprinted with permission of Merrill Publishing Co.)*

complex. The idea emerged to serve two purposes: (1) to assert the exclusive right of a nation claiming its territorial sea to fish the claimed area, and (2) to define in war the extent of a neutral country's territory. That eighteenth-century cannons could fire about three nautical miles offshore had much to do with the width of the earliest territorial seas—the "three-mile limit." Thomas Jefferson, in fact, referred to the United State's three-mile limit as the "cannon-shot" limit.

Of course, in today's world, this simple "cannon-shot" method of defining offshore territorial limits is not practical. Since World War II, claims have extended out to 6, 12, or even 200 nautical miles. (A nautical mile is about 15 per-

cent longer than an ordinary land mile.) A nation having valuable fisheries would want to extend its limits for the purpose of excluding other nations from fishing in its seas. The extreme example of this is Peru, which currently claims sovereignty offshore to 200 miles. Mineral rights have also made maritime territorial limits a significant issue in recent decades.

Another related change in the postwar era has been development of technology to exploit ocean resources worldwide. For example, Japan and Russia now have huge fishing fleets capable of traversing the globe; North American companies drill for oil in the North Sea and other offshore regions. At present, over 20 percent of the world's oil and

ELEMENTS OF A STATE: CRITERIA FOR EVALUATION OF POLITICAL UNITS

Almanacs and atlases provide information about the countries of the world, but a geographic framework is necessary for meaningful comparison. The following checklist (with examples of countries) often is used by political geographers for analysis and prediction of geopolitically related events. How many of these critical elements of information are given by the news media when they report world events?

Location.   What are the limits of absolute location? Are there any "outliers" away from the major territorial limits (for example, Hawaii and Alaska in the case of USA)? What are the effects of latitudinal location (Russia)? What are the locational characteristics with reference to land and sea (Bolivia)? How about strategic location (Panama)? What are the specific site and situation of a place?

Size.   What advantages or disadvantages has a state with a large area (Brazil, China, Canada, Russia, USA)? Consider resources, defensibility, transportation costs, or environmental diversity. What are common characteristics of exceptionally small nation-states?

Shape.   A country with a compact shape is normally easier to plan for development and provision of services. Elongated nation-states (Chile, Thailand, Vietnam) have special problems, especially for communication and transportation. Likewise, island countries (Philippines, Japan, Cuba) must consider these factors.

Relief.   Mountain barriers within and between countries are a historic limitation to national unification. Although less important in the age of air travel, some countries have exhibited less national unity than others (Balkan countries, nations in central Asia). Mountain environments limit agricultural opportunities but often provide valuable mineral resources. Of course, there are exceptions (Switzerland).

Climate.   Countries in high latitudes (Iceland) and in the centers of continents (Mongolia) tend to have greater limits upon what can be produced. Most large-population groups depend upon grain crops, so the mid-latitude countries produce the most food and have the largest variety of products.

Natural Vegetation and Soils.   Many of the best soils for farming were formally grasslands (Argentina). A country's endowment of forests and agriculturally productive soils provides primary sector employment and the opportunity for exports (Canada).

Mineral Resources.   Possession of rare minerals (South Africa) and large deposits of energy resources (Saudi Arabia) strengthen a state's position in global affairs. Handicaps of having fewer such resources can be overcome (Japan).

Hydrology.   Larger concentrations of population in the modern world lie mostly in river valleys or on coasts, so it is no surprise that lack of these features can limit the economic and demographic unity of a state (Chad or Botswana). Water for irrigation can control settlement patterns (Egypt). Water flooding can be a natural hazard (Bangladesh).

Culture.   Ethnicity, language, religion, and traditional folkways all have been important factors in the causes of international conflicts especially in the Middle East and in Southeast Asia. Generally, the less culturally homogeneous a state, the more likely is the possibility for internal conflict (see *centrifugal forces* below).

Population.   Extremely large populations or uneven population distributions (India, Brazil) may pose problems for national planning and development.

Economy.   The more diversified a nation-state's economy, the more competitive advantages it has in the global marketplace. Countries dependent upon primary extractive industries are more subject to economic conditions beyond their control, such as Bolivia and Saudi Arabia. Nations with large service and manufacturing sectors in their economy tend to have higher standards of living and higher gross national products (USA).

Political Systems.   A country's political system often affects which nations it can trade with, which nations it may be allied with through military agreements, or which nations may pose a threat to it. A few countries (Switzerland, Sweden) have prospered through longstanding declarations of neutrality in global disputes.

Raison d'Être.   A nation-state's "reason for being" can be a strong unifying factor politically, economically, or militarily. For example, much internal unity can be generated by the continual prospect of aggression from neighboring countries (Israel, Croatia). Religion can be a national unifying factor (Pakistan, Eire).

Centripetal Forces.   As noted, certain forces tend to unify a nation-state from the inside. Political and economic processes can unify a country when meeting a crisis. Other internal unifying factors could be environmental conditions, a strong sense of nationalism, and a long common history (China and Iran).

Centrifugal Forces.   Centrifugal forces are those that tend toward disunity. Differences in language, religion, ethnicity, and historical roots have commonly led to dissention and internal political difficulties (former Yugoslavia).

   This list or a similar structure for comparison of nation-states may aid in an understanding of global events and may assist teachers in comparing the world's countries. For up-to-date information about the nation-states of the world, consult John Paxton, ed., *The Statesman's Yearbook* (New York: St. Martin's Press, published annually).

gas comes from offshore areas. The near future will see nations turning to undeveloped mineral resources in seabeds worldwide.

This dependence on ocean resources will further confuse the demarcation of territorial ocean boundaries. Over the last two decades representatives of seafaring countries have held "Law of the Sea" conferences to find agreement on maritime boundaries, extraction of ocean resources, and the rights of passage for ships. To date, total agreement is illusive, but work toward resolving this complicated issue is ongoing.

The technologies of the twentieth century have brought the concept of the territorial sea to an abrupt end. The dual concepts of sovereignty and freedom no longer can be balanced to cope with new environmental and resource problems, because humans can now effectively control and even inhabit remote ocean areas.

Political Geography in Action: A Case Study on the Former Nation of Yugoslavia

Southeast Europe has long been one of the most ethnically diverse regions of the world. A variety of language divisions, religious groups, and ethnic factions have made this area an especially difficult place in which to draw political boundaries. According to one Serbian commentator: "Yugoslavia was a state bordering on seven states on the outside and another eight on the inside." This viewpoint was echoed by India's Prime Minister Nehru when he said, "Yugoslavia is one land with two alphabets, three religions, four languages, five peoples, six republics, and seven neighbors."

The former nation of Yugoslavia is a good example of the many problems that occur in a physically and culturally complex multinational state. Most of the area's topography is extensive hills and mountains. Over the centuries these natural barriers have created intricate divisions of languages and religions. Of the 20 million people who live in republics carved out of the former larger nation of Yugoslavia, almost nine-tenths speak a South Slavic language (Figure 7-20).

The official nation-state of Yugoslavia, "Kingdom of Serbs, Croats, and Slovenes," was first created at the end of World War I when the South Slavic territories were added to Serbia. Governmental control came from the Serbs who tried to centralize strong control over the diverse people of the new nation.

At the end of World War II, the northern and western regions of Yugoslavia (Slovenia, Croatia, and Vojvodina), which for centuries had been integrated with Western European culture via Hapsburg or Venetian rulers, were merged with the southern and eastern regions, which had lived for centuries under Turkish rule. In addition to these cultural differences, the ethnic groups in the north were predominantly Catholic, used the Latin alphabet, and had a high level of economic development. The southern and eastern Yugoslavs, however, were mainly members of the Eastern Orthodox or Muslim faiths, used the Cyrillic alphabet, and had a lower level of economic development.

During the war, Joseph Tito, a charismatic leader of the Communist party (and who was half Slovene and half Croat), united various Yugoslav factions to drive the Germans, who invaded in 1941, out of the region. This created the prospect of a united country. After the war, when political boundaries were realigned, an attempt was made to draw lines of demarcation according to language groupings. The result was still imperfect (Figure 7-21). The new division left a large number of ethnic Serbs inside other newly formed republics, especially Bosnia. Conscious that both Croatia and Serbia laid historical claim to Bosnia, Tito declared during the war that its future would be "neither Serbian nor Croatian nor Muslum but rather Serbian *and* Croatian *and* Muslum." The result was a disaster.

In the period after World War II, the state was reorganized in the Soviet style. Six "Peoples Republics" were formed, each of which encompassed an autonomous region and an autonomous province. This division of power seemed to help ease tensions.

The original reason for Yugoslavia's existence as a unified nation was to keep out the Germans. Once that threat was gone, there has been little to bind the diverse ethnic groups together except the need to keep out the Russians during the Cold War era. The end of the Cold War and the collapse of Soviet-, style communism took the threat of Soviet invasion away. Since the late 1980s, the area's strong opposing political parties have been competing for control. Until its collapse in 1991, Yugoslavia was basically an unhappy marriage of its two largest nationalities, with little or no influence from the other smaller groups. In early 1991, Croatia and Slovenia separated from Yugoslavia and a civil war began. Bosnia then claimed its independence in 1992 prompting an invasion by Serbia. Yugoslavia no longer existed and neither the intervention of the United Nations nor the threat of NATO action have quelled the fighting. By 1995 the war claimed over 200,000 lives and has displaced over a million refugees.

The Balkan Peninsula is a mosaic of different groups, and even the best political geographer cannot do them justice. It serves as a fascinating example of the difficulty of analyzing political patterns in the world today. The drawing

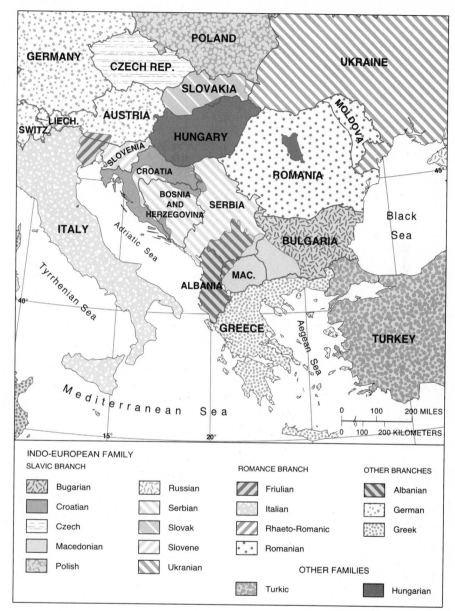

FIGURE 7-20
Slavic languages in the Balkans. (From James M. Rubenstein, The Cultural Landscape: An Introduction to Human Geography, *4th ed., New York: Macmillan, 1994, Fig. 7-16, p. 287. Reprinted with permission of Macmillan Publishing Co.)*

of political boundaries, the defining of nation-states, and the understanding of the forces that hold them together are the subjects of the majority of the global news stories appearing in our media everyday.

Numerous other topics of interest to political geographers are beyond the scope of this text but warrant further study in future course work. They include analyzing differences between rimland (coastal) and heartland (interior) countries, studying the electoral college system of voting in the United States, and other fascinating topics.

THE CULTURAL LANDSCAPE

This chapter has presented several topics of interest to human geographers. We have seen how languages, religions, political patterns, and population distribution affect Earth's human patterns. The resultant features on the land are visible evidence of cultural and historical patterns and processes.

FIGURE 7-21
New ethnic nations of former Yugoslavia.

A.

B.

FIGURE 7-22
*A. Amish farm in western Pennsylvania. (Photo by author.)   B. Chinese farm south of Canton.
(Photo courtesy of Steven Herman.)*

The **cultural landscape** is best defined as the surface of Earth as modified by humans. North American geographers have been interested in studying the cultural landscape ever since Carl Sauer and his students stressed the importance of understanding present-day landscapes as evidence of historic processes. These **sequent occupancies** have left their mark.

Most of our planet, except for the most inhospitable places, has been occupied by people at one time or another. Human transformation of Earth depends on human decision making, objectives, attitudes, and technology. Figure 7-22 compares the varying impact humans can have upon the land. Compare the differences and similarities of land use patterns on the Amish farm in western Pennsylvania with the intensity of land use on a Chinese collective farm.

Clearly, the cultural landscape varies with time and place. J. B. Jackson, founder of *Landscape* magazine, a publication focusing upon the importance of understanding landscapes of all types, had this to say about culture change and landscape studies: "every ruin, whether in Asia Minor or in the American Southwest reveals a fragment of a rural or urban landscape which became obsolete..."[9]

A more recent editor of *Landscape,* Bonnie Loyd, is an avid defender and describer of the cultural landscape and place. Her popular public radio commentary, "The Urban Geographer," made California urban cultural landscapes known to a wide audience (see the box on page 188).

Social patterns and processes are not static but change through time. Cultural change often is reflected in the cultural landscape as each group leaves its signature on the land. The more complex the culture, the greater is its impact on the environment.

We have seen that people are able to modify their environment through the material culture they develop. Cities, roads, cemeteries, field patterns, and airports are all expressions of human occupance. The cultural landscape remains a tangible, physical record of a given culture.

Geographers have a practical interest in studying the cultural landscape. Regional and city planners try to bring harmony to our environment by balancing the landscape of the city. The prerequisite for any urban or rural plan is full knowledge of the existing cultural landscape so that much of value may be preserved.

Pierce Lewis, a North American geographer known for his astute observations of ordinary landscapes, suggested these four axioms for studying and teaching about the cultural landscape:

1. *The axiom of landscape as a clue to culture:* Ordinary things that people create on Earth say something about the kind of people we are, and about what we are in the process of becoming.

2. *The axiom of cultural unity and cultural ecology:* Almost all landscapes reflect culture in some way. All items are important in landscape studies.

3. *The axiom of common things:* Common landscapes are difficult to study.

4. *The axiom of history:* The present-day cultural landscape says something about the past.[10]

Understanding how people and their cultural baggage influence the landscape leads us directly into further study about how people make their living from the land. These economic features of the theme of Place are explored in the following chapter, as we continue our study of human geography.

MEET THE GEOGRAPHER
Sharing Landscapes

In high school Bonnie Loyd wondered how her interests would fit together into a career. Then she went to New Zealand as an exchange student. There, as in other countries of the British Commonwealth, geography is an important part of the curriculum. On weekend hikes her geography teacher could both point out earthquake faults and discuss current events in China. Bonnie recalls, "I realized I could use geography every day."

At the University of Wisconsin, after much thought, Bonnie chose geography as a major instead of journalism. As a graduate student at Syracuse University, she was encouraged by some of the best writers in geography, while outside the classroom she worked on a student social science journal and began a national newsletter for women in geography.

After graduate school Bonnie moved to San Francisco and became a freelance editor of college geography textbooks, doing everything from assembling glossaries to planning illustrations. She also worked as a project writer and researcher on urban planning reports for a major consulting firm. She notes that the ability to write clearly and the ability to understand information about local environments are the most valuable skills for planners.

After writing an article for *Landscape* magazine in Berkeley, California, she was asked to be the editor of this unusual publication on the human geography of places. In this role, she finds new authors, edits articles, writes news items, chooses photographs, designs the layout, and supervises marketing. About 40 percent of the articles are written by geographers and the others come from authors in kindred disciplines such as architecture, landscape architecture, art history, literature, photography, and history. Recent issues have contained articles on French cellars, the Nevada Nuclear Test Site, billboards, the first skyscrapers, and landscape painting. The magazine, which has been published for nearly 40 years, has subscribers around the world. In addition, Bonnie writes book reviews for newspapers and magazines such as the *Los Angeles Times* and the *Professional Geographer*.

When asked about the fascination of her work, Bonnie says, "I like exploring, so I feel lucky that I can be a tourist every day, not just when I'm on vacation."

KEY CONCEPTS

Birth and death rates	Demography	Language family	Rimland
Boundary	Enclave	Malthusian doctrine	Sequent occupance
Centrifugal forces	Ethnic religion	National cohesion	Toponyms
Centripetal forces	Exclave	Nation-state	Universalizing religion
Cultural landscape	Geopolitics	Population distribution	
Culture	Heartland	Population pyramid	
Demographic transition	Iconography	Rate of natural increase	

A spinoff of Bonnie Loyd's experience in print is her weekly series for public radio, entitled "The Urban Geographer." Bonnie writes and presents short commentaries about familiar things in the city, such as parking meters, motels, graffiti, tree houses, elevators, and even roller coasters. The series has won a national broadcasting award.

Here is the script for one of Bonnie's typical broadcasts on public radio station KQED-FM in San Francisco. Her topic is "Drive-in Restaurants."

I'm always amazed by how much the architecture in this country has been shaped by the automobile. As soon as Americans began to drive, they began to design new buildings. The automobile brought us gas stations, drive-in movies, motels, garages, even shopping malls. But my favorite artifact of car culture is the drive-in restaurant.

Drive-in restaurants are totally unnecessary to civilized life, but wonderful nevertheless. Now I'm not talking about the fast-food franchise where you stand at the counter and they shove an emaciated burger at you. No, no, no. I mean the kind of place where you cruised in on a warm night in your '57 Chevy, and a pretty carhop in a sort of majorette uniform sidled up to the car to take your order. While you waited for some serious burgers and industrial-strength malts, you basked in fluorescent light and checked out the occupants of other cars. The food arrived on special aluminum trays that clamped to the side of your car. If you needed some salt for the fries, you flashed your headlights to beckon the waitress. Blowing the paper wrapper off your straw was optional.

The first drive-in restaurant, the Pig Stand, opened in 1921 on the Dallas–Fort Worth Highway. Texas was first, but California wasn't far behind. In 1923, the A&W people opened a drive-in root-beer stand in Sacramento.

I was delighted to discover a book on the unlikely topic of the architecture of chain restaurants. The author, Philip Langdon, claims that "the finest drive-ins were reserved for California." Well, that's not surprising. We have great weather, plenty of automobiles, and a sense of humor.

One reason California drive-ins were special is that many of them were round. Drive-ins present one of the few situations where a round building makes sense. All the customers who pull up to the restaurant want to face the front, and a round building lets them do it. Carhops like the arrangement because they have to walk less. And the circular plan also mimics the curve of a car as it swings around the building.

Drive-in restaurants haven't survived very well. A combination of things pushed them out of business—rising real-estate prices, fears of juvenile delinquents, and self-service restaurants. But there are still a few drive-ins scattered around Northern California. If you spot one, pull in and order a strawberry milkshake for me.

FURTHER READING

The study of culture is shared by geographers, anthropologists, and other social scientists and humanists. The geographer's interest usually is based on the cultural and societal relationships of people to Earth. A recommended and classic survey of research topics pursued by cultural geographers is to be found in the various chapters contained in Philip L. Wagner and Marvin W. Mikesell, eds., *Readings in Cultural Geography* (Chicago: University of Chicago Press, 1962).

Three highly recommended survey texts, each with a different approach to the topic of cultural geography, are Jerome Fellmann, Arthur Getis, and Judith Getis, *Human Geography:* *Landscapes of Human Activities* (Dubuque, IA: William. C. Brown, 1990); Terry Jordan, Mona Domosh, and Lester Rowntree, *The Human Mosaic: A Thematic Introduction to Cultural Geography,* 6th ed. (New York: Harper & Row, 1994); and James M. Rubenstein, *The Cultural Landscape: An Introduction to Human Geography* (New York: Macmillan, 1994).

Cultural geography in the United States is summarized in Wilbur Zelinsky's book, *The Cultural Geography of the United States,* 2nd ed. (Englewood Cliffs, NJ: Prentice-Hall, 1993). A collection of ideas on how cultural geography may be mapped is in John G. Rooney, Jr., Wilbur Zelinsky, and Dean R. Louder, eds., *This Remarkable Continent: An Atlas of United States and Canadian Society and Culture* (College Station: Texas A & M University Press, 1982).

The subfield of geography that deals with population is well represented in Nathan Keyfitz and Wilhelm Flieger, *World Population Growth and Aging: Demographic Trends in the Late Twentieth Century* (Chicago: University of Chicago Press, 1990); Paul and Anne Ehrlich, *The Population Explosion* (New York: Simon & Schuster, 1990); Gary L. Peters and Robert P. Larkin, *Population Geography: Problems, Concepts, and Prospects* (Dubuque, IA: Kendall/Hunt, 1989); and the World Bank, *World Development Report* (New York: Oxford University Press, published annually). Also useful in teaching geography is the annual *World Population Data Sheet,* published by the Population Reference Bureau, 777 14th Street, NW, Suite 800, Washington, DC 20005. PRB has also published some very useful materials for teaching about population issues. Some of the most creative and comprehensive of these publications are entitled *Connections: Linking Population and the Environment* by Kimberly A. Crews (Washington, DC: Population Reference Bureau, 1990); *Making Connections: Linking Population and the Environment,* by Kimberly A. Crews (Washington, DC: Population Reference Bureau, 1992), and *The Future of World Population* by Wolfgang Lutz (Washington, DC: Population Reference Bureau, 1994).

The general topic of the geography of languages is discussed more by linguists than by geographers. A recommended survey book is Grover S. Krantz's *Geographical Development of European Languages* (New York: Peter Lang, 1988). A geographic approach to appreciating the spatial patterns and diffusion of the English language is Robert McCrum, William Cran, and Robert MacNeil, *The Story of English* (New York: Viking Penguin Books, 1986). Other useful sources on languages include the classic "The Geography of Languages" by C. M. Delgado de Carvalho, in *Readings in Cultural Geography,* Philip L. Wagner and Marvin W. Mikesell, eds. (Chicago: University of Chicago Press, 1962); *Language in Culture and Society* by Dell H. Hymes (New York: Harper & Row, 1964); and Colin H. Williams, ed., *Language in Geographic Context* (Clevedon, Avon: Multilingual Matters, 1988).

The cultural geography of religions is effectively summarized in *Sacred Worlds: An Introduction to Geography and Religion* by Chris C. Park (New York: Routledge, 1994), and David E. Sopher, *Geography of Religions* (Englewood Cliffs, NJ: Prentice-Hall, 1967). A study of how religious beliefs may effect food and diet restrictions and choices is Frederick J. Simoons, *Eat Not This Flesh: Food Avoidances in the Old World* (Madison: University of Wisconsin Press, 1961). Other sources on the geography of religion are Adrian Cooper, "New Directions in the Geography of Religion," *Area* 24 (June 1992): 123–129; *The Mormon Landscape*

by Richard V. Francaviglia (New York: AMS Press, 1978); and Trevor Ling's *A History of Religions East and West* (London: Macmillan, 1968).

Ethnic geography has become an increasingly important subfield in the discipline of geography in recent decades. With the establishment of a new American Ethnic Geography Specialty Group in the Association of American Geographers and the publications of numerous new books and articles on the subject, this issues-based focus of cultural geography has found a focus in the multicultural climate of the 1990s. The following sources are offered as a beginning to your study of this important topic: James P. Allen and Eugene J. Turner, *We The People: An Atlas of America's Ethnic Diversity* (New York: Macmillan, 1987); Susan Wiley Hardwick, *Russian Refuge: Religion, Migration, and Settlement on the North American Pacific Rim* (Chicago: University of Chicago Press, 1993); "Readaptation and European Colonization in Rural North America," *Annals of the Association of American Geographers* 79 (1989): 489–500; Jesse O. McKee, ed., *Ethnicity in Contemporary America* (Dubuque, IA: Kendall/Hunt, 1985); Richard L. Nostrand, *The Hispano Homeland* (Norman: University of Oklahoma Press, 1992); and Robert Ostergren, *A Community Transplanted: The Trans-Atlantic Experience of a Swedish Immigrant Settlement in the Upper Middle West, 1835–1915* (Madison: University of Wisconsin Press, 1988).

The diversity of political geography can be seen by surveying *Political Geography of the Twentieth Century: A Global Analysis,* Peter J. Taylor, ed. (London: Belhaven Press, and New York: Halstead Press, 1993); *Nation, State, and Territory: A Political Geography* by Roy E. H. Mellor (London and New York: Routledge, 1989); Stanley D. Brunn's classic, *Geography and Politics in America* (New York: Harper & Row, 1974); Peter Slowe's *Geography and Political Power: The Geography of Nations and States* (London: Routledge, 1990); and *Systematic Political Geography* by Harm DeBlij (New York: John Wiley, 1989). An excellent geopolitical reference is by economists Michael Kidron and Ronald Segal, *The New State of the World Political Atlas* (New York: Simon & Schuster, 1987).

A world systems theory approach to the study of political geography is discussed in Immanuel Wallerstein's *Geopolitics and Geoculture: Essays on the Changing World System* (New York: Cambridge University Press, 1991).

A collection of readings introducing the study of cultural landscapes is found in J. B. Jackson's *Discovering the Vernacular Landscape* (New Haven, CT: Yale University Press, 1984) and Christopher L. Salter, ed., *The Cultural Landscape* (Belmont, CA: Duxbury Press, 1971). Another collection of essays about the landscape experience is found in Donald W. Meinig, ed., *An Interpretation of Ordinary Landscapes: Geographical Essays* (New York: Oxford University Press, 1979). Other sources on the cultural landscape are discussed in Chapter 4.

ENDNOTES

1 Wilbur Zelinsky, "The Use of Cultural Concepts in Geographical Teaching: Some Conspiratorial Notes for a Quiet Insurrection," in *Introductory Geography: Viewpoints and Themes* (Washington, DC: AAG, 1967), AAG publication No. 5, p. 75.

2 Paul Vidal de la Blache, *Tableau de la Géographie de la France* (Paris, 1903).

3 A. L. Kroeber and C. Kluckhohn, "Culture: A Critical Review of Concepts and Definitions," *Papers of the Peabody Museum of Archaeology and Ethnology* (New York: Peabody Museum, 1952).

4 Zelinsky, "The Uses of Cultural Concepts," p. 75.

5 Donald Steila, Douglas C. Wilms, and Edward P. Leahy, *Earth and Man* (New York: John Wiley, 1981): 229.

6 David Sopher, *Geography of Religions* (Englewood Cliffs, NJ: Prentice-Hall, 1967): 34.

7 Richard Hartshorne, "Political Geography," in P. E. James and C. F. Jones, eds., *American Geography: Inventory and Prospect* (Syracuse, NY: Syracuse University Press, 1954): 192–193.

8 Julian Minghi, "Boundary Studies in Political Geography," in Roger E. Kasperson and Julian Minghi, eds., *The Structure of Political Geography* (Hawthorne, NY: Aldine, 1969): 147.

9 J. B. Jackson, *Discovering the Vernacular Landscape* (New Haven, CT: Yale University Press, 1984): 68.

10 Pierce Lewis, in D. W. Meinig, *An Interpretation of Ordinary Landscapes* (New York: Oxford University Press, 1979).

8

THE THEME OF PLACE: ECONOMIC SYSTEMS

Bringing in the goats in Pamir Mountains, China. (Photo courtesy of Stephen F. Cunha.)

The economic characteristics of place focus upon the production, distribution, and consumption of **resources.** A resource base is finite, but our perception and use of resources varies. Resource location is a prime factor in economic advantage or disadvantage for a country or region.

Places also differ in their productivity and consumption patterns. For this reason, humankind has allocated land for various uses. Economic activities based on land use are thus spread across Earth unevenly (but not randomly).

It is the predictability of these economic patterns that forms the core of economic geography. Knowledge of these predictable patterns is the essence of this chapter. (All places have a location and those locations can be very important. For more about the locational aspects of economic geography we refer you to Chapter 3.) The *National Geography Standards* related to the contents of this chapter include:

Standard 11: *The geographically informed person knows and understands the patterns and networks of economic interdependence on Earth's surface.*

Standard 16: *The geographically informed person knows and understands the idea of "Resource" and the changes that occur in the use, distribution, and importance of resources.*

ECONOMIC PATTERNS IN GEOGRAPHY

The exchange of goods and services is based on availability, accessibility, wants, and needs. The economic geographer seeks patterns and consistencies, so generalizations or concepts are necessary to explain economic patterns and processes. In this chapter, we broadly overview the most important generalizations. Some concerns of the economic geographer are discussed in other chapters; for example, economic factors in urban geography are presented in Chapter 9, and the economic concepts of trade and transportation are discussed in Chapter 11 in the theme of Movement.

Economic geography emphasizes the *location* of economic activity and the *character* of places that result from these activities. Our main focus here is upon the production, distribution, and **consumption** of goods and services. For analysis it is useful to divide resource utilization and other economic activities into four categories:

1. Primary—basic forms of livelihood on Earth, including agriculture, mining, fishing, and forestry (activities dealing with nature).

2. Secondary—processing of primary products, an important basis for industrial societies (value is added during the production process).

3. Tertiary—distribution of goods and other servicing industries; dominates highly industrialized countries.

4. Quaternary—dissemination of information by superskilled and technically trained "workers." The 1990s are a time of great change in the economic sector, particularly the globalization of many activities and the growing importance of transnational companies. Our "information society" is a reflection of rapid changes in technology and significant culture change.

Goods are classified as either *producer goods* (used to produce other goods, for example, a machine tool) or *consumer goods* (used personally by individuals, for example, food, cars, sports equipment). The geography of consumption is, of course, quite uneven in the world; economic wants and needs and resulting consumption through purchases are the driving forces behind economic activity.

It is important to keep in mind that the disparity between consumption in the industrialized world and consumption in many other countries is increasing, not shrinking. Implications of this imbalance may be seen in decision making, both political and economic.

PRIMARY ECONOMIC ACTIVITIES

Primary production occupies most of Earth's surface. Spatially, the activities of agriculture, mining, fishing, and forestry cover large parts of the planet. Temporally, these basic economic systems came first. Numerically, primary production involves well over half of the world's people.

Traditional Hunting, Fishing, Gathering

The "most primary" of the many types of primary production is rudimentary hunting, fishing, and gathering. Before humans developed agriculture, all practiced one or more of these basic economic activities. In today's world, they are economies of exception—they are the oldest occupation on Earth, they involve the fewest people, and they demand more land area than any other economic system.

In areas having cultures that are poorly adapted to climatic limitations, gathering is still practiced by native peoples. Tools are simple and limited, the domestication of plants and animals is not practiced, and trade is minimal. Hunters and gatherers usually live in small groups, often smaller than 50 people, because more would quickly exhaust local resources. They survive by collecting food daily and so must move frequently to locate edible products. These groups are located in widely scattered parts of the world:

- Pygmies of central Africa
- Semang and Sakai of the Malay Peninsula
- Kubu of Sumatra
- Ge tribes of eastern Brazil
- Hottentot (Khoikhoi) of South Africa
- Aborigines of Australia
- Inuit (Eskimo) of northernmost North America (see the box on page 194)

AGRICULTURAL SYSTEMS

Farming is the most universal economic activity on Earth. Highly mechanized "agribusiness" in the United States contrasts with the basic subsistence farm common in many other parts of the world. Farmers in much of the world grow only enough food to survive, with little or no surplus. In developed countries, farmers rely on machine technology to feed thousands of people with their labor. Since the origin of agriculture at least 15,000 years ago, the population of Earth has expanded at least a thousand times because farmers have taken on the responsibility for others (Figure 8-1).

The idea of agriculture was certainly one of Earth's

CATCHING A SEAL

The following historical description of seal hunting offer an exciting glimpse of the influence of economic systems on place.

The whole principle of successfully stalking a seal is just in realizing from the first that he is bound to see you and that your only hope is in pretending that you are also a seal. If you act and look so as to convince him from the first that you are a brother seal, he will regard you with unconcern. Imitating a seal well enough to deceive a seal is not difficult, for, to begin with, we know from experience that his eyesight is poor. You can walk up without taking any special precautions until … you are within two hundred and fifty or three hundred yards. Then you have to begin to be careful. You move ahead while he is asleep, and when he wakes up you stop motionless. You can safely proceed on all fours until within something less than two hundred yards, but after that you will have to play seal more carefully. Your method of locomotion will then have to be that of the seal which does not differ very materially from that of a snake, and which therefore has its disadvantages at a season of the year when the surface of the ice is covered with puddles of water anywhere from one to twenty inches in depth as it is in spring and early summer. You must not crawl ahead, seal fashion, but you must be careful to always present a side view of your body to the seal, for a man coming head-on does not look particularly like a seal.

Until you are within a hundred yards or so the seal is not likely to notice you, but somewhere between the hundred yards and the seventy-five yard mark his attention will suddenly be attracted to you, and instead of going to sleep at the end of his ordinary short period of wakefulness, he will remain awake and stare at you steadily. The seal knows, exactly as well as the seal hunter knows, that no seal in this world will sleep continuously for as much as four minutes at a time. If you lie still that long, he will know you are no seal, and up will go his tail and down he will slide into the water in a twinkling of an eye.

When the seal … has been watching you carefully for twenty or thirty seconds, you must raise your head twelve or fifteen inches above the ice, look around seal-fashion, so that your eyes will sweep the whole circle of the horizon, and drop your head again upon the ice. By the time he has seen you repeat this process two or three times in the space of five or six minutes, he will be convinced that you're a seal, and all his worries will be gone. From then on you can proceed more rapidly, crawling ahead while he sleeps and stopping while he remains awake, never doing anything unbecoming to a seal. In this way, you can crawl to within five or ten yards of him if you like, and as a matter of fact I have known of expert seal hunters who under emergencies would go after a seal without any ordinary weapon and crawl so near him that they could seize him by a flipper, pull him away from his hole, and club or stab him.

This description of place and economics was written in 1913. Today, the Inuit people of Alaska and Canada no longer practice seal hunting in this traditional way. Completion of the Alaskan pipeline, depletion of biotic resources, and cultural modification due to increased contact with the outside world have all contributed to cultural and economic change in the region as "modernization" has occurred.

Adapted from My Life with the Eskimo _by Vilhjalmur Stefansson (New York: Macmillan, 1913), reprinted in John R. Coleman,_ Comparative Economic Systems _(New York: Holt, Rinehart & Winston, 1968). Used with permission._

most radical revolutions in human cultural and economic development. Today, significant variations in types of agricultural production may be observed on Earth as climatic and cultural opportunities combine to produce unique patterns such as those shown in Figure 8-2. (The origin and diffusion of agriculture is detailed in Chapter 11.)

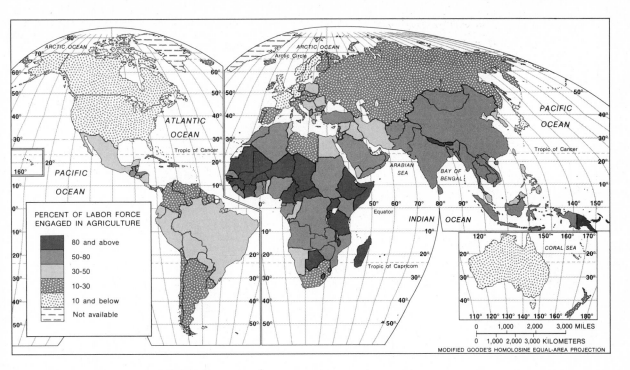

FIGURE 8-1

Percent of labor force engaged in agriculture worldwide. (From James M. Rubenstein, The Cultural
Landscape: An Introduction to Human Geography, *2nd ed., Columbus, OH: Merrill Publishing Co.,
1989, Fig. 8-3, p. 281. Reprinted with permission of Merrill Publishing Co.)*

Shifting Cultivation ("Slash and Burn Agriculture")

The largest percentage of people on Earth are subsistence
farmers. Most of these "primary producers" live in Asia,
Africa, and Latin America. Extensive subsistence agriculturalists often practice shifting cultivation, or "slash and
burn" agriculture. This type of agriculture is most commonly called *swidden,* from a word meaning "to singe".
Small patches of land are cleared with machetes and other
bladed tools, trees are stripped and killed, and fires are set
to clear the land for planting. With the use of a simple dig-
ging stick or hoe, various subsistence crops are planted for
use by the producer and kinship groups.

Swidden cultivation requires very large areas of land
because small fields are rotated from one place to another.
It is the least labor-intensive of any type of farming.
Although techniques do not require much capital, this system of agriculture is actually quite successful in the tropical world. Formerly viewed as a "primitive" form of production because of its simplicity, swidden is now
considered a sophisticated production system requiring
knowledge of the physical environment and crop sensitivi-
ties. Because of swidden farmer contact with the outside
world as well as increased population growth, however, this

type of farming is currently coming under scrutiny as a negative influence on the fragile tropical world.

Nomadic Herding

Other types of extensive subsistence production include nomadic herding, or the raising of animals that are completely dependent on natural vegetation (Figure 8–2). A relatively small number of people are still involved in this
activity. Nomadic herding occurs in many different physical environments and in places with a strong tribal spirit.
Sheep, goats, and camels are the most important animals
raised, but cattle, horses, reindeer, and yaks also may be important locally.

Mobility is one answer to meager, slow-growing vegetation. Families, animals, and other possessions are
moved regularly to another location as forage and water
are exhausted in the original settlement area. Movement
is seasonal between summer and winter pastures that usually are remote from each other. This seasonal movement
between pastures is known as **transhumance.** Transhumance is carried out between the northern and southern steppes in interior Asia, or between different types of
terrain, that is, between plains and mountains, between

FIGURE 8-2
Areas of subsistence and commercial agriculture in the world. (From James M. Rubenstein, The Cultural Landscape: An Introduction to Human Geography, *2nd ed., Columbus, OH: Merrill Publishing Co., 1989, Fig. 9-3, pp. 314–315. Reprinted with permission of Merrill Publishing Co.)*

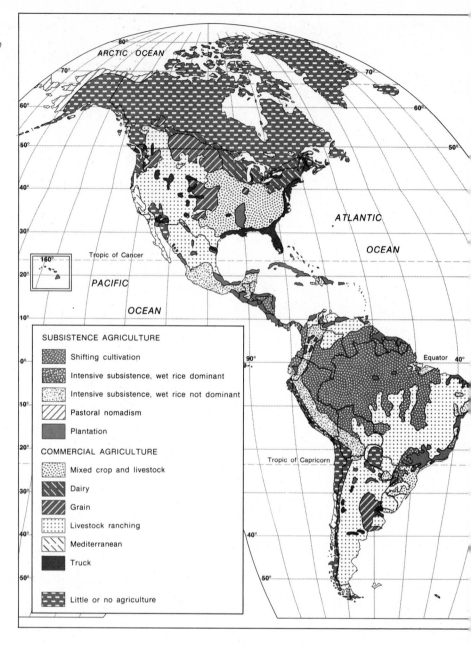

SUBSISTENCE AGRICULTURE

Shifting cultivation

Intensive subsistence, wet rice dominant

Intensive subsistence, wet rice not dominant

Pastoral nomadism

Plantation

COMMERCIAL AGRICULTURE

Mixed crop and livestock

Dairy

Grain

Livestock ranching

Mediterranean

Truck

Little or no agriculture

plains and plateaus, between valley bottoms and mountain slopes, or between opposing mountain slopes with strongly contrasting climatic conditions. (The best example of the last type of transhumance occurs in the Himalaya Mountains where snow-covered north-facing slopes are abandoned in winter for snow-free south-facing slopes.)

A typical cultural group practicing nomadic herding subsistence production is the Kirghiz people in central Asia (Figure 8-3). Although the majority of Kirghiz are no longer completely nomadic, they are still very much dependent on goat meat for survival. Other nomadic herders are the Lapps in northern Finland. In 1986, contamination by nuclear fallout from the Soviet Union's Chernobyl nuclear reactor incident severely contaminated their reindeer stock. Many contaminated animals were destroyed. Radiation contamination will continue to be a problem for at least the next decade.

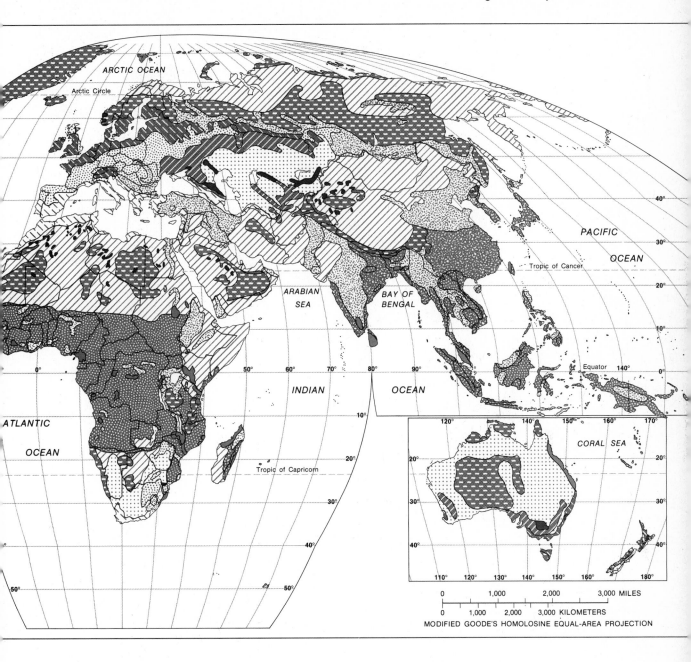

MODIFIED GOODE'S HOMOLOSINE EQUAL-AREA PROJECTION

Intensive Subsistence Farming

Intensive **subsistence** agriculture, practiced for over 8000 years, continues to support over half of the world's population. It is characterized by:

1. A high level of local knowledge concerning planting times, soil fertility, and weather patterns

2. A variety of implements and tools

3. A technique that greatly alters the landscape

4. The ability to support large populations

5. The production of only a very small exportable surplus

One of the most densely populated places in the world, southern China, has depended upon intensive subsistence farming for centuries. Here, many individual crops are grown at the same time and in the same field. Attention is given to each plant, with planting and harvesting done primarily by hand. Large draft animals are almost completely absent in this system. The most impressive evidence of in-

FIGURE 8-3

Kirghiz, a nomadic people in the Pamir Mountains of central Asia, milking their goats. (Photo courtesy of Stephen F. Cunha.)

tensiveness, however, is in the local landscape. Everything seems subordinated to the need for cropland. Roads are narrow and confined whenever possible to the edges of rough or forested land. Paths follow the tops of dikes that separate rice paddies. Houses and villages are located on hillsides and other unproductive land.

Terracing is the most noticeable evidence of local efforts to expand cropland in many intensive subsistence areas. In China, over one-fourth of all cropland is terraced. Terraces minimize slope erosion and the consequent deposition of rubble on top of fertile soils at the foot of the slope. They also reduce water runoff so that the moisture supply for crops is increased. However, poorly built or maintained terraces may actually hasten deterioration and fragmentation of a region.

Through terracing and other methods, intensive subsistence agriculture has fed millions of people for centuries. Recent advances include intercropping and multiple cropping (harvesting more than one crop from the same patch of land each year). These innovations have doubled and tripled crop yields in many parts of the world in recent years. The effects of the "green revolution" also have increased yields significantly in intensive subsistence agricultural regions. Details of the almost unbelievable expansion of crop yields from this modern-day "revolution" are discussed in Chapter 10.

Growing Rice: China versus California

Most of the world's population depends upon grain crops

as the staple of diet. The most frequently grown grain is rice, and there are many ways to produce it. A good example of how agricultural methods vary due to a contrast in cultures and economies can be seen in the cases of intensive subsistence rice production in Asia and in commercial rice production in California's Central Valley.

China. This nation has a land mass of over 3.7 million square miles (9.6 million km^2) and is by far the largest country in Asia. Yet, in this huge country, only 29 percent of the land can be used for agriculture. Despite this lack of arable land, China produces a variety of staple crops for its home market. Wheat, cotton, and maize are the main crops in the north; rice is the primary crop in the south. The rice-producing areas of southern China are the most densely populated parts of the country.

Chinese farmers usually do not live on their rice farms. Instead they live in small villages near the fields or on large communes. Few draft animals and no modern machinery are used to help cultivate the land for rice production. Tiny acreages are intensively planted, cared for, and harvested generally by one Chinese family, with attention focused upon each plant. After gathering, rice usually is removed from the hull by beating it on the ground or by having animals trample it. Grain often is available in local villages as brown rice, which has had only the hull removed. The typical rice farmer in southern China is poor—often destitute—because he must depend on only a few acres of land to support an entire family. Taxes and necessary consumer goods are paid for from the family rice supply.

FIGURE 8-4
Plowing the rice fields, Indus River plain, Pakistan. (Photo courtesy of Stephen F. Cunha.)

Other nations in Southeast Asia also grow rice as a labor-intensive crop. Figure 8-4 illustrates this type of subsistence farming in Pakistan.

California. In dramatic contrast to China's labor-intensive rice production, the Central Valley in California produces rice on a massive scale using highly technical machinery on very large acreages. About 85 percent of California's rice is harvested in the Sacramento Valley, which extends about 150 miles (240 km) north of Sacramento in northern California (Figure 8-5). In formerly unproductive swampy areas between river levees and the foothills of nearby mountains, rice is planted by airplane in fields graded by laser-guided earthmovers.

An abundance of level land, the availability of low-cost water, and a warm climate have been cited as major reasons for the success of rice-growing in this part of the American West. As shown in Figure 8-6, total rice acreage and yield per acre have increased with improved technology. Since the advent of planting and fertilizing by airplane, pesticide application for weed and insect control, implementation of advanced methods for disease control, and other scientific solutions to rice-growing problems, California's production has increased dramatically.

From the earliest days of commercial rice production in California, innovative farmers have continued to increase rice yields per acre. In recent years, this effort has been assisted by scientific research at the University of California at Davis and at the U.S. Department of Agriculture. In fact, there has been a 200 percent increase in average yield per acre between 1912 and 1975. These efforts have resulted in cultural and economic practices that favor commercial production of rice in a competitive

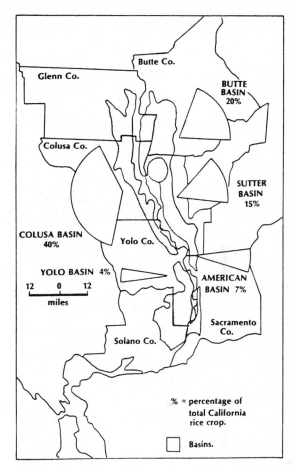

FIGURE 8-5
Relative importance of the Sacramento Valley's basins as rice growers. (From Milton D. Miller, Rice in California, Richvale, CA: Butte County Rice Growers Association, 1979, Fig. 1, p. 79. Used with permission of Butte County Rice Growers Association.)

world market. Today, a large percentage of California rice is shipped to Asia!

However, the dependence upon fossil fuel–based agriculture and manufactured chemical sprays has caused a buildup of fertilizers and hazardous wastes in drainage channels. This has reminded growers that bigger or faster may not always be environmentally better. In recent years these residues, such as selenium, have been observed in downstream reservoirs and rivers at dangerous concentrations for waterfowl and fish. Disposal of agriculturally generated toxic wastes is an unsolved environmental problem that reminds us of the first law of ecology: everything has to go somewhere.

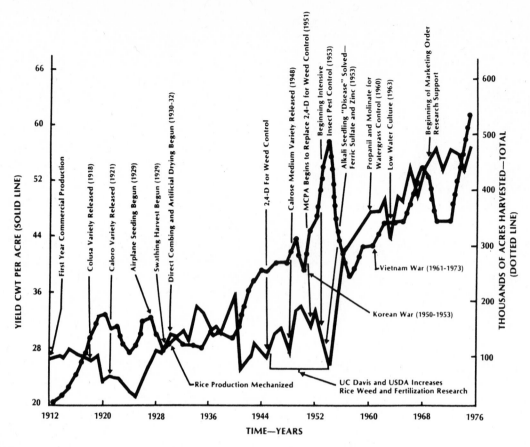

FIGURE 8-6

California rice acreage and yield trends vs. technological developments. (From Milton D. Miller,
Rice in California, *Richvale, CA: Butte County Rice Growers Association, 1979, Fig. 8.5, p. 134.*
Used with permission of Butte County Rice Growers Association.)

Commercial Agriculture

California's rice production is only one of many different types of **commercial** farming around the globe. Commercial agriculture operates under many natural and economic controls; as outlined by economic geographer Richard Morrill, these include:[1]

- Environment, including landforms, soil, temperature, moisture, and growing season, all of which influence the cost to produce various crops

- Location relative to major markets and the resulting transport costs for different crops

- Consumer demand for various products

- Inherent characteristics of the crop (productivity and the labor required, for example)

- Productivity of the crop in response to inputs such as fertilizer or machinery

- Regional differences in labor quality, costs of land, form of ownership, population pressures on the land, and the presence of alternate employment opportunities

Commercial agricultural systems have interconnections and specializations that do not exist in subsistence systems. These spatial connections, production and consumption patterns, and the overall complexity of commercial agriculture have been studied by geographers in great detail. Land use theories and location models have been developed to explain patterns of agricultural production. Modern location theory in economic geography is based on a model developed in 1826 by a German landowner, Johan Heinrich von Thünen.

Von Thünen's Isolated State

Von Thünen's book *The Isolated State* may be summarized as follows:[2]

If environmental variables are held constant, then the farm product that achieves the highest profit will out-bid all other products in the competition for location.

In other words, the most intensive farming will be located nearest the market, where the owner's investment will be entirely in labor and capital instead of in transportation. Farther away from the market, land use becomes more expensive as transportation costs increase but rent for land decreases. Thus market gardening and dairying would be located closest to markets, and cattle ranching would be farthest away (Figure 8-7).

Von Thünen's theory was geography's first formal spatial model. As with most models, it generalizes *patterns* of location but does not explain or predict *individual decisions* regarding location. Von Thünen's model idealizes a situation in which the basic assumptions are that the following will exist:

1. A flat land surface with homogeneous climate and soil conditions

2. A uniform transportation system

3. A single market at the center; its needs being supplied entirely by the surrounding agricultural area, which is in turn surrounded by a forested area that completely cuts off trade with the outside world

The effect of von Thünen's model is most easily visualized with a series of concentric circles surrounding the market center. As shown in Figure 8-8, intensive land use nearest the market will be a zone of market gardens and feed lots, surrounded progressively by dairying, livestock fattening, commercial grain farming, livestock ranching, and ultimately by nonagricultural land. Of course, variations from these ideal conditions will affect the mapped locations of land use patterns.

It is easier to visualize von Thünen's ideas if they are applied to actual places. Numerous geographers have, in fact, mapped and tested the model in various locations worldwide.[3] Figure 8-9 illustrates the approximate locations of each zone of his model in generalized land use regions of the United States. "Rent" referred to in the figure is the cost of using land, whether it is owned, leased, borrowed, or rented (in the common sense of that term) by the user. "Location rent" assumes that the highest cost for use of land

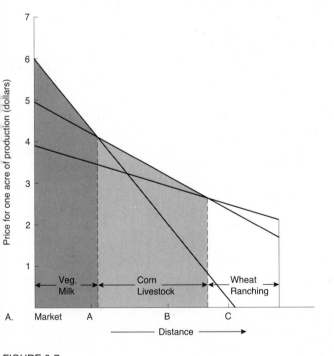

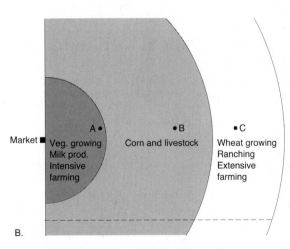

FIGURE 8-7

Location rent for competing land uses. A, Side view, and B, overhead view of the area affected by a selected market. Small vegetable farms, with their intense applications of labor and capital, predominate adjacent to the market. In contrast, large wheat ranches with their low per-acre applications of labor and capital utilize remote areas. (From Gordon F. Fielding, Geography as Social Science, *New York: Harper & Row, 1974, Fig. 5.22, p. 145. Reprinted with permission of Harper & Row, Publishers, Inc.)*

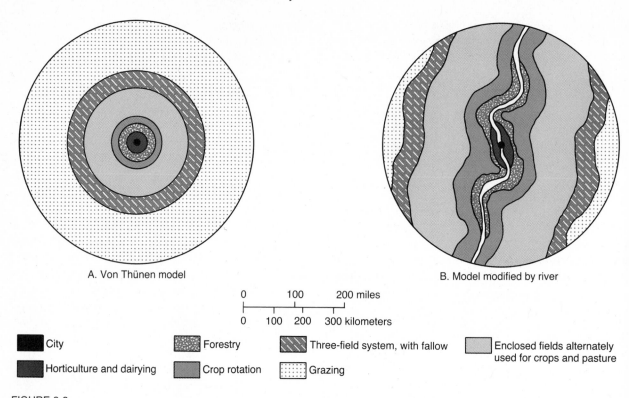

A. Von Thünen model

B. Model modified by river

| 0 | 100 | 200 miles |
| 0 | 100 | 200 | 300 kilometers |

■ City

□ Horticulture and dairying

▨ Forestry

▦ Crop rotation

▨ Three-field system, with fallow

▦ Grazing

□ Enclosed fields alternately used for crops and pasture

FIGURE 8-8
Von Thünen's isolated state model in three stages of formation. (From James M. Rubenstein, The Cultural Landscape: An Introduction to Human Geography, 2nd ed., Columbus, OH: Merrill Publishing Co., 1989, Fig. 9-5, p. 326. Reprinted with permission of Merrill Publishing Co.)

is at or near the market. "Economic rent," in this illustration, could also be called "environmental rent."

Let us examine several of the land use categories in von Thünen's model in some detail as illustrations of types of commercial agriculture and further examples of primary production.

Dairy Farming. Until about 1850, fresh milk was rarely found outside of the local farms and villages where it was produced. However, during the past century, specialized dairy farms have developed in response to urban markets, environmental conditions, and cultural food preferences in certain parts of the world. Dairy farming has developed a well-defined spatial pattern in northeastern North America, northwestern Europe, New Zealand, southeastern Australia, and in other scattered, tucked-away valleys near big cities.

Around large urban centers there are easily observable milk supply areas called "milksheds" (think of "watersheds") governed by demand in the city. The rapid increase in urban populations has provided an expanded market for milk and milk products. As with other forms of commer-

cial farming, dairying has grown in response to market demand in developed countries.

Environmental conditions also encourage dairying where market demand exists. Year-round rainfall, cool summers, and nutrient-weak podzolic soils are favorable to hay and pasture grasses used to feed cows. In most dairying regions, cows graze in open fields during the spring and summer but depend upon locally grown hay in winter. Rugged or poorly drained land often is used for permanent dairy pasturing because it is unfit for much else.

Consumption of dairy products also depends upon the distance over which they must be transported. For example, butter and cheese, which are less perishable and more expensive, may be produced farther away from market centers (although specific locations of these dairy products is governed by local cultural and economic conditions). Remote New Zealand, for instance, is known for butter production. Similarly, the "western" edge of the American dairy belt in Wisconsin and Michigan is known for fine butter and cheese products. Culturally induced food and diet preferences in this part of the world also have encouraged cheese

FIGURE 8-9

The Von Thünen model in action: theoretical land use zones based on location rent and economic rent. (From John F. Kolars and John D. Nystuen, Physical Geography: Environment and Man, *New York: McGraw-Hill Book Co., 1975, Fig. 10-14, p. 239. Used with permission of McGraw-Hill Book Co.)*

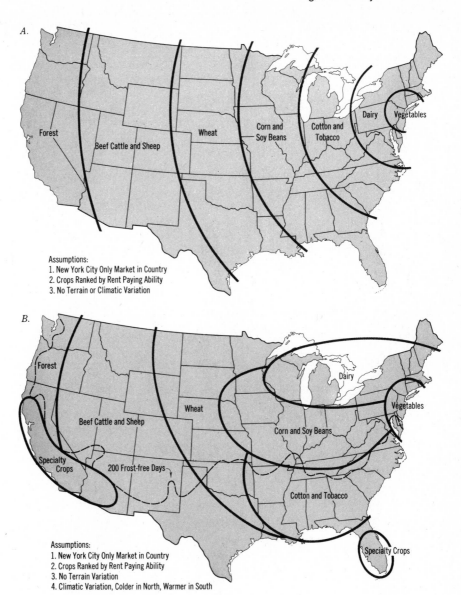

A.

Forest

Beef Cattle and Sheep

Wheat

Corn and Soy Beans

Cotton and Tobacco

Dairy

Vegetables

Assumptions:
1. New York City Only Market in Country
2. Crops Ranked by Rent Paying Ability
3. No Terrain or Climatic Variation

B.

Forest

Beef Cattle and Sheep

Specialty Crops

200 Frost-free Days

Wheat

Corn and Soy Beans

Dairy

Vegetables

Cotton and Tobacco

Specialty Crops

Assumptions:
1. New York City Only Market in Country
2. Crops Ranked by Rent Paying Ability
3. No Terrain Variation
4. Climatic Variation, Colder in North, Warmer in South

production. Dutch and Scandinavian immigrants near the Great Lakes region have played an important role in the expansion of the American dairy belt (Figure 8-10).

Mixed Crop and Livestock Farming. The classic image of a "family farm" in the minds of most people in North America is the commonest type of commercial agricultural system: the mixed crop and livestock enterprise. The economic plight of these farms has recently received much publicity and public support ("Farm Aid" concerts, films, and videos such as *Heartland* and *The River*). After decades of government subsidies and bank loans, decreasing exports and foreign markets, increasing agricultural imports, and

changing domestic policies, the classic crop and livestock farm faces many changes in the latter part of the twentieth century.

Despite financial and environmental problems, the mixed farm remains the most self-sufficient commercial agricultural system on Earth. Farmers produce most of their own seed and specialized stock. Many also produce enough vegetables, dairy products, and poultry for their own needs as well. A high percentage of the land is cropped, although animal sales bring in the most revenue. Livestock are supplied the cheapest feed available (pasture grass such as hay is most common), and in turn their manure is used to improve soil fertility. Geographers Cotton Mather and John

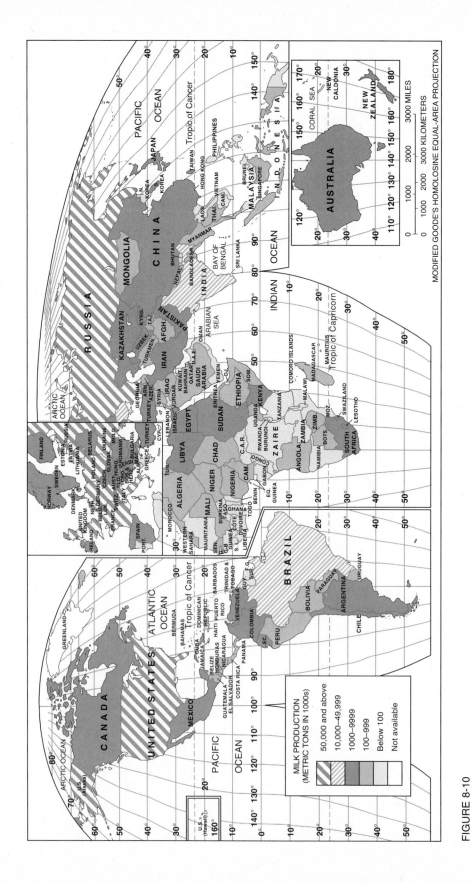

FIGURE 8-10

World milk production. Distribution of milk production closely matches the division of the world into relatively developed and developing countries. (From Anthony R. de Souza and Frederick P. Stutz, The World Economy: Resources, Location, Trade, and Development, 2nd ed., New York: Macmillan, 1994, Figure 7-12, p. 288. Reprinted with permission of Macmillan Publishing Co.)

F. Hart investigated the use of manure in their classic study of the American Corn Belt and found it to be an independent part of the integrated system of production.[4]

Commercial Grain Farming. Grain is the seed of cereal grasses such as wheat, oats, rye, and barley. It is grown on mixed crop and livestock farms such as the ones in the American Corn Belt, largely to feed animals. Grain for commercial sale (human consumption) rather than for animal feed is grown on a large scale in only six countries: the United States, Russia, Ukraine, Argentina, Canada, and Australia. Generally located on dry land unfit for most other agricultural uses, grain farms usually have very large acreages, are highly mechanized, and are low in labor intensity. Commercial grain farms usually are located in the dry margins of humid climatic zones where drought and temperature extremes are common environmental hazards. However, fertile soils (such as the prized chernozems) and large expanses of level land are advantages that encourage the planting of huge acreages of wheat even in climatically sensitive areas.

Wheat is the most important crop grown in commercial grain regions of the world. Figure 8-11 shows the wide distribution of world wheat production. In this highly specialized land use system, a large percentage of the land is planted in wheat (often over 80 percent). In North America, "spring wheat" is planted in spring and harvested in autumn in the Dakotas and in the prairie provinces of Canada; "winter wheat" is planted in fall and harvested in midsummer in Kansas and surrounding states (Figure 8-11).

Wheat farms are among the largest of all farms worldwide and require a huge capital investment for land and machinery. Labor requirements are most intense during the summer months when some farmers known as "suitcase farmers" or "sidewalk farmers" who work in town during the winter months visit their fields in time for the harvest. Local fulltime farmers often resent these part-timers who seem to lack commitment to their often-neglected fields.

Large-scale wheat production is a risky operation. Tempted by inexpensive level land and the lure of profits, early farmers often planted crops too far west in North America and found themselves in a temperamental climate. Frequent drought, lack of groundwater or surface water, easily eroded soils, and seemingly constant wind were some of the environmental hazards of grain farming. The Dust Bowl of the 1930s and periodic droughts in the decades since have been the most dramatic lessons of this climatic variability. Today's agricultural techniques offset these problems where possible: "dry farming" allows fields to lie fallow during alternate years so they may accumulate moisture; windbreaks planted along field boundaries retard wind damage; and, most recently, large-scale irrigation has been installed where water is scarce.

Center-pivot irrigation is now used in much of the winter wheat belt along the driest western edge of the wheat-growing region. This system maximizes field acreage by watering circular fields from their centers with motorized sprinklers. Airline passengers flying over the region often remark on the distinctive landscape created by these curious round fields.

Large-scale wheat production is significant in the global economy because much of this wheat is grown for international trade. Canada and the United States are the world's largest wheat exporters. This one product has provided these nations with dependent trade partners in many parts of the world and given economic strength to North America. The opening up of the former Communist block in eastern Europe and the former Soviet Union has added further interdependence to world trade networks. This has had the long-term result of easing international political tensions.

Livestock Ranching. The von Thünen ring that lies farthest from markets is the grazing zone beyond the grain belt. Here, ranching seems to be the only agricultural possibility. A climate too dry for crops, rugged land too difficult for mechanization, and a natural grass vegetation cover— all contribute to the economic potential for livestock grazing. However, ranching is not simply a response to environmental conditions and von Thünen expectations. It also is culturally controlled because ranching exists primarily in areas settled by Europeans. An efficient transportation system is also necessary for ranching success. Extremely large acreages are required because productivity per acre is so low; ranches in the western United States average 2000 acres and may be 20,000 acres or more.

As irrigation methods are improved and applied to larger areas of the arid and semiarid American West, land available for ranching is decreasing. Much like a hundred years ago, cattlemen and farmers are competing for land and water. Today it is most common for young cattle to be raised on open land and then shipped to massive feed lots nearer to markets for fattening.

Land-extensive systems such as livestock ranching often create the cultural perception of a region. The distinct image of the American West, for example, has been shaped in part by cattle ranching and the American cowboy. The box on page 208 is taken from the first book ever published by a "real" cowboy, Andy Adams. The images evoked in

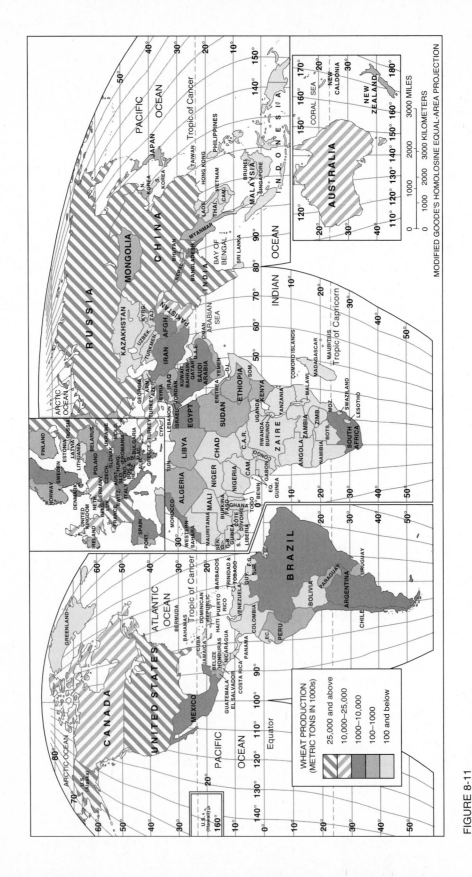

FIGURE 8-11

Distribution of world wheat production. (From Anthony R. de Souza and Frederick P. Stutz, The World Economy: Resources, Location, Trade, and Development, 2nd ed., New York: Macmillan, 1994, Figure 7-13, p. 289. Reprinted with permission of Macmillan Publishing Co.)

this excerpt will be familiar to most readers who have seen countless western movies and television programs over the years.

Specialized Commercial Farming Systems

Mediterranean Agriculture. Some of the most productive agricultural lands in the world are found in limited areas having a *Mediterranean climate.* Here, as in lands surrounding the Mediterranean Sea, mild temperatures with rainy winters and dry summers allow a wide variety of crops to be grown. Although traditional farms near the Mediterranean Sea were planted in wheat and barley, today's Mediterranean farming is known more for its specialty crops. Oranges, figs, olives, and grapes are the primary export crops of **Mediterranean agriculture.** In fact, the majority of the world's highest quality wines come from such Mediterranean-climate regions as the Loire Valley in France and the Napa Valley in northern California. Figure 8-12 shows other wine-producing areas of the world; most are located in Mediterranean-climate regions. (In Chapter 13 we pursue in depth an exciting geographic topic: the geography of world wine production.)

Plantation Agriculture. We have seen that Mediterranean agriculture depends upon one particular type of climate. Another specialized farming system that is linked with a certain climatic zone is plantation agriculture. According to agricultural geographer Howard Gregor, the term *plantation* has undergone many changes of meaning in the past several decades.[5] First associated with greedy European farming in the tropics, recent years have seen an expansion outside the tropics of farms that resemble plantations. These rural production areas involve massive investment, crop specialization, intensive labor, central management, and advanced farm technology, much as the old plantations did. Gregor suggests that this agribusiness trend is another, newer type of plantation in the modern sense of the word.

Traditionally, "plantations" have implied the introduction of foreign capital and power onto an indigenous people for profit. Also, "plantation" crops were often foreign to the area being developed. Sugar, cacao, coffee, tea, rubber, and tobacco are examples of plantation crops grown outside their native area. The most important plantation crops today are coffee, tea, sugar cane, rubber, tobacco, and peanuts. Coffee is today's most important plantation crop in terms of value and countries involved in production.

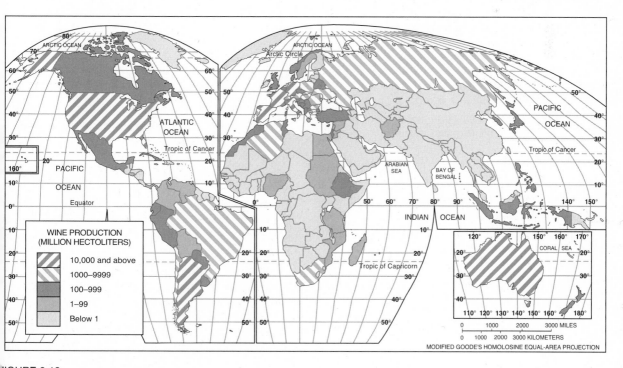

FIGURE 8-12
Distribution of the world's vineyards and world wine production. (From James M. Rubenstein, The Cultural Landscape: An Introduction to Human Geography, *4th ed., New York: Macmillan, 1994, Figure 6-14, p. 243. Reprinted with permission of Macmillan Publishing Co.)*

A COWBOY'S PERCEPTION OF OPEN SPACE IN THE WEST

In the first book ever published by a real cowboy, Andy Adams recalls his feelings about the open expanses of the American West:

> After leaving the country tributary to the Solomon River [in northern Kansas], we crossed a wide tableland for nearly a hundred miles, and with the exception of the Kansas Pacific Railroad, without a landmark worthy of a name. Western Kansas was then classified, worthily too, as belonging to the Great American Desert, and most of the country for the last five hundred miles of our course was entitled to a similar description. Once the freshness of spring had passed, the plain took on her natural sunburnt color, and day after day, as far as the eye could reach, the monotony was unbroken, save for the variations of the mirages on every hand. Except at morning and evening, we were never out of sight of these optical illusions, sometimes miles away, and then again close up, when an antelope standing half a mile distant looked as tall as a giraffe. Frequently the lead of the herd would be in eclipse from these illusions, when to the men in the rear, the horsemen and cattle in the lead would appear like giants in an old fairy story. If the monotony of the sea can be charged with dulling men's sensibilities until they become pirates, surely this desolate arid plain might be equally charged with the wrongdoing of not a few of our craft.

From Andy Adams, The Log of a Cowboy (Boston: Houghton Mifflin, 1903), pp. 232–233.

Other Types of Primary Production

Other primary extractive industries—fishing, forestry, and mining—also contribute substantially to the world's economies in the secondary sector. Even if you never see a tree fall, you still are reading a book made from it. Even if you never witness a mining operation personally, you may still wear a ring from a country dependent upon mineral production.

Due to the nature of the nineteenth-century colonial systems that established primary production economies throughout the world, most primary production continues to be located in developing nations. In fact, many of the socioeconomic and political problems of Third World countries in this century hinge upon primary-resource management and production. It is impossible to understand the politics of many nations without understanding their primary production, for example, oil production in the Middle East or mining in South Africa.

Let us examine these three other types of primary production, with a special focus upon selected regions and their particular extractive industries.

Fishing. As diminishing food supplies and rising prices arouse concern throughout the world, many nations are turning to ocean and river resources. However, the search for food from the hydrosphere has long-range ecological implications.

Figure 8-13 shows worldwide distribution of fisheries. Although fishing provides much less of the world's food supply than does agricultural production, some countries (including Iceland, Finland, and Japan) rely on ocean resources for the majority of their dietary protein. For other countries, fresh fish is a luxury but not an essential part of their diet. In recent decades, as human consumption has increased and capital has become more available to selected nations of the world, there has been a trend toward greater use of technology in the fishing industry. Japan, for example, has major fleets of factory ships to catch, process, and freeze fish. Using radar and satellite imagery, schools of fish are targeted by these fleets. Aboard the ships are specialists in biology, oceanography, and related fields. Some governments such as Norway and Great Britain provide subsidies to develop and outfit these huge operations.

Complete international agreement has yet to be reached regarding the sustainable yield of each species and region. For example, Japanese vessels off the Pacific Northwest coast are catching salmon spawned in hatcheries in Canada and the United States. How much right does each nation have to the resource, and how much of it may be caught before the viability of the fishery is threatened?

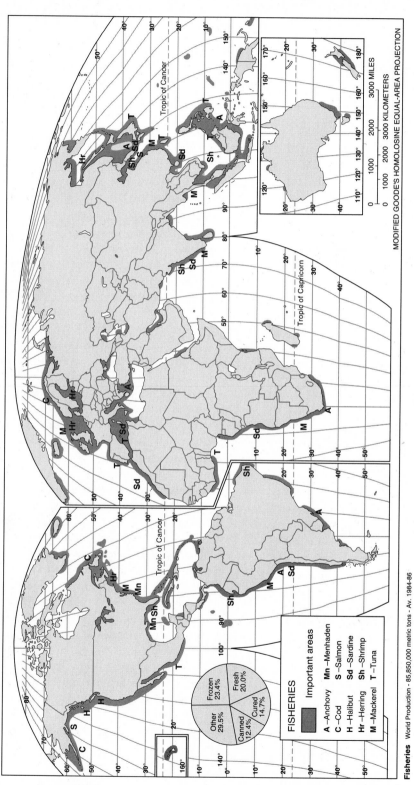

Fisheries World Production - 85,850,000 metric tons - Av. 1984-86

FISHERIES

Important areas

A –Anchovy   Mn –Menhaden
C –Cod   S –Salmon
H –Halibut   Sd –Sardine
Hr –Herring   Sh –Shrimp
M –Mackerel   T –Tuna

Other 29.5%
Frozen 23.4%
Fresh 20.0%
Cured 14.7%
Canned 12.4%

MODIFIED GOODE'S HOMOLOSINE EQUAL-AREA PROJECTION

3000 MILES
3000 KILOMETERS
1000   2000
0   1000   2000

Disposition of World Catch, 1986
Marine Catch – 87.9%
Inland Catch – 12.1%

Japan 13.7	China 8.0	India 3.3	S.Kor. 3.2	Indon. 2.8	Thai. 2.5	Phil. 2.2	N.Kor. 2.0	Other 5.5	Soviet Union 12.6	Chile 5.8	Peru 5.1	Other 3.7	United States 5.6	Other 3.7	Norway 2.5	Den. 2.7	Other 10.2	Africa 4.5
←				ASIA				→		← S. AMER. →			← N. AM. →		← EUROPE →			

0   10   20   30   40   50   60   70   80   90   100%

FIGURE 8-13

Distribution of the world fisheries. (After Edward B. Espenshade, Jr., ed., Goode's World Atlas, Chicago: Rand McNally Co., 1986, p. 36. Used with permission of Rand McNally Co.)

All of the world's fisheries are now in a state of decline after reaching a peak of production between 1973 and 1991. The most impacted areas of the Pacific and Atlantic oceans have declined by 30 percent! Of the world's 17 major fisheries, 13 are either in decline or are commercially depleted. These include red snapper off of Mexico, orange roughies in New Zealand, cod from the Canadian Grand Banks, Atlantic blufin tuna, and Pacific Coast salmon.

When fishermen depended upon luck, catches took 10 to 20 percent of available stock. Fish finders, loran charting, spotter planes, driftnets, and satellite imaging have contributed to more efficient operations leaving fewer fish for reproduction of the resource stock.

Despite some international limits on fishing areas and fishing methods, cultivation is probably the only way that the industry can become sustainable. **Aquaculture,** the raising of sea animals and plants for commercial purposes, has become an important economic activity throughout the world. Commodities produced on freshwater and marine farms include shellfish, tropical fish, pearls, and various types of food fish for human consumption. One example of aquaculture important in today's economy is the Scottish salmon industry.

Ten years ago the Scottish salmon business was started as a high-risk investment. At that time it was believed impossible to raise salmon in a captive environment. Today, however, there are over 100 fishfarms in Scotland that collectively produce over 3000 tons of salmon annually. This is three times the yield produced under natural conditions in Scotland.[6]

This new industry has brought prosperity to a number of small villages in areas of formerly high unemployment. The hatcheries themselves consist of neat rows of circular fish tanks hidden beneath trees, and sea tanks that float out in the lochs. This process for growing salmon was first tried in Norway and has become remarkably successful in many sea-based countries of northern Europe. Most managers of these fish farms are young, and many have degrees in an environmental science. Some salmon "ranches" are now operating in the Pacific Northwest of North America, and in Maine and maritime Canada.

In many other parts of the world, fish farming has become a mainstay of the local economy. In Japan, fish have been raised commercially for decades. China and Israel are developing this cash crop production method. Norway has specialized in cage production. In many parts of the Mediterranean, aquaculture methods are raising sea bass and sea bream. In central North America catfish are produced on hundreds of small farms, which claim to produce more protein per acre than any agricultural (land-based) crop.

The manager of Marine Harvest, Scotland's largest salmon farm, estimates that the 1000 tons of salmon produced each year from their farm is the equivalent of 6 million meals. Peace Corps and other local-development agencies often encourage production of tilapia and carp in tropical zones as sources of protein. It is becoming evident that a partial solution to the problem of world hunger is aquatic farming as a supplement to traditional agriculture.

Forestry. The most versatile resource used by people other than water is probably wood. Life without lumber, paper, and other wood products is almost inconceivable, even in this age of plastics, light-weight metals, and electronics. The cutting of wood is almost as old as human history, and deforestation of vast wooded lands has become a much-publicized environmental problem. Indeed, extreme clear-cutting of forests is occurring in all major countries worldwide as demand increases for fuel wood, building lumber, paper products, and other tree materials. Chapter 10 expands on deforestation as a major global concern in the 1990s.

The relationship between humans and their forests has varied with location. Most of the world's forests grow in the mid-latitudes, and historically these have been the most severely clear-cut. In Western Europe, early humans colonized the land by clear-cutting both deciduous and coniferous forests. In the Mediterranean region, lowland forests already had disappeared in Greek and Roman times as construction and fuel needs multiplied with increasing populations. As humans took permanent control of their land, forests were pushed back to allow for expanded settlement.

North America has had a similar story. Early settlers from Europe cleared the forests to make room for farms, towns, and cities. Trees were viewed as an obstacle to civilization and were removed as a first step in settlement. The ax was as important as the plow to early settlers.

In tropical regions, much of the virgin forest has been cleared by shifting cultivation. Second-growth forests are now most common. Commercial trees grown in the tropics now are usually grown on tree farms because natural tropical forests have a large variety of species mixed together, making exploitation difficult and expensive. Until recently, plantation farming of trees by colonial powers was the most common technique. Although lumbering techniques vary in different types of forests, the dominant theme in the world today is the removal of trees for human use and to make way for human habitation.

Mining. Extraction of minerals from Earth is another type of primary production. Minerals play a major role in the lo cation of manufacturing or secondary production. They are of obvious economic importance to many developing countries.

Less than 1 percent of Earth's minable surface has any significant concentration of minerals (with the exception of the coal fields). The most obvious political effect of this concentration of minerals is the concentration of political power in the hands of only a few mineral-rich nations. It is an interesting and somewhat tragic coincidence that minerals have helped both to build and to destroy economies.

A corollary to the concentrated location of minerals is that, although most minerals are located in a few areas, not all minerals are found together in the same location. This erratic distribution is very important economically (1) because countries need a full complement of minerals to develop their secondary industries to capacity and (2) because continual mining depletes mineral deposits through time.

The natural dispersal of minerals has led to an intricate transportation system connecting mineral-producing areas to processing sites. Water is the cheapest form of transportation, so rivers and ocean routes most often are used to transport ore to factories. This has favored centralized control, especially when a country has a monopoly on one mineral or when the resource is a major export item. Even so, most mineral production in today's capitalistic world is done by a very few large private organizations, each concentrating on a particular mineral. Companies that produce tin, copper, and oil are among the world's largest corporations. Largeness begets largeness: large operations, whether state-owned or private, can afford the high cost of exploring for new deposits.

Mineral depletion and overuse has led to shortages and subsequent economic decline in certain parts of the world. For example, in the nineteenth century Great Britain was the world's leading producer of lead, copper, tin, iron, and coal, and access to minerals outside of the British Empire was unnecessary. Today, much of Britain's economy faces a shortage of these essential minerals. The iron and steel industry especially is affected by the shortage of local coal and iron ore. Germany, long a self-sufficient steel producer, now must import iron ore from at least seventeen different countries for its iron and steel industries.

Perhaps no other resource depends so much upon human attitudes as minerals. Our value judgments about minerals are rooted both in culture and technology. Early people valued flint, obsidian, and jade as resources for weapon-making and for their beauty. Much later, gold and silver became our most prized metals. In fact, in the early history of the American West, gold and silver mining carried the economy of the region and became the reason for

settling many parts of this vast region. The following case study illustrates the importance of gold mining in the economy of the American West and its contribution to the evolution of California's culture.

The California Gold Rush: Impact on the Culture of the West? In one of the ironies of history, it was not the Spanish seventeenth-century treasure-seeking conquistadors who discovered gold in California. Nor was it southern California rancher Francisco Lopez, who in 1842 pulled some onions from his garden in the mountains behind Mission San Fernando and found some gold clinging to their roots.

The major gold strike in California happened in the northern part of the state. After John Sutter built his fort in the Sacramento Valley and founded the pioneer town of New Helvetia (today's city of Sacramento), he opened a lumber mill east of the fort to supply wood for his growing settlement. Choosing a small valley on the south fork of the American River 45 miles (72 km) east of the fort, Sutter sent a staff carpenter, James Marshall, to run the operation. On the morning of January 24, 1848, Marshall noticed a few glittering particles on the bottom of the millrace. Putting the particles in the brim of his hat, Marshall quickly traveled to Sutter's Fort to test them.

The rest of the story is history. In 1848 and 1849, people from all over the world rushed to northern California to seek their fortunes in the gold fields. By early June 1848, half of San Francisco's population had disappeared into the hills. Business stopped, real estate sales halted, farmers left their fields unfinished, army recruits deserted—everyone rushed to the mines.

This exodus soon created a lifestyle of extreme informality. California became a place to go when one had to escape problems at home. A rough, masculine lifestyle soon developed in the mining towns, creating an image of the West that has been retold many times on film, in books, and on television. Prostitution, gambling, rough jokes, and a lack of bathing facilities emerged in this challenging environment of quick opportunity.

Many have called the California Gold Rush one of the most significant events in history. In the long run, California probably would have evolved into a prosperous, populous place without the Gold Rush, but it is hard to deny the enormous effects of this single event on California's culture and development. Let us examine several possibilities for the region if the Gold Rush had never happened:

1. The first transcontinental railroad might not have been linked to Sacramento but might have been routed through Oregon to Portland. Oregon was settled and

developed earlier than California, so the builders of the railroad might have considered it a better risk. The completion of the transcontinental railroad encouraged settlement and economic development of the region.

2. Cities like San Francisco and Sacramento might eventually have developed into major urban centers, but their growth would probably have occurred later in the history of the state. When gold was discovered, San Francisco was a small town with two hotels, two unfinished wharfs, and 812 people. By the end of 1849, the city had a population of over 40,000.

3. California was admitted as a state in 1850, long before many other states that had been settled earlier. Without the Gold Rush, its admission probably would have happened much later.

4. A major effect of the Gold Rush was the in-migration of people into the United States. New immigrants not only migrated to California but were welcomed to deserted farms and factories in the East and Midwest that had been vacated by enthusiastic would-be miners. This expanded the American market and labor supply and encouraged its ethnic diversity, both east and west. Without the Gold Rush, migration westward would have been much more gradual.

5. Most important, the Gold Rush set the stage for a culture based on mass hysteria and greed. The image of California in many outsider's minds to this day is one of a glamorous, wealthy, somewhat manic place where "unique" people find their homes! Perhaps because of the importance of the Gold Rush, this image of the state remains a dominant theme in the media and in popular culture.

SECONDARY ECONOMIC ACTIVITIES

Since the late nineteenth century, the most significant measure of a nation's wealth has not been its level of primary production but rather the pace of its manufacturing growth, or secondary production. The cultural transformation set in motion by the diffusion of the Industrial Revolution from Great Britain to diverse places in the world dramatically changed Earth's socioeconomic and physical patterns. Industrialization is a process that involves transforming raw materials into a more useful form. These materials are brought together at a manufacturing plant where labor and energy are applied to produce a finished product. This final product is then shipped to market.

In this section, we will first examine the location patterns of industries, and then turn our attention to more specific types of manufacturing at a regional level. Why do certain plants locate at a particular place? What site location factors encourage manufacturing firms to locate where they do? Answers to these questions are frequently quite complicated because site requirements for different industries vary in both time and space.

INDUSTRIAL LOCATION FACTORS

Categorizing types of primary production, as we did in the preceding sections of this chapter, is relatively simple compared with categorizing location patterns of industries. When evaluating sites for manufacturing we must consider the location of both resources and markets, changing market demand, transportation costs, techniques and scale of production, and even amenity factors such as climate, cultural amenities, and nearness to a university.

Industrial location theory, first developed by Alfred Weber and others, provides us with a model for a general understanding of this complicated problem. Weber's simplified industrial landscape and his explanation of the location of industries is presented in his book *Theory of the Location of Industries.*[7] Weber's theories actually are quite general because he designed them with a wide range of political, social, cultural, and economic systems in mind. However, because virtually all early industrial location theory is based on Weberian analysis, it is important to discuss his ideas in some detail here.

Weber's Location Triangle

Weber's intention was to establish the lowest-cost location for a manufacturing plant. He first laid out five basic ground rules for his theory:

1. Isotropic Plane Assumption: the model of this theory would apply to a single country assumed to have a uniform topography, climate, technological level, and economic system.

2. He considered one finished product at a time, not a variety of different products.

3. He assumed that raw materials and the market were fixed at certain locations.

4. He assumed that labor was also fixed in location, but was available in unlimited quantities at any production site.

5. Transport costs were a direct function of the weight of the product and the distance shipped.[8]

Several basic terms must be defined to clarify Weber's model:

Ubiquitous—raw materials available everywhere in the isotropic plane at the same cost (for example, sand, gravel)

Localized—raw materials available only at specific locations (for example, mineral ores; the opposite of *ubiquitous*)

Pure materials—finished products that lose essentially no weight in processing (for example, petroleum)

Weight-losing materials—those that impart only some of their weight to the finished product (that is, the weight of steel is much less than the combined weight or iron and coal)

Weber's model may be analyzed most easily by using a series of locational triangles. The triangles provide us with a visual tool for any analysis of industrial location. In the simplest case—two raw material sources and one market—the ideal location for an industry will be within the triangle and closest to the location that involves the highest transport cost. Figure 8-14 depicts this thinking—the cost of transport from P to M is here assumed to be higher than 1 to P or 2 to P, and thus P is the optimum economic location for the plant.

More complex location questions may be answered by using Weber's isodapane method. The goal is the same as in the triangle—to seek the optimum location for an industrial plant. First, one draws lines of equal transportation cost for any raw material or finished product; these lines are called **isotims**. Any point on the map will then show transport costs for all products involved. The isodapane is then found by totaling all the isotims at one location. *Isodapanes* are lines that show constant total costs. These lines make it possible to locate the least expensive site for optimum production (Figure 8-15).

In the early days of iron and steel production and other industrial development, Weber's principles applied. Although his theories set the pattern in the nineteenth cen-

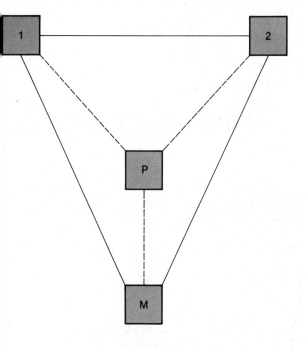

M—Market

1—Localized gross raw material

2—Localized gross raw material

P—Production center

---Transportation route followed

FIGURE 8-14

Weber's location triangle in the simple case of a triangular arrangement of two material sources and one market (1, 2, and M). (From Roy J. Sampson and Martin T. Farris, Domestic Transportation Practice, Theory, and Policy, *5th ed., Boston: Houghton Mifflin, Fig. 14-2. Copyright © 1985 by Houghton Mifflin. Used with permission.)*

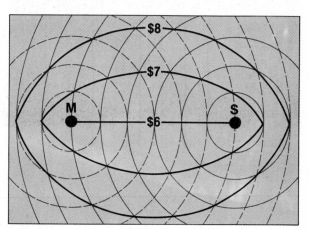

FIGURE 8-15

Weber's theory: isodapanes (black lines) superimposed onto isotims (white lines). (From James O. Wheeler and Peter O. Muller, Economic Geography, *New York: John Wiley & Sons, 1986, Fig. 9-8, p. 207. Reprinted with permission.)*

tury and remain visible through historical inertia today, present industrial-location decision making has changed. Due to the generalized nature of location theory, Weber's ideas necessarily had certain limitations. He failed to consider variations of market demand; he oversimplified transportation costs; he overlooked the variability of labor supply due to migration patterns; and he considered manufacturing plants that produced only one product. Many industrial firms actually produce a wide range of products. In addition, developments in technology after the 1960s have lowered transportation costs and have encouraged location based upon amenities and a concern for quality of life. Despite these limitations, however, Weber's ideas continue to be tested and retested by many economic geographers, and his ideas remain fundamental in the spatial analysis of industry.

Other Industrial Location Theories

Since Weber's 1909 publication, other theorists have made contributions to our understanding of secondary production. August Lösch, writing in 1939, developed a theory based on maximum profit.[9] Lösch almost exclusively focused upon the importance of spatial variation in the sales potential of various manufacturing plants. Nazi persecution forced Lösch from his native Germany to do much of his applied study in the United States, specifically in Iowa in the 1930s. Thus, many of his publications are about the American Midwest.

Numerous industrial theorists such as economist Walter Isard also have made important contributions to our understanding of location theory. Isard's 1956 book *Location and Space Economy* expanded the ideas of Weber,

von Thünen, and Lösch and attempted to link them into a more general economic theory.[10]

These theories of industrial location are only several of the many that have been tested and retested in the past century. Most have considered the location of resources, markets, or labor supply. One way to understand this theoretical base in economic geography and to make it more "real" is to examine case studies of industries in various manufacturing regions in the world. Figure 8-16 maps manufacturing regions of North America as a guide for our location analysis.

Market Dependency: Printing.   Some industries must be located near their markets so they can deliver their products as rapidly as possible. Industries such as bakeries, dairies, and daily newspaper printers depend upon speed and rapid delivery for their success.

An examination of the printing industry in the 1990s discloses a booming enterprise. From speedy "copy" shops in small towns and cities across the developed world to extensive publishing companies, printing has become an essential industry in the "information society" of our times. Quick-print shops usually are located near universities and colleges, or in close proximity to busy office complexes. A personal relationship often develops between supplier and purchaser in these settings.

Labor is more important than raw materials as a localizing factor for print shops, so these industries often are located downtown or at the edge of the central business district at the focus of local transportation. Printing also is increasingly done in more suburban locations. As business firms are drawn toward the outer city by land availability, tax advantages, and other amenities, print shops also are lo-

FIGURE 8-16
Manufacturing regions of North America. (From James M. Rubenstein, The Cultural Landscape: An Introduction to Human Geography, *2nd ed., Columbus, OH: Merrill Publishing Co., 1989, Fig. 10-3, p. 352. Reprinted with permission of Merrill Publishing Co.)*

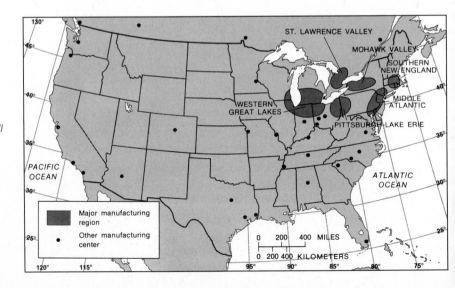

cating in suburban business parks and along commercial ribbon streets.

Other reasons that printing industries locate near cities and busy suburbs are "perishability" and convenience. Your daily newspaper is highly "perishable" in that it is out-of-date almost before publication. Because of the need to deliver newspapers immediately, they are manufactured in their primary market area. The newspaper *USA Today* demonstrates that nationwide distribution of daily newspapers is possible via satellite transmission of text and pictures, but each regional edition is still printed in its market area.

It is important to realize that market-oriented industries such as printing may not be located in the lowest-cost places. Central cities are often the most expensive places to produce a product due to high land costs, high taxes, and high wage rates. However, the market advantages outweigh these cost disadvantages. It is obvious that if sales revenues are high enough, a firm can accommodate these higher production costs (Figure 8-17). Print shops located in more expensive prime location points are responding to market pressure rather than to other locational considerations.

Resource Dependency: Commercial Canning and Freezing.

We have seen that print shops depend upon market demand for their location in or near settlements. However, other industries respond to forces other than market pressures. For example, commercial canning and freezing plants process ripe, nontransportable products, so they most often are located close to their raw materials, rather than to their market. As these types of industries often are small and localized, output per plant is limited because of the high cost of raw material transport. Thus, processors of fruits, vegetables, and related products are usually located in remote, sparsely populated places.

Commercial canning and freezing plants in the United States tend to be located in several clusters rather than at one site. The primary region is the Northeast, with scattered plants in central Florida, southern Texas, northern Utah, southern and central California, and western Oregon and Washington. These locations are all dependent upon the perishability of raw materials and so lie near food-production regions.

A good example of a food processing plant located near raw materials is a sugar cane factory. Unlike sugar beets, sugar cane requires immediate processing, so it is seldom transported very far to a processing plant. The radius of an area within which sugar cane may be transported to the factory is much smaller than the radius for a beet sugar mill. After the cane has been processed to produce raw sugar, the sugar is taken to a refinery.

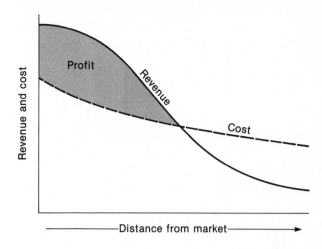

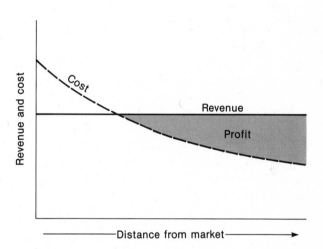

FIGURE 8-17

Relationship between revenue and cost with distance from the market. Top diagram represents a market-oriented industry. Although costs are high, revenue is also high at the optimum market-oriented location. Lower chart represents an industry having constant revenues regardless of location, but it incurs higher production costs in the market area, perhaps as higher rent. (From James O. Wheeler and Peter O. Muller, Economic Geography, *New York: John Wiley & Sons, 1986, p. 196. Reprinted with permission.)*

These secondary processing plants often are located some distance from the original processing site. In fact, refineries are often located in intermediate places between the sugar fields and the market. Cuba formerly supplied most of the raw sugar for U.S. refineries. Today, people in the United States use cane sugar from Louisiana, Florida, or from plantations in Hawaii, Puerto Rico, the Dominican Republic, and the Philippines. U.S. refineries also use sugar beets grown in scattered locations across the country. For example, Hawaiian cane sugar is transported to San Francisco East Bay refineries such as Antioch, California, for final processing.

Multiple Dependencies: The Iron and Steel Industry. Psychologically, our dependency on iron and steel production in the world's economy has remained vital. Iron and steel traditionally have been among the first industries developed during the economic growth of a nation, and so in many countries there may be an exaggerated perception of their importance. The almost universal use of steel lies in its strength, flexible properties, and low cost, which make it useful for a wide variety of products. In fact, the per capita steel consumption in an economy is a measure of that country's level of development.

Iron-making technology began in the Middle East about 2000 B.C. and rapidly spread to other regions having access to iron ore and charcoal. Preindustrial steel production was carried on in early Spain and other places around the Mediterranean Sea, although production methods in local areas frequently were a well-guarded secret.

Raw material requirements for pig iron production include coking coal, iron ore, and limestone. Historically, because such great quantities of coal were needed, coal availability was the dominant control over the location of iron and steel plants. The best location was a place having large amounts of coal that was also relatively close to markets.

Modern steelmaking technology began in the late eighteenth century with blast furnaces using coal in the form of coke. Later, the Bessemer converter, open hearth furnace, and electric furnaces made large-scale steel production possible.

Location patterns for iron and steel production show a progression from early concentration in areas having local ores, to much more concentrated large-scale production in places like Pittsburgh, Pennsylvania, to widely scattered areas, with this dispersal occuring since 1940. In North America, the Pittsburgh-Cleveland area remains the most important region of steel production, even though its production levels have fallen drastically in the past decade. With its excellent local supplies of high-quality coking coal, with iron ore brought in from the Great Lakes system, with nearness to eastern markets and Detroit's automobile manufacturing, and with three large rivers to bring in raw materials and ship out finished steel, Pittsburgh has had multiple early advantages for becoming an iron and steel giant. Today, the industry hangs on in western Pennsylvania but with much reduced production, primarily because of inertia that resists change and competition

from foreign production. No new plants have been added for over 80 years.

Steel production in North America has dispersed beyond the confines of this early center. Places like the Great Lakes region (Chicago, Detroit, and Gary in Indiana), Baltimore and Philadelphia, Birmingham in Alabama, and even western cities, such as Los Angeles and Pueblo in Colorado, are steel-production sites important in today's American economy. A recent Japanese steel-production venture in Pittsburg, California, typifies this dispersal.

Other iron and steel manufacturing regions in the world include Germany's Ruhr region, Ukraine, and the Southern Ural Mountains and the remote Siberian Kuznetsk Basin in Russia.

Location patterns of iron and steel manufacturing in the world today reflect both the inertia of remaining in original locations and the more recent advantages of availability of new raw materials or markets. In fact, nearness to markets has become a more important force in the past century (Figure 8-18). As the cost of bulk transport of coal and iron ore was lowered by cheap railroad transportation, iron and steel production became less tied to the nearness of raw materials and more oriented toward growing urban markets. After World War II, when the iron ore supply shifted to overseas sources, new steel mills were established near the

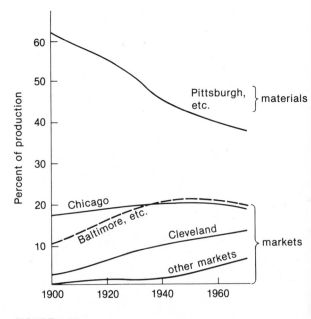

FIGURE 8-18

Change in locations of steel production. Locations near markets have become more important for steel production over time. (From Richard L. Morrill, The Spatial Organization of Society, *Belmont, CA: Wadsworth Publishing Company, Inc., © 1970, p. 104. Reprinted by permission of the publisher.)*

eastern port cities of Baltimore and Philadelphia. These mills are also closer to large urban markets.

Transportation Dependency. Diversified manufacturing plants tend to locate at the junctures of transportation routes. These "break-of-bulk" points (places where modes of transportation change, such as ports and railyards) tend to become centers of production. It is to the advantage of many manufacturers to locate at terminals where raw materials or finished products are loaded and unloaded, or at junctions where the mode of transport changes (for example, from rail to truck). As goods must be loaded and unloaded at these locations, processing might as well occur there, too.

It is interesting to note the number of cities in the world that developed at these break-of-bulk locations because of transportation advantages. North American cities along the "fall line" between the Appalachian Mountains and the Atlantic seaboard are classic historical examples of this locational consideration for industry and energy production. Likewise, Chicago served as a focus of railroad lines in the 1860s and 1870s when railroads had varying gauges (track widths). Goods and raw materials had to be unloaded and reloaded onto different railroad lines at this location, thus strengthening Chicago's early economic position. Today's billionaire cities often owe their success to their break-in-bulk function.

Today many of the world's port cities are the focus of the greatest petrochemical refineries. These break-in-bulk points between land and water transportation illustrate the importance of transportation in industrial location patterns. Break-in-bulk points often become "hub cities" and are used by airlines and overnight express companies. For example, US Air uses Pittsburgh and United Airlines uses Chicago as their corporate headquarters.

Footloose Industries. Some industries, especially in the tertiary and quaternary sectors, do not depend upon resource availability, transportation costs, or market demand. Their major concern is an adequate supply of skilled labor. Such endeavors often are called "clean industries" because their main products are low-weight electronic goods or no-weight information. Research-and-development parks have evolved, particularly near universities, in areas such as the Research Triangle area of North Carolina, the greater Boston area, and Santa Clara County, California. Today's automobile industry is much less dependent on the locations of raw materials, and it now locates manufacturing plants close to cost-efficient labor pools.

An example of the globalization of the auto manufacturing business is detailed in Peter Dicken's *Global Shift: The Internationalization of Economic Activity.*[11] In this book, the author points out that Ford Motor Company was the first automobile producer to take advantage of the development of the European Community. In 1967 it reorganized its entire European operation into a transnationally integrated operation with each plant performing a specialized role in order to achieve economies of scale. Figure 8-19 shows how the production of the Fiesta model and later the Escort model was more market- and labor-oriented than resource-based. The result is a highly complex network of cross-border flows of finished vehicles and components, a network made even more complex by the involvement of hundreds of outside suppliers.

ISIC ECONOMIC CLASSIFICATION

In 1958, the Statistical Office of the United Nations published an *International Standard Industrial Classification of All Economic Activities* (ISIC), which recognized the following ten divisions:

- 0 Agriculture, forestry, hunting, fishing
- 1 Mining and quarrying
- 2, 3 Manufacturing
- 4 Construction
- 5 Electricity, gas, water, and sanitary services
- 6 Trade, banking, insurance, real estate
- 7 Transport, storage, communication
- 8 Services
- 9 Other activities

In our discussion thus far we have examined primary production (Categories 0 and 1) and secondary production (Categories 2 and 3). Now we will look at tertiary production (Categories 4, 5, 6, 7, and 8) and at quaternary communication of information (Category 9).

TERTIARY ECONOMIC ACTIVITIES

Most textbooks in economic geography spend voluminous pages discussing primary and secondary production but find little room for much detail on tertiary activity. Yet, according to recent statistics, nearly three-quarters of all U.S. nonfarm employment is service oriented. Tertiary activities include all types of services—trade, marketing, education,

FIGURE 8-19
The Ford Fiesta production network in Europe. (From Peter Dicken, Global Shift: The Internationalization of Economic Activity, *New York: Guilford Press, 1992, p. 300. Reprinted by permission of Guilford Press.)*

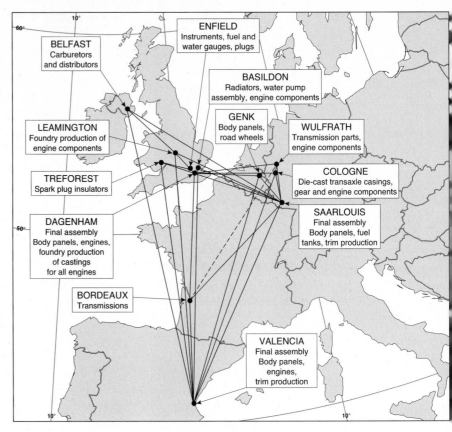

insurance, and banking, to name but a few. Indeed, most of what goes on in your local shopping mall may be classified as tertiary production.

The retailing component of this activity is detailed in Chapter 9. "Marketing geography" has recently become well known as an applied field, and market geographers have won prestigious positions as site analysts and geomarketing experts (see the box on page 219). The location of service industries is of vital concern in an economy that is ever-more dependent upon the tertiary sector.

This third type of production strongly depends upon people. In fact, people are both the raw material and the finished product. It is thus logical that tertiary activity is usually located near clusters of population. The entire focus is upon consumers. Is it any wonder that marketing geographers are finding employment in the 1990s, the "age of the consumer"?

As metropolitan areas continue to grow, the need for an expanded tertiary sector is obvious. Health care services, shopping facilities, educational institutions, and many other types of services must be enlarged to meet the growing needs of sprawling urban regions. (The next chapter takes

a closer look at this emerging, dominant settlement form.) Tertiary economic activities are increasingly international in scope, as surplus capital flows from some countries to others having available land or resources for development.

In recent years, Japanese investment in Hong Kong, Thailand, and North America has intensified. Japanese investment firms have focused upon acquiring banks and other financial institutions, large manufacturing plants, downtown office buildings, and raw land. Investments in commodities and the retail sector have not interested the Japanese as much. Japanese investment in the United States totaled $135 billion in 1988, and it has been predicted that it will increase 30 percent annually thereafter. It is estimated that about $5 billion per year is invested in real estate varying from downtown high-risers to suburban apartment buildings. Some Japanese firms are entering land-development opportunities.

The first wave of Chinese and Japanese investments in the early 1980s was primarily in San Francisco, Los Angeles, New York, and Boston. In the 1990s, all parts of the United States and Canada have been viewed as investment opportunities by enterprising Asian investors.

MEET THE GEOGRAPHER
Geography at Work: Real Estate Geographer Analyzes the Market

Real estate professionals and geographers have at least three things in common: location, location, and location. Real estate professionals of all types traditionally have been concerned with the financial worth of land and structures; geographers have looked at the land, the structures, and the people that interact with them. It would seem to be a natural marriage of business and discipline, but the two remained separate job markets until the growth of applied geography brought geographers into real estate.

Jeanette Rice is senior vice-president of Research and Analysis for Holliday, Fenoglio, Dockerty, & Gibson, a national commercial mortgage banking firm in Houston, Texas. A mortgage banking firm arranges financing for commercial real estate property transactions. She came to the real estate industry with a master's in geography. The analytical and research skills learned in her study of geography, her general location and spatial understanding, and her knowledge of urban and economic geographic principles, were easily transferred to her present responsibilities in real estate market analysis. She answers questions about the regional economy and real estate market from a geographic perspective. Jeanette, one of the first practitioners in this combined area of expertise, asserts that more geographers will develop careers in real estate as the industry becomes more complex and requires more in-depth analysis for good decision making.

Jeanette's position is relatively unusual, and her geography background is a plus for her firm. She demonstrates that geography can provide a path to business success and that it is a legitimate, respected discipline outside academia.

Jeanette encourages geography students to pursue careers in real estate, and feels that professors should teach students to "think geography." Students need a notion of what to do with geography, what jobs are available, and how to sell themselves in the business world.

From Association of American Geographers Newsletter *23 (November 1988), p. 16, and personal correspondence with the authors. Used with permission of Association of American Geographers.*

QUATERNARY ECONOMIC ACTIVITIES

A fourth level of activity was recently recognized to be an integral part of the economic system. Using a term suggested by geographer Jean Gottmann, *quaternary activity* has been defined as economic development based on the exchange of **information** or on "superservices." The 1990s are certainly an "age of information," as personal computers become commonplace in many homes in the developed world (for example, this textbook was both written and edited on personal computers). Internet and other global computer connections now link tens of thousands of workers at universities, businesses, and in government offices.

Business firms are becoming ever more closely linked into a global communications system. When the U.S. stock market fluctuates, repercussions are felt almost immediately in Japan, Western Europe, and many other parts of the world. Transworld communication has become virtually instantaneous; a telephone message travels nearly at the speed of light (186,000 miles a second), while sound through the air travels much slower (1100 feet a second). For example, a Minnesotan who yells at a person 1500 feet away while talking to a New Zealander on the phone will be heard by the New Zealander first! Indeed, the globe seems to be growing smaller as quaternary activity grows, along with its relation to international banking and Third World debt.

The now well-known "Quaternary Revolution", as it has evolved in the United States in the 1980s, and early 1990s, is perhaps most clearly illustrated by one region in one state, a former farmland between the California cities of San Francisco and San Jose in the Santa Clara Valley.

Known as "Silicon Valley" because of the region's dependence upon semiconductor computer chips made of silicon, this "valley" has gone from pastoral to postindustrial in less than a generation. Known as the "Prune Capital of America" as recently as the 1950s, almost all of Silicon Valley lies within Santa Clara County. Silicon Valley is now the ninth largest manufacturing center in the nation and is the birthplace of countless computer chips, pocket calculators, arcade games, cordless telephones, laser technology, microprocessors, and digital watches. The "Information Society" depicted in Naisbitt's trend-spotting book is clearly visible in the economic landscape of this part of the American West.

An economic geographer interested in quaternary production might ask, "Why Santa Clara County?" What is significant about the location of this place? Undoubtedly military and defense contracts played a large part in the early days of Silicon Valley's evolution. Attracted by skilled labor, affordable land, and low taxes, the Lockheed Corporation moved to northern California in 1956. The U.S. Department of Defense purchased about 40 percent of the semiconductors produced by this large company. In addition, the area has had an unusually active scientific and intellectual "community of scholars" because of its close association with Stanford University. These factors, coupled with innovative computer technology developed by a rare group of inventive "native sons" (Tandem, Apple, Hewlett-Packard), have helped create this nexus of quaternary economic activity.

The evolution of this economic region has not always been so smooth. Silicon Valley's dominance in the world of electronics and information has been challenged by the overseas production of computer chips and other electronic equipment. Following the boom years in the early 1980s, the region suffered a decline after 1984 due to a saturation of the market for computers, an oversupply of domestically produced computer chips, and competition from Japan and Korea.

Recent years have seen a renewal of this high-tech land of promise as industries are finding specialized niches or expanding their geographical scope of business. Apple Computer, for example, long almost exclusively dependent upon the home and educational markets, has turned its attention to the corporate market.[12] Other firms are developing products in robotics, and still others are entering biotechnology.

Just as world economies have gone through evolutionary stages, so have their settings. Cities were at first assemblages of dwelling units for people who worked in the surrounding fields, woods, and mines. As manufacturing became important, cities took on the role of housing those workplaces. Tertiary economic activities are also predominantly urban, as centers for the exchange of services and goods became important functions in the city. Likewise, the beginnings of the quaternary economy thus far have taken place in cities and their suburbs. Our next chapter views cities as dynamic economic and cultural environments for human activity through the ages.

KEY CONCEPTS

Agribusiness	Mediterranean agriculture	Resources	Transhumance
Aquaculture	Neolithic revolution	Secondary production	Von Thünen
Consumption	Plantation	Subsistence	Weber's location triangle
Domestication	Primary production	Swidden	
Footloose industry	Production	Terracing	
Information society	Quaternary production	Tertiary production	

FURTHER READING

There are many excellent surveys of economic geography, each with a slightly different emphasis but all containing a central body of essential information. One of the most useful is Brian J. L. Berry, Edgar C. Conkling, and D. Michael Ray's *The Global Economy: Resource Use, Locational Choice, and International Trade* (Englewood Cliffs, NJ: Prentice-Hall, 1993). Another excellent survey with emphasis on location and quality of life is Thomas J. Wilbanks, *Location and Well-Being: An Introduction to Economic Geography* (New York: Harper & Row, 1980). A similar work with a strong emphasis on transportation is James O. Wheeler and Peter O. Muller, *Economic Geography* (New York: John Wiley, 1986). Other more recent books include Peter Dickon's

Global Shift: The Internationalization of Economic Activity (New York and London: Guilford Press, 1992) and *The World Economy: Resources, Location, Trade, and Development,* by Anthony R. deSouza and Frederick P. Stutz (New York: Macmillan, 1994).

More theoretical approaches include "Economic Geography in the 1990s: The Perplexing Geography of Uneven Development," in *The Student's Companion to Geography,* edited by Alisdair Rogers, Heather Viles, and Andrew Goudie (Cambridge, MA: Blackwell, 1992), and Joseph H. Butler, *Economic Geography: Spatial and Environmental Aspects of Economic Activity* (New York: John Wiley, 1980).

A comprehensive work cited in Chapter 11 is Ronald F. Abler, John S. Adams, and Peter R. Gould, *Spatial Organization: The Geographer's View of the World* (Englewood Cliffs, NJ: Prentice-Hall, 1971). Also cited is a useful collection of case studies, most of which are still valid, which was compiled by Richard S. Thoman and Donald J. Patton in *Focus on Geographic Activity: A Collection of Original Case Studies* (New York: McGraw-Hill, 1964).

Our recommended survey of the geography of agriculture is a book written for the general public by a scholar: John Fraser Hart's *The Land That Feeds Us* (New York: W. W. Norton, 1991). Often cited in this chapter is a classic by Howard Gregor, *Geography of Agriculture: Themes in Research* (Englewood Cliffs, NJ: Prentice-Hall, 1970). Michael Chisholm's *Rural Settlement and Land Use: An Essay in Location* (Atlantic Highlands, NJ: Humanities Press, 1979) focuses upon global agricultural patterns. Agricultural development in the United States from social and environmental viewpoints is offered in Walter Ebeling, *The Fruited Plain: The Story of American Agriculture* (Berkeley: University of California Press, 1979).

Sustainable agriculture, an increasingly important topic of inquiry in geography, is investigated in *Smallholders, Householders: Farm Families and the Ecology of Intensive, Sustainable Agriculture* by Robert MacNetting (Stanford, CA: Stanford University Press, 1993). Also useful for teaching about agriculture is L. R. Brown and J. E. Young's "Feeding the World in the Nineties," in *State of the World* (New York: W. W. Norton, 1990): 101–141, and "Sustainable Agriculture" by I. P. Reganold, R. I. Papendick, and J. E. Parr in *Scientific American* (June 1990): 88–95.

Industrial geography is changing at a rate that makes it difficult for survey books to keep abreast of the changes. A provocative attempt at this, however, is Barry Bluestone and Bennett Harrison, *The Deindustrialization of America: Closings, Community Abandonment, and the Dismantling of Basic Industry* (New York: Basic Books, 1982). An encyclopedic evaluation of industrial location analysis is the three-volume work by F. E. Ian Hamilton and Godfrey J. R. Linge, eds., *Spatial Analysis, Industry, and the Industrial Environment: Progress in Research and Applications* (New York: John Wiley, 1979–1984).

Other useful sources on secondary production include *The Politics of International Economic Relations,* by J. E. Spero (London: Allen and Unwin, 1990); UNIDO's *Industry in a Changing World* (New York: United Nations, 1983); "Occupational and Industrial Diversification in the Metropolitan Space Economy in the United States, 1985–1990," by S. Bagchi-Sen and B. M. Pigozzi in *The Professional Geographer* 45 (1993): 44–54); and B. J. Cohen's "The Political Economy of International Trade," in *International Organisation* 44 (1990): 261–281.

ENDNOTES

1. Summarized from Richard L. Morrill, *The Spatial Organization of Society* (Belmont, CA: Duxbury Press, 1970): 40.
2. Johan Heinrich von Thünen, *Der Isolierte Staat in Beziehung auf Landwirthschaft und Nationalo Konomie* (Hamburg, 1826); translated version: Peter Hall, ed., C. M. Wartenburg, trans., *Von Thünen's Isolated State* (Oxford: Pergamon Press, 1966).
3. Other studies of the von Thünen theory abound in the geographic literature. For example, see U. Ewald, "The von Thünen Principle and Agricultural Zonation in Colonial Mexico," *Journal of Historical Geography* 3 (1977): 123–133; Ernst Griffin, "Testing the von Thünen Theory in Uruguay," *Geographical Review* 63 (Oct. 1973): 500–516; Ronald Horvath, "Von Thünen's Isolated State and the Area Around Addis Ababa, Ethiopia," *Annals of the Association of American Geographers* 59 (June 1969): 308–323; and Robert Sinclair, "Von Thünen and Urban Sprawl," *Annals of the Association of American Geographers* 57 (March 1967): 72–87.

4. Cotton Mather and John Fraser Hart, "The Geography of Manure," *Land Economics* 32 (Feb. 1956): 25–28.
5. Howard Gregor, "The Changing Plantation," *Annals of the Association of American Geographers* 55 (1965): 221–238.
6. Gillian Wainwright, "Scottish Salmon Aquaculture Now Three Times Wild Catch," *Aquaculture* 11 (Jan.-Feb. 1985): 26–29.
7. Alfred Weber, *Theory of the Location of Industries* (Chicago: University of Chicago Press, 1909).
8. Summarized from James O. Wheeler and Peter O. Muller, *Economic Geography* (New York: John Wiley, 1986).
9. August Lösch, *Economics of Location* (New Haven, CT: Yale University Press, 1954).
10. Walter Isard, *Location and Space Economy* (Cambridge, MA: MIT Press, 1956).
11. Peter Dicken, *Global Shift: The Internationalization of Economic Activity,* 2nd ed. (New York: Guilford Press, 1992).
12. Pamela Nakaso, "The Valley of Silicon Lives Again," *Sacramento Bee* (May 23, 1988): C-1.

9 THE THEME OF PLACE: URBAN SYSTEMS

Helsinki, Finland, market on a Saturday morning. (Photo by author.)

Earth is rapidly becoming an urban place. Most people who live in the economically developed countries of the world today live in cities. In most nations of the world, the rate of **urbanization** exceeds that for all previous decades. In the United States, almost 80 percent of the population is urban; in Canada and Japan, about 75 percent; and in Australia, over 85 percent of the population lives in cities. Figure 9-1 illustrates the spatial distribution of urbanization in the world today.

With its focus on urban settlement patterns and processes, this chapter addresses geography content contained in the following *National Geography Standard:*

Standard 12: *The geographically informed person knows*

and understands the processes, patterns, and functions of human settlement.

Two primary innovations enabled urban places to come into being: the innovation of agriculture, coupled with greater opportunities for storage and distribution of resources. As hierarchical social structures and the complexities of rural life intensified, cities evolved as **nodes** or centers of human activity. Cities have become the dominant, most important form of twentieth-century settlement. They are distinctive places with distinctive landscapes, because urban places reflect local cultural beliefs, resource uses, levels of technology, and population densities.

For some, the word *city* brings to mind images of air and water pollution, overcrowding, traffic problems, and other negative issues. For others, *city* means ethnic diversity and cultural advantages. Cities are often thought of as places of excitement and action. They are social regions, places where people interact and cultures evolve rapidly in contrast to the gradual change of rural areas.

All cities perform economic functions. As a result of these economic and cultural factors, distinct and observable patterns of land use exist within cities. In fact, the distribution of commercial and industrial districts, residential neighborhoods, and other land use districts have created similar patterns inside many different cities. There also are links between cities; economic linkages, social ties, and transportation networks create systems of cities that interrelate and are interdependent.

This chapter focuses upon the similarities and differences among cities. We begin by examining the evolution and history of the urbanization process to discover why cities developed in the first place. Then we look at land use inside the city as we discover patterns and processes inside neighborhoods and inside residential, commercial, and industrial districts. Next we explore the exciting diversity of the social geography of the city, with a look at the city as a microcosm. Finally, we study relationships between urban places and their spheres of influence by examining the network of cities across the globe.

CITIES OF THE PAST

Cities are a recent cultural **innovation**, representing the best and worst of our human condition. For the first several million years of human existence, we lived in dispersed rural settlement sites. Before 1850 no society could be considered "urban," and by the year 1900 only the United Kingdom was primarily urban. Today, but a few decades later, all industrialized nations are "urbanized." (Note: we use the term *urbanized* to mean a nonagricultural, agglomerated place in or near cities.) The process is continuing at a rapid rate.

As far as is known, the world's earliest cities developed over 9000 years ago in the Fertile Crescent in the valleys of the Tigris and Euphrates rivers in the Near East (Figure 9-2). At least three conditions were necessary for these early cities to develop:[1]

1. Ample natural resources to support a large population (water, good soils, and level land)

2. An agricultural system that could provide enough food and storage facilities for a concentrated sedentary population

3. A social system for the distribution of food and resources and for the organization of the society

Other early cities developed later in the Nile Valley in Egypt (3200 B.C.), along the Indus River in India (2500 B.C.), in the Huang Ho Valley in China (1500 B.C.), and later in the New World in Peru, Guatemala, and Mexico. Early cities favored hill-tops or other protected landforms for defense, or river junctions and riverside sites for ease of transportation (Figure 9-3). For a brief tour through the ancient city of Anyang, China, see the box on page 227. Also, for a reassuring (?) account that human nature has changed little over the millennia, see the box entitled "'Generation Gap' in Nippur and Ur" on page 228.

The process of urbanization was a long and complicated transition that continued for thousands of years. Scholars have suggested four models of urban evolution to explain the development of cities:[2]

- *The hydraulic civilization model* suggests that the development of large-scale irrigation systems and the subsequent development of centralized authority were necessary for urbanization to occur. Irrigation brought about not only increased crop yields but centralized power in the hands of those who controlled the water system.

- *The innovation model,* on the other hand, suggests that urbanization depended on some new invention to trigger city growth. One group would gain an advantage over another because they had some new technology.

- *The environmental stress model* looks at change in the physical environment as the trigger for urbanization. As one group gains control over the environment in some new way, they assume a dominant position.

- *The coercion and warfare model* explains urban growth as a reaction to danger. Towns are more secure than the countryside, so people flocked to them in wartime. Supporters of this model argue that cooperation, not competition, was necessary for development of urban places.

It is obvious that no single model can provide all the answers. Undoubtedly some combination of factors encouraged early cities to grow and prosper in different parts of the world.

These early urban centers differed from nonurban societies in many ways. According to V. Gordon Childe, not-

FIGURE 9-1
Distribution of the world's cities having a population of 2 million or more. (From James M. Rubenstein, The Cultural Landscape: An Introduction to Human Geography, *4th ed., New York: Macmillan, 1994, Fig. 11-3, p. 458. Reprinted with permission of Macmillan Publishing Co.)*

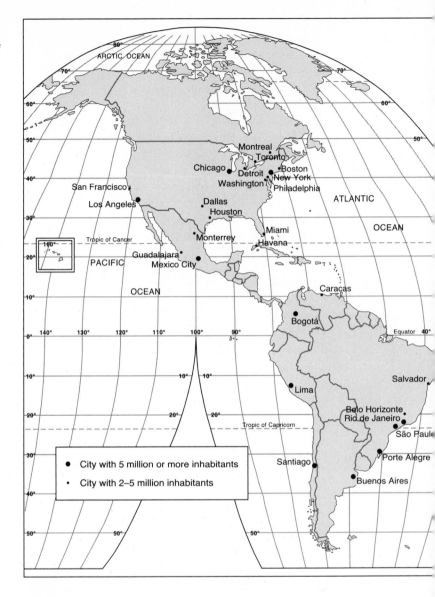

City with 5 million or more inhabitants

City with 2–5 million inhabitants

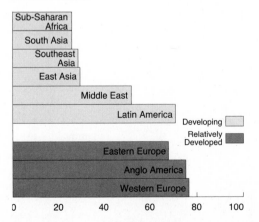

PERCENT URBAN

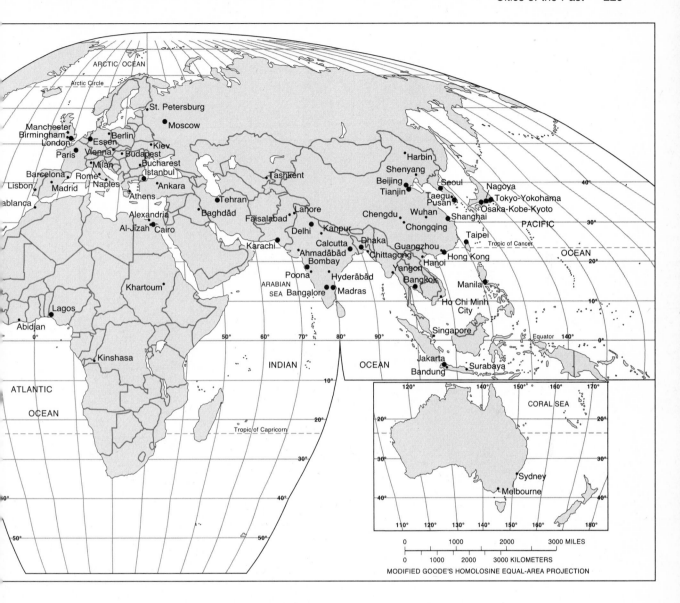

ed urban historian, the following characteristics distinguished urban places from rural ones:[3]

1. Urban settlements were large and more densely populated.

2. Cities supported full-time craftsmen.

3. Farmers produced a surplus, some of which they turned back to the city society.

4. Cities built monumental public buildings.

5. Food surpluses were kept in central granaries owned or managed by priests.

6. Cities were centers for recording and applying science.

7. Cities furthered development of the predictive sciences.

8. Cities were centers for artistic advances.

9. Cities used imported raw materials.

10. Urban social organization was based on residence, not kinship.

Urban Development Around the Mediterranean Sea

The world's earliest cities fulfilled what are thought to be primarily religious and political functions, and their economic role in early centuries was limited. Markets were lo-

FIGURE 9-2
*Where and when early cities arose.
(From E. B. Phillips and R. T.
LeGates,* City Lights, *New York:
Oxford University Press, p. 84.
Copyright © 1981 by E. Barbara
Phillips and Richard T. LeGates.
Reprinted by permission of Oxford
University Press, Inc.)*

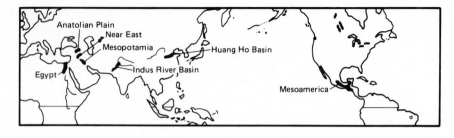

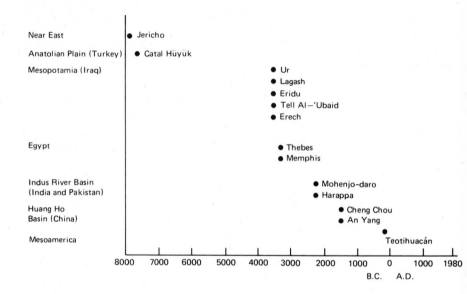

FIGURE 9-3
*Roman ruins are visible in down-
town Rome in the 1990s. (Photo by
author.)*

ANYANG: AN EARLY CHINESE CITY

The city of Anyang, located along the rich fertile soil of the Yellow River (Huang He) in China, is brimming with life as the residents go about their business. Little matter that the date is about 1390 B.C. On the outskirts of the walled part of town are rich farmlands. Farmers are tending their fields of millet, rice, and barley, made fertile by the loess soil or "yellow earth." Domesticated pigs, dogs, sheep, oxen, and horses are a part of the landscape, providing food and transportation in an economic system that uses the cowrie shell as a medium of exchange.

Upon entering the walled section of Anyang, we see numerous buildings of various shapes and sizes, although round houses are the most common. Artisans are busy at their jobs. Luxury items must be in high demand because of the variety of goods being produced by different craftspeople. Wandering through the streets of the city, we see cloth being produced from both hemp and silk. Metal tools and bronze statues are being created as well as beautiful pottery. Other artisans are carving jade and marble into items of beauty.

Anyang society is hierarchical, so an intelligentsia may be found. These literate officials have devised a written language to keep track of government records. They also perform future predictions on oracle bones upon which questions have been inscribed. The bones are burned and the resultant cracks determine the answer. In addition, this class has devised a calendric system to keep track of the solar year and lunar months.

The elite of Anyang are the king and his nobility. These individuals rule the lower classes. Their homes are elaborate palaces set away from the commoner parts of the city. Their primary job is to lead the city in warfare, but most of their time is actually spent in hunting wild animals to supplement their diet of domesticated plants and animals, and in partaking of the pleasures of the good life.

Anyang, China, is a city with a hierarchical governing body, an agricultural base, and a system of commerce. This early urban center along the Huang He may be called a city in the truest sense of the word.

cal, and crafts operated on a small-scale, restricted basis. However, with establishment of cities around the Mediterranean Sea, trade became increasingly important as navigation and ship-building technology improved. First in trade were the Phoenicians at Tyre, then the Greek city-states, and finally Roman cities. New patterns expanded to accommodate ever-increasing population densities and economic opportunities. By 600 B.C. there were over 500 towns and villages on the Greek islands. Subsequent diffusion around the Mediterranean followed the expanding Greek Empire. Athens soon developed into the ultimate expression of Greek urban life, with the Acropolis in the center of the city, the Agora for community meetings, and a concern for aesthetics. The Greeks were, in fact, the first to formulate ideas for city planning. Greek cities were used as models for urban planning during the Renaissance many centuries later.

Roman cities also used Greek cities as their model. The idea of Greek grid-pattern streets (a concept probably borrowed from cities in the Indus Valley) was expanded and refined with ideas from Etruscan cities north of Rome. The Greek Acropolis evolved into the Roman forum as numerous monuments and beautiful public buildings were erected. Roman cities had well-planned residential, commercial, and industrial districts (Figure 9-3). These cities soon spread beyond the Mediterranean as the Roman Empire dominated lands as far away as England, the Alps, and Eastern Europe. Indeed, imperialism may be credited with the spread of urban patterns around the globe.

Medieval Towns

Following the decline of Roman power, cities played a much less central role in European society. Hundreds of

"GENERATION GAP" IN NIPPUR AND UR

The following text, assembled from fragments found at the early Middle Eastern cities of Nippur and Ur, has parallels in today's society:

You who wander about in the public square, would you achieve success? Go to school, it will be of benefit to you. What I am about to relate to you turns the fool into a wise man.

Because my heart had been sated with weariness of you, I kept away from you. I kept away from you and heeded not your fears and grumblings. Because of your clamorings, I was angry with you. Because you do not look to your humanity, my heart was carried off as if by an evil wind. Your grumblings have put an end to me, you have brought me to the point of death.

I, never in all my life, did I make you carry reeds to the canebrake. The reed rushes that the young and little carry, you, never in your life did you carry them. I never sent you to work as a laborer. "Go, work, and support me" I never in my life said to you. Others like you support their parents by working.

If you spoke to your kin and appreciated them, you would emulate them. They provide 10 gur of barley each—even the young ones provided their fathers with 10 gur each. They multiplied barley for their father, maintained him in barley, oil, and wool. But you, you are a man when it comes to perverseness, but compared to them you are not a man at all. You certainly don't labor like them—they are the sons of fathers who make their sons labor, but me, I didn't make you work like them.

Night and day am I tortured because of you. Night and day you waste in pleasures. You have expanded far and wide, have become fat and puffed. But your kin waits expectantly for your misfortune and will rejoice in it because you looked not to your humanity.

From Samuel Noah Kramer, The Sumerians *(Chicago: University of Chicago Press, 1963). Reprinted with permission of the University of Chicago Press.*

years elapsed between the great cities of Greek and Roman times and the modern cities of Europe. The nonurban character of rural Europe during the early Middle Ages discouraged city growth, erased earlier urban patterns from the landscape, and set the stage for an entirely new urban form.

Kingsley Davis suggests two reasons for the new urban age that was to come.[4] First, the low productivity of medieval agriculture made it necessary for farmers to trade (and to manufacture something to trade). Second, the feudal social system meant that people could not gain political dominance over their hinterlands and thus become warring city-states. These nonurban people, then, began to specialize in commerce and manufacture. Craftsmen soon were housed in towns where commercial exchange was maximized. Small villages and towns soon developed, centered around the marketplace and church, and surrounded by a great wall for protection from attack. Cities became economic centers as well as a place of safety.

These medieval towns remained small, and much of the population of Europe stayed in rural areas. For exam-

ple, in the year 1600, considering only cities with populations of 100,000 or more, only 1.6 percent of the population was urban. Using the same measure, by 1700, only 1.9 percent lived in urban places, and in 1800, only 2.2 percent were urban. From these statistics, it is obvious that at the beginning of the Industrial Revolution, Europe was an extremely nonurban place.

Despite the small size of these preindustrial cities, many of their characteristics still remain in modern Europe. The narrow, winding streets so typical of this period may be seen in most European cities today, and numerous church buildings reflect the old ways. The real tradition of modern urban life began during this period, and a great deal of the medieval urban landscape may still be seen on European travels (Figure 9-4).

The Industrial City

Until the Industrial Revolution, cities could be thought of as being essentially "parasitic." Economically, they produced little and depended almost entirely upon agricultur-

FIGURE 9-4
Medieval towns such as this one at Assisi in central Italy often were built on hilltops for security. (Photo by author.)

al surpluses from rural areas. This pattern changed slightly during medieval times, but it was the development of large-scale industry that made cities the centers of economic production after the Industrial Revolution.

Following the development of centralized industry and power, cities began to grow rapidly. Old marketplaces were soon replaced by banks, merchants, and speculators. A new form emerged as cities became a focus for the assembly of raw materials, their processing, and the distribution of products. Massive rural-to-urban migration brought in labor from the surrounding countryside, with an unprecedented growth in urban population. City planning was replaced by what Lewis Mumford has called "laissez-faire utilitarianism" as chaotic land use patterns developed in the industrial city.[5]

For a view of London as the prototypical industrial city at the end of the nineteenth century, see the box on page 230.

Development of Cities in North America

Anglo-American cities differed from European cities in that their pattern was not based on preexisting, historical city sites but rather on brand-new, undeveloped sites. Geographer J. R. Borchert analyzed American cities in a recent article and defined five factors that encouraged city growth in the United States:[6]

1. *Resources*—cities grew because they were near resources, such as agriculture or minerals.

2. *Population migration*—most cities grew larger because a population migrated to them. This was important in the nineteenth century when people were coming from Europe, and it remains important in the twentieth century as growth has shifted to the western and southwestern states with immigrants from Asia and Latin America.

3. *Transportation and technology*—cities frequently grew because of advantageous locations along transportation routes. (This was especially true at break-of-bulk points where one form of transport was replaced by another.)

4. *Functional interrelationships between cities*—some cities grew because they assumed important roles in already existing urban systems.

5. *Changes in Economic and Social Systems*—in the nineteenth century, for example, resettlement in the West was encouraged.

Borchert then recognized five periods, or epochs, in American urban growth. Geographers commonly use this classification system to analyze North American historical patterns. Borchert's five stages include the following:

1. *The colonial heritage (to 1790)*—Cities were mainly the creation of European mercantilism, with products such as timber and food shipped to Europe. The largest cities during this early period were those with the best access to the interior. By 1790, New York was the largest city (population 33,000), with Philadelphia and Boston its two biggest rivals.

2. *Era of sail and wagon (1790–1830)*—This was a period of major agricultural settlement of the Northeast and the Old Northwest Territories (now the Midwest). This period contains the only decade when American urban populations dropped, due to an increased focus upon agriculture and decreased linkages with Great Britain. During this period all the major cities of the Midwest were settled, including Pittsburgh, Cincinnati, Louisville, and St. Louis.

3. *Epoch of rail and steam (1830–1870)*—This period showed increasing transportation efficiency with the appearance of the steamboat and railroad. The entire continent became a market for goods as East Coast cities continued to grow with an increasing manufacturing function. The first industrial area was centered around Boston. In addition, Chicago rapidly became a railroad center and rose from obscurity in 1830 to become one of the largest cities in the United States.

4. *Steel-rail epoch (1870–1920)*—This period saw rapid extension of national accessibility to the remaining isolated regions of the West and South. By the end of this period, most major urban centers were joined in a standardized nationwide rail network, and the modern period of major urban centers was beginning to emerge. New growth cities were usually associated with the exploitation of remaining agricultural lands in the West or

LONDON: THE PROTOTYPE INDUSTRIAL CITY

Imagine a world filled with industrial filth, crime, and crowding, where living conditions were so bad that individuals spent the bulk of their wages for a one-room apartment. This room, without electricity, heat, or sanitation, was shared with seven or eight other people and hoards of vermin. Consider a world where a child searched the muddy streets for bits of food lying among the scattered debris and refuse. This was a world in which adults performed two days of heavy labor in a workhouse only to receive inadequate food, but nevertheless considered themselves lucky. Imagine a world where jobs were so scarce that entire families, including small children, worked in sweat shops to scrape out a meager existence. This was the East End of London in 1902, in the midst of the Industrial Revolution.

Development of large-scale industry and the centralization of power created a weekly invasion by thousands of in-migrants into cities such as London. These people came from the surrounding countryside in search of jobs and a better way of life. What they found instead were few jobs and many workers, so that wages were kept at a minimum. In fact, to subsist (to obtain food and rented housing only), a family of five had to make at least $5.25 a week, but in reality, few families made close to $4.00 a week. Not only were wages inadequate; there was no job security. Those missing work because of injury or illness often found that their job was unavailable upon their return. Further, there were no old-age benefits; the elderly were simply left in the streets to survive as best they could when they could no longer work.

City planning in industrial London was practically nonexistent. Workers were forced to live in East End ghettos because of low wages. These neighborhoods consisted of street after street of houses converted into one-room hovels. Entire families lived in these crowded rooms and often sublet floor space to outsiders to make ends meet. The number of individuals sharing a room could be extended by sleeping in shifts; a day worker could occupy the floor space during the night and vice versa. These rooms lacked sanitation or garbage facilities. To get rid of refuse, garbage was tossed from windows onto the roofs of cardboard shacks that surrounded the tenement buildings.

Death rates were high as a direct result of these difficult living conditions in the industrial city, and the increasingly dense populations. The average life span of the East End resident was only 30 years. Conditions in industrial London in 1902 did not allow for long lives.

Consider how much of the above description could be applied to some Third World cities currently reported in the news media, such as the "Tomb Cities" in Egypt or the Paracaidistas* of Mexico City.

Paracaidista is Spanish for "parachutist." In Mexico City, *paracaidistas* are "tent city" people who live in makeshift fabric-covered dwellings.

with isolated mineral deposits. In the East, growth was related to the iron and steel boom and other heavy manufacturing.

5. *Auto-air-amenity epoch (1920–present)*—This is primarily a period of changes within the pattern established by

the end of the previous epoch. Much of this change is associated with the rise of vehicles powered by the internal-combustion engine as a major mode of transport. Changes in the epoch include:

- The decline of the Appalachian coal and railroad centers,

as petroleum-fueled vehicles with internal–combustion engines replaced steam transport

- The growth of automobile-related centers
- The rising importance of cities as service centers
- The growth of the West and Southwest
- The increasing mobility of populations that leads to sub-urbanization

This historical development of urban places on Earth sets the stage for understanding the form of the modern city. Cities are divided into a number of land use districts, each with recognizable and distinct functions.

INTERNAL STRUCTURE OF THE CITY

Housing, Home, and Neighborhood

A view of the city can be macroscopic or microscopic, and in this section we will take the microscopic approach. A close-up view at its most extreme would be an analysis of a household as it interacts with its larger urban context. Chapter 4 introduced the concept of home in various settings; here we examine "home" in the urban place, multiplied by the number of nearby neighbors to create an urban neighborhood.

Neighborhood is not always easy to define. Neighborhoods are usually set apart from other parts of a city by boundaries. These may be physical barriers such as freeways, major streets, shopping districts, or topographic boundaries (ravines, hills). Neighborhood boundaries may also be based on human expressions. Different social or ethnic groups may remain isolated from each other in distinct and observable neighborhoods. It is fascinating to notice the cultural differences in modern residential districts—a Mexican barrio, for example, looks very different from a Polish neighborhood.

An upper-income neighborhood offers a different landscape expression than a lower-income district. Age and gender also create varying patterns in neighborhoods. A district with predominantly upwardly mobile career-oriented residents differs from an area of primarily working-class people. Cultural groups often express themselves very differently in terms of their living space.

Neighborhoods change dramatically through time. North Beach in San Francisco, for example, has experienced many changes since Asian immigrants began flooding into the area in the 1970s, replacing the earlier Italian residents. Although the Italian flavor of North Beach still remains in its commercial districts, the residential charac-

ter is rapidly transitioning into a mixture of Italian and Asian people (see the box on page 232).

A neighborhood, then, is an area in the city where an individual lives his or her everyday life. Geographers are interested in the locations and patterns of neighborhoods, and in neighborhood change. To depict these changing patterns, maps of social variation may be constructed. A most comprehensive analysis of urban neighborhoods was conducted by Toronto geographer R. A. Murdie,[7] and two of his maps of Toronto neighborhoods are shown in Figure 9-5.

Murdie's Toronto study was based largely on census figures and statistical analysis. However, more recent studies have gone farther and have begun to study the **perception** *of place* revealed through interviews and questionnaires. This "perception of neighborhood" theme is a recent

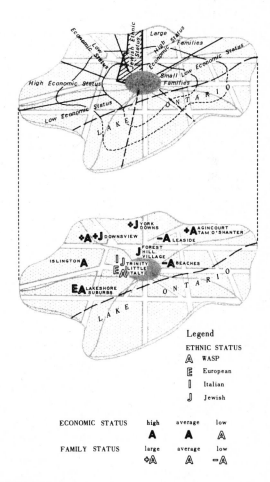

FIGURE 9-5
Murdie's "communities" in metropolitan Toronto. (From R. A. Murdie, "Factorial Ecology of Metropolitan Toronto, 1951–1961," University of Chicago Department of Geography Research Paper No. 16, 1969, p. 169. Used with permission.)

NORTH BEACH: NEIGHBORHOOD IN TRANSITION

Asians now are moving into the old Italian neighborhood of North Beach in San Francisco. North Beach still retains much of its past exciting ethnic character, however. Even though half of the businesses in the area are now Asian owned, the smell of garlic and olive oil lingers as a reminder of the neighborhood's early history. Pasta, pizza baked in a wood-fired oven, cappuccino cafes, and gelato shops still abound along the winding, hilly streets of this quaint district. Many of these Italian specialty shops now are Chinese owned!

The first settlers to arrive in North Beach were Russians. Then in turn came the Germans, Irish, Italians, and now the Asians to enrich this rapidly transitioning neighborhood. Today the most striking difference in demographic changes may be observed at Saints Peter and Paul Church in the heart of North Beach. Since the 1920s this sand-colored, twin-steepled church has been famous throughout northern California as the "Italian Cathedral." Now, however, over 25 percent of its membership is Chinese. About 70 percent of the children in the church school are Chinese, and over 80 percent in the child-care center are Asian. Italian-language classes are no longer taught at the school, but Cantonese classes are held daily after school for Asian children from the neighborhood.

Washington Square, formerly full of Italian men in black suits smoking pungent cigars, is now full of Chinese men practicing t'ai chi, a meditative, slow-motion exercise. Public space as well as private space is transitioning as more Asians move across Columbus Avenue into North Beach. This busy commercial street formerly served as a visible boundary between Chinese and Italian space in San Francisco. Today Columbus Avenue is visible as a cultural mixture of Asian and Italian businesses and residences as the two ethnic groups merge their neighborhood space.

In San Francisco's North Beach, commercial development represents a mixture of Italian and Chinese cultures in the 1990s. (Photo by author.)

attempt to capture the heart of the city as seen through its residents' eyes. Kevin Lynch, in his book *The Image of the City,* was one of the first to look at the "legibility" of the city in this way (the ability to "read" a landscape).[8] After studying the subjective experiences of neighborhood residents, Lynch suggested that the image of the city is based on physical forms, including:

1. *Paths*—channels of movement (sidewalks, roads)
2. *Edges*—linear boundaries (rivers, railroads)
3. *Districts*—large city areas (central business district)
4. *Nodes*—strategic focal points (shopping areas, office complexes)
5. *Landmarks*—point references in the city (monuments)

A neighborhood, then, is actually a multiple reality. It is a place with many different meanings to different people. Residents of a neighborhood perceive their space differently from the way visitors might. Demarcating a neighborhood's boundaries depends on the gender, ethnicity, and socioeconomic status of the residents. These varying perceptions of neighborhoods make analysis and planning of the city a challenging task.

This "feel" of the neighborhood has been studied by other social scientists in an attempt to understand more about urban patterns. Herbert Gans's book about Boston's West End neighborhood, *The Urban Villagers,* emphasized that the neighborhood is more than an ecological or statistical concept and is often based upon the perceptions of the residents themselves.[9] He found that the idea of the West End as a single neighborhood was entirely foreign to residents of the area! Each ethnic group saw their turf as a distinct subregion, with social differences between ethnic areas that were often much greater than their actual geographic distance would indicate. Until the beginning of redevelopment in the West End, no one perceived it to be a distinct neighborhood. After redevelopment, the residents were drawn together in a common cause. In many other cities around the world, political decisions often set the stage in much the same way and encourage neighborhood cohesion.

Sociologist Walter Firey also studied Boston neighborhoods.[10] He defined two conditions of ethnic and socioeconomic neighborhoods that are perceived by their residents—*sentiment* and *symbolism.* Symbolic features are especially important in historic neighborhoods such as Boston's Beacon Hill. Sentiments are also often attached to a locale. Firey described this condition in Boston's North End Italian neighborhood, which significantly resembles San Francisco's North Beach and other Italian neighborhoods in North America. Sentiments in these ethnic neighborhoods are often tied to family attachments and a sense of community.

A neighborhood, although often difficult to define from the outside, is clearly defined by its residents and forms communities in the city. These spatially defined areas of the city, in turn, form larger land use districts observable in urban areas in many parts of the world. These land use patterns are frequently observable from city to city and offer exciting field trip opportunities for geography teachers.

URBAN LAND USE PATTERNS

The city, viewed from the window of your car or from a bus or perhaps a trolley, shows busy people not only at home but in transit to work, school, entertainment, and so on. Where we go is usually based on **land use.** We often travel from our residential districts to commercial districts or industrial districts. Land use is, in fact, the flesh of the city. Transportation lines are its circulation system. Both are needed for a healthy "organism".

Land use in cities usually conforms to certain predictable patterns. Land owners have a tendency to develop land to the best advantage, and this encourages observable land use districts to develop within urban areas. These districts may be residential (about 50 percent of land area in a city), transportation (about 20 percent), commercial, office, and industrial use (about 25 percent), and all others uses—such as education, parks, and public use (about 5 percent). All land use theory is based on **accessibility** to a place. Accessibility includes the relative advantages in time, cost, or physical entry over other similar places.

Residential land use takes up the greatest amount of space in the city and was discussed in the preceding section of this chapter. Commercial land is also widely distributed and is the key to much of the traffic flow in cities. Commercial activities are highly clustered and tend to concentrate where land values are highest. The image of the city is often created in its central business district, while strip commercial streets give suburbs their individual image. These commercial zones also serve as employment nodes for residential districts.

Geographer Brian J. L. Berry has developed a distinct typology of commercial land use in the city.[11] Pointing out that cities have different types of commercial districts, Berry calls one type "specialized areas." These clusters—"auto rows," medical districts, or printing districts—often share advertising and customers.

A.

B.

FIGURE 9-6

*A. Shopping district in downtown Stockholm. (Photo by author.) B. Shopping district in the sub-
urbs of Stockholm. (Photo by author.)*

Berry called his second type of commercial area a "strip commercial district." Here, businesses are scattered along an arterial street, rather than being grouped or clustered. Three types of strip commercial developments are highways (motels, restaurants), shopping streets (clothing stores, groceries), and neighborhood arterials.

Berry's third type of commercial land use in the city is known as a "nucleation." This clustered commercial district is a concentration of stores and offices. The best-known nucleation is the *central business district* (CBD) in the oldest part of the city's downtown. A hierarchy of nucleations outside the CBD include:

1. Regional shopping centers (50–100 shops with a large department store as an anchor)

2. Commercial shopping center (30–70 shops with a discount drug store as anchor)

3. Neighborhood shopping center (10–40 shops with a supermarket as anchor)

4. Isolated clusters (1–3 stores with a convenience store or "mom-and-pop" grocery as anchor)

In the United States, Berry's typologies of commercial land use are easily observable in the commercial landscape of modern cities. In Sweden, other types of commercial districts have developed. Careful planning in Stockholm, for example, has created tightly clustered commercial districts around many metro stations. These nucleations are located in both central city and suburban areas and efficiently serve local residents. The centralized districts act as nodes for parking, offices, and shopping in a convenient location, with access to many types of transit (trains, automobiles, bicycles, and pedestrian pathways). The photographs in Figure 9-6 illustrate the multiuse advantages of two types of commercial districts in Sweden.

In modern cities, then, a typology of commercial land use exists. The central business district (CBD) is at the top of the commercial hierarchy throughout the world. "Downtown" is usually the oldest part of town where land values are highest. *The peak-value intersection,* called the "100 percent location," has the highest land values in town and is the center of the city. This area is traditionally dominated by banking and retailing, but other specialized areas also develop in the CBD. Office complexes, governmental centers, and financial districts surround the peak-value intersection. Warehousing, commonly carried on around the edges of the CBD, has now been cleared away in many large American cities, and the former "warehouse district" now may be occupied by low-income residential housing projects. There is often a vertical zonation in the CBD as well, with first-floor retailing locations topped by offices and factories (or sweat shops) in the upper floors of tall buildings.

This downtown commercial district forms the core of the city and is the centerpoint for twentieth-century land use theories. Residential, commercial, and industrial land are all considered in these models of urban expansion.

The earliest theory to predict land use within cities was developed in the Chicago "school of sociology" in the 1920s.[12] "Human ecologists" from the Chicago school patterned their ideas after earlier land use models developed for rural land by von Thünen in the early 1800s (see Chapter 8 for a detailed discussion of this early theory). The first rural theories stated that items with the highest demand and transportation costs would be produced closest to the city. This von Thünen model was thus a *land rent* or **bid rent** expression of land values around the city. Burgess refined the von Thünen model and applied it to urban rather than rural areas.

The Burgess **concentric zone theory** model (Figure 9-7) is based on continual immigration into cities and the

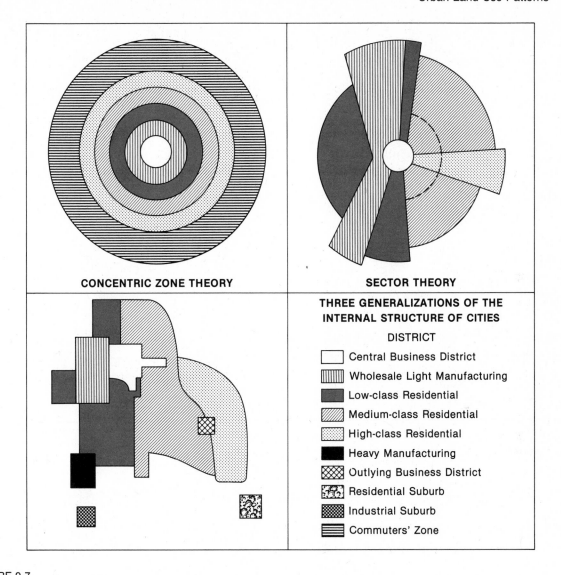

FIGURE 9-7

Three models describing the internal structure of cities.   A. Concentric zone model, in which a city grows in a series of rings around the central business district.   B. Sector model, in which a city grows in a series of sectors radiating from the CBD.   C. Multiple nuclei model, in which a city grows in individual activity nodes. (From Arthur Getis, Judith Getis, and Jerome Fellman, Introduction to Geography, *Dubuque, IA: William C. Brown Publishers. Reprinted with permission of William C. Brown.)*

subsequent assimilation of these new residents into mainstream American neighborhoods. Today's cities, however, are different from those of the nineteenth and early twentieth centuries because most new inmigration consists of African-Americans, Mexicans, Asians, and other groups rather than European immigrants. Assimilation of these groups is slower than the assimilation of earlier European immigrants, so the concentric zone model is no longer ap-

plicable in most cases. The classic invasion-succession model just does not usually fit today's multiracial, multicultural urban experience.

In 1939, Homer Hoyt refined Burgess's earlier ideas into the **sector theory**.[13] As shown in Figure 9-7, his model is based on residential neighborhoods and their expansion through time into sectoral patterns. This descriptive statement about land use in cities considers the importance of transportation lines and other amenities in shaping urban patterns.

Hoyt's diagram may be understood more easily by listing his observations:

1. The highest rent (land cost) area is located in one or more sectors on one side of the city.

2. High rent often takes the form of wedges that extend out from sectors along radial lines leading outward from the center to the edge of the city.

3. Middle-range rental areas tend to be located on either side of the high-rent zones.

4. In some cities, middle-rent units tend to be found on the edge of low-rent residential districts as well as high-rent areas.

5. All cities have low-rent districts. These are usually found opposite the high-rent areas and are in more central locations.

Geographers C. Harris and E. L. Ullman (1945) suggested that both earlier land use models, which were based on concentric zones and sectors in the city, had major shortcomings.[14] Their **multiple nucleii theory** (Figure 9-7) indicated that contemporary American cities have a number of areas that serve as centers of retailing, manufacturing, residential, and other land uses. These multiple centers tend to act as nodes within the urban region, drawing similar activities inward.

An extension of the multiple nuclei concept for explaining the layout of a large city is to carry it further in order to explain the relationships of several cities in a metropolitan area. Joel Garreau suggests that outlying areas of a metropolitan region are the nodes for new completely self-sufficient cities with a complete mix of residential, employment-generating, and service-providing land uses. Examples of these **edge cities** are Columbia, Maryland; Tysons Corner, Virginia; Crystal City, Virginia; Bellevue, Washington; Walnut Creek, California; Clayton, Missouri; Scottsdale, Arizona; and King of Prussia, Pennsylvania. Garreau lists over 190 places that qualify as edge cities. Sixteen such places were counted in the Washington DC area alone and more are "on the drawing board" (in the computer).

All these descriptive land use models explain different processes and decision making in the city. Taken separately, however, each is too simplistic to explain the complexities of urban patterns in the 1990s. Urban geographers and sociologists have written countless pages in countless journals, books, and dissertations attempting to correlate these models with reality in modern North American cities.

CLASSIFICATION OF CITIES

This chapter has illustrated some of the similarities among cities in the world. All cities have distinguishable patterns, within and without, that link them and that create spatial interaction. Like most North American cities with their heavily franchised commercial districts, cities throughout the world have more similarities than differences among them.

Each city is also unique and has a unique physical setting, however. Each has a personality or sense of place that makes it special. Some cities have songs written about them or movies that have made them famous. The song "I Left My Heart in San Francisco" may be heard in bars and restaurants all around the Pacific Rim. Similarly, images of cities, such as in the song "New York, New York" or as depicted in art and films, are clear in the minds of many who have never actually been there.

Geographers usually classify cities according to their economic function—manufacturing, trade, educational, resort, or other—that is, according to their dominant economic activity. Singapore, for example, is a city that developed largely through trade and commerce (see the box on page 237). Of course, most cities have several functions and fall into overlapping categories.

Cities also may be classified by size. Such categorization allows comparison of cities by the number of functions they perform. Such functions may include operating a professional sports complex, an international airport, downtown movie theaters, muffler shops, and so on. Generally, the larger the population, the more numerous functions are supported. Comparison of urban places by size also makes possible determination of the minimum population threshold necessary to support a specific function. For example, roughly 12,000 persons are required to support a supermarket. An urban area must have over half a million persons to draw 50,000 spectators to a professional football game. At the other extreme, many video rental stores survive with a population base of only 500 people.

Places are classified in size by various governmental agencies, particularly by the U.S. Bureau of the Census or by its equivalents in other countries. In the United States, an "urban place" is a population concentration of 2500 persons or more. "Metropolitan areas" consist of a city and adjacent built-up areas, whether incorporated or not. In 1959 the "metropolitan area" was changed to *Standard Metropolitan Statistical Area* (SMSA). Some SMSAs may contain more than one large city, such as Minneapolis–St.

SINGAPORE: A CITY BUILT ON COMMERCE

Singapore, the large island in a series of small islets, is located at the southern end of the Malay Peninsula. The island is 26 miles (41.8 km) in length, 14 miles (22.5 km) in breadth, and has a coastline that extends over 120 miles (193 km). This long, flat, sandy coast has been a tremendous economic asset in the development of the city into a major center of commerce and trade. In addition, the island boasts a well-sheltered harbor that has long been recognized as an excellent port of call.

The Chinese were the first to appreciate the attributes of Singapore's harbor in A.D. 231. Even at that early date, merchants and seamen concluded that Singapore would be a fine location for trade. From that early date on, Singapore has passed through the hands of many colonizing countries, all with the idea of developing this island as a trade center. Finally, in 1965, Singapore became an independent country and city.

Singapore today is a contemporary, modern island, built upon trade networks via shipping lines and storage facilities. Singapore trades products from places as far away as Japan and the United States. In 1978 alone, Singapore participated in 11 trade fairs in the United States, Australia, Japan, and East Asia. The city also is a trading center for developing nations; among the major products handled from Third World countries, Singapore has traded rubber, petroleum, textiles, wood products, pigs, poultry, fish, fruits, orchids, and tobacco.

The excellent economic location of Singapore has led to the establishment of urban centers such as Jurong. Jurong Town Corporation was established in June of 1968 as a large industrial center. It now boasts 955 factories with a labor force of 82,000, most of whom are considered skilled labor. A modern port has been established, with bulk cargo–handling facilities. A marine base has also been built in recent years to provide access for an offshore oil-exploration industry in Southeast Asia.

The Jurong population of 106,000 is well served by a wide range of amenities such as markets, schools, clinics, hospitals, banks, post offices, community centers, and day care centers. Every neighborhood has a swimming pool and a sports center. Jurong also has a bird park, a Chinese garden, a drive-in movie theater, and a sports stadium.

Because of its location as a major trade center, the island of Singapore has evolved into a modern urban region. Have you examined a globe or world map to understand more fully why Singapore is in such an advantageous location for trade? How many cities do you know that also are countries?

Paul in Minnesota. The largest contiguous SMSAs in the United States are designated *Standard Consolidated Areas* (SCAs); these include major centers such as Chicago-Gary (Illinois and Indiana), Houston-Galveston (Texas), Seattle-Tacoma (Washington), and several others. "Megalopolis" is not a census term but often is used in reference to the contiguous built-up area from Boston to Washington ("Boswash") or other such agglomerations.

Canadian census subdivisions are similar to those in the United States, but their terminology is different. An "urban area" is the basic unit, consisting of 1000 persons or more and a population density of 1000 persons per square mile. A *Census Agglomeration* (CA) unit belongs to an urbanized core having a population of 2000 to 100,000 persons. A *Census Metropolitan Area* (CMA) is the commuting area of an urbanized core populated with 100,000 or more.

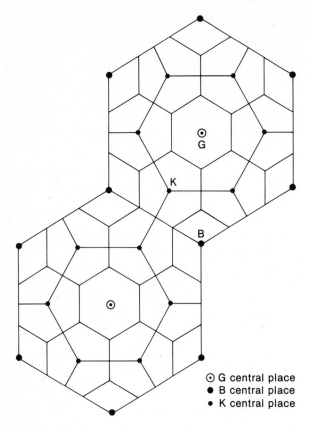

⊙ G central place
● B central place
• K central place

FIGURE 9-8
Christaller's central place theory is a logical hexagon-shaped model of urban location. Central places (G) are the largest. Smaller central places (B) offer fewer goods and services and serve only the areas of the intermediate-sized hexagons. The many still-smaller central places (K) are closely spaced and serve still smaller hinterlands. Goods offered in (K) places are also offered in (B) and (G) places, but the latter offer more and higher-order goods. Note that places of the same size are equally spaced. (From James M. Rubenstein, The Cultural Landscape: An Introduction to Human Geography, 2nd ed., Columbus, OH: Merrill Publishing Co., 1989, Fig. 11-4, p. 390. Reprinted with permission of Merrill Publishing Co.)

INTERACTIONS BETWEEN CITIES

Central Place Theory

Whatever their main economic functions, towns and cities also provide goods and services for outlying areas. These tertiary activities often dominate the economic scene in urban places. Urban centers form patterns of central places, usually focused upon a regional node. Relationships be-

tween city size and distance from other cities were first proposed by German regional economist Walter Christaller in his book *The Central Places of Southern Germany,* published in 1933.[15] Christaller's **central place theory** is concerned with spacing between service centers.

His idea is based upon a series of assumptions and concepts that do not apply to all places but that do provide an equal footing for comparisons.

Christaller assumed that the region under consideration would be an unbounded plain with soil of equal fertility everywhere, that there would be an equal distribution of resources, that there would be a uniform transportation network in all directions, and that there would be an even distribution of population and purchasing power. His basic concepts include:

- *Central place*—a place where goods are purchased or services provided
- *Range*—how far people will travel for a good or service
- *Threshold*—the market area required for a good; that is, the physical area enclosing the number of people necessary for a business to make a profit possible
- *Hinterland*—area around a central place where people use the goods and services provided by that central place
- *Order of places*—having a certain set of functions; that is, "low-order places" have only a few functions (for example, gas station, restaurant, grocery store) and "high-order places" have many functions not found in smaller places (for example, art galleries)

Christaller linked these concepts into a theory. He reasoned that while some service centers are small and offer a limited number of goods for consumers, other centers offer extensive goods and services, and that therefore an orderly hierarchy of centers emerges on the landscape. At the top of this hierarchy are large cities that act as regional metropolises. At the bottom of the hierarchy are villages and hamlets that may contain nothing more than a tavern or small grocery store. Each higher-order center contains all the functions of lower-ranked centers along with at least one other service or good.

Using this hierarchy of centers, Christaller then developed a model depicting the hypothetical spacing of urban centers. This model is a form of the gravity model discussed in Chapter 3. In his model, if lines between service centers were drawn between sets of equidistant places, they would form a hexagonal pattern (Figure 9-8). The hexagon is the most efficient shape and encloses the greatest area in the shortest distance while still completely interlocking with other similar shapes.

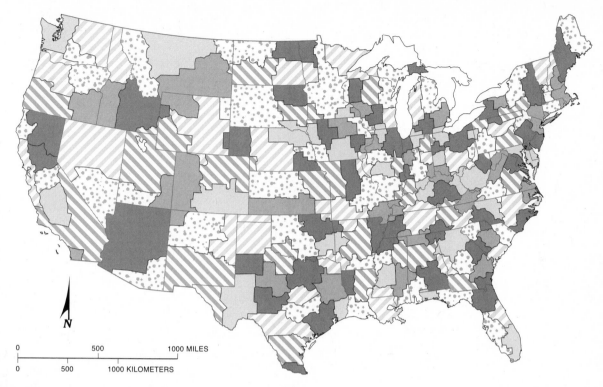

FIGURE 9-9

Daily urban systems, 1994. The metropolitan trade areas of the United States, based on commuting to the central city, on newspaper circulation, and on a number of other patronization/marketing entices, are shown on the map. (From Anthony R. de Souza and Frederick P. Stutz, The World Economy: Resources, Location, Trade, and Development, *2nd ed., New York: Macmillan, 1994, Figure 9-4, p. 365. Used with permission of Macmillan Publishing Co.)*

The central place theory is highly idealized, but it does conform to the real world in a number of places. For example, if you look closely at a map of the Midwest in the United States, or fly over it at night when the pattern of city lights is visible, you may notice that towns and cities in southern Illinois, southern Indiana, and surrounding regions approximate the model described above, because conditions here most closely approximate Christaller's assumptions. In this region, Chicago as a regional center is clearly the first-order city. (It is noteworthy that the Chicago-centered pattern was created during the nineteenth-century period of railroad expansion, and that most of the smaller places are on rail lines that converge at Chicago.)

The assumptions in the central place theory have been useful in areas where regional planning has been employed, such as in Israel (the kibbutz system is hierarchical), the Netherlands (in the Zuider Zee "polder" area—the area reclaimed from the sea), and in Sweden (for locating health care delivery systems).

The urban characteristics of a place, then, are often orderly and hierarchical, especially when based upon a ranking of functions, goods, and services. Cities are economic entities connected to each other in predictable ways.

INTERRELATIONSHIPS BETWEEN CITIES

The Urban Region

All places are necessarily connected to other places. Inhabitants of a place are not just a biological phenomenon; they are linked to the physical environment, their cultural background, and economic and political realities. Similarly, cities are not just economic phenomena; they also are environmental, cultural, and political centers with a wide range of influence on the surrounding region.

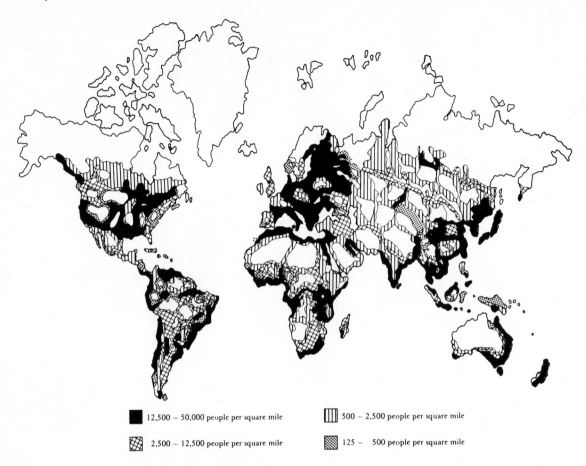

■	12,500 – 50,000 people per square mile	|||	500 – 2,500 people per square mile
▨	2,500 – 12,500 people per square mile	▦	125 – 500 people per square mile

FIGURE 9-10

Ecumenopolis, 100–150 years from now. (From Michael E. Eliot Hurst, I Came to the City, 1975, p. 316. Reprinted with permission of M. E. Eliot Hurst, Simon Fraser University, Burnaby, B.C.)

These urban **spheres of influence** exert a pull on their **hinterlands** similar to a magnet. This influence weakens as the distance from the city increases, a phenomenon known as **distance decay**. Eventually a point is reached where another city exerts more influence upon the region. The size of a nodal region depends on the number of goods and services produced in the primary city. Ultimately a hierarchy of cities may emerge in a region, with a national metropolis as most important, regional centers as second-order cities, and metropolitan centers as third-order cities.

The clearest way to observe this urban sphere of influence is by studying information flows between cities and their hinterlands. Fads, as expressed in music videos, hairstyles, or slang expressions, may be mapped for distribution and diffusion as innovations move outward from large cities to smaller towns. The volume of telephone messages may also be mapped to show spheres of influence.

A 1994 map showing "daily urban systems" in the United States is based on commuting patterns into the city, newspaper circulation, and several other variables. The map shown in Figure 9-9 dramatically illustrates the interconnections between cities and within cities.

PLANNING FOR FUTURE URBAN GROWTH

The Global City

In the late twentieth century, the most powerful urban sphere of influence to date has begun to emerge. This **global city** has become the prototype megacity with its worldwide economic, political, and cultural influence. As telecommunication linkages, trade networks, and increasingly larger urban regions begin to dominate the world's economies, we are gradually becoming a global urban soci-

FIGURE 9-11
Megalopolis on the sprawling north-eastern seaboard of the United States in the 1990s.

ty. Today's sprawling megalopolis may ultimately connect into an even larger "world city" or Ecumenopolis (Figure 9-10).[16] The photograph of downtown New York City, an integral part of the elongated urban region known as "Boswash," North America's largest megalopolis shown in Figure 9-11, depicts what may be only the beginning of future, expanded, and overlapping urban growth.

Changes in economic conditions on Earth are the primary reason for the massive growth rate of our cities. According to recent estimates, the world population of 7 billion in the year 2000 will have at least 5 billion people living in cities.[17] Sensitive environmental planning for this future Ecumenopolis will aid in our future adjustment to an urban world.

Futurist Buckminster Fuller suggested a plan that departs drastically from classifications used by previous city planners. He recommended a typical town plan that locates cities according to great circle routes, prevailing winds for transport, and isotherms of temperature for residence. Fuller's ideas center around the house of the future as an individual shelter that is easily transported to new sites as needed. He supported the concept of mass services to people in cities, including mass-produced houses.[18]

As residents of planet Earth, we are all qualified to speculate about what we think the city of the future should contain. It is not unrealistic to plan future cities with completely recycled water systems, inexpensive rapid transit systems, comfortable and airy housing, access to recreation and wilderness areas, and proximity to places of employment, school, or entertainment.

Whatever solutions we adopt as world cities evolve, some plan must be visualized that seeks a balance between human needs and services on the one hand, and environmental issues on the other. Our next chapter deals with human-environment interaction at varying levels of intensity. Understanding humans in the context of their environment is critical to our future on Earth.

KEY CONCEPTS

Accessibility	Edge city	Node	Suburb
Agglomeration	Hinterland	Peak value intersection	Threshold
Central business district	Innovation	Perception	Urban planning
Central place theory	Land use	Range	
Concentric zone model	Multiple nucleii model	Sectoral model	
Distance decay	Neighborhood	Sphere of influence	

FURTHER READING

As more of the world's population becomes urban, it is ever more important to understand the reasons for the growth of cities. The classic work on the form, function, and beginnings of cities throughout history is Lewis Mumford, *The City in History: Its Origins, Its Transformations, and Its Prospects* (New York: Harcourt, Brace & World, 1961). James E. Vance, Jr., analyzes industrialization and the growth of Western cities in *This Scene of Man: The Role and Structure of the City in the Geography of Western Civilization* (New York: Harper's College Press, 1977).

A more recent book that brings together the important work of a dozen researchers in the field of urban geography is John Fraser Hart's *Our Changing Cities* (Baltimore and London: The Johns Hopkins University Press, 1991). These well-written essays effectively capture the flavor of ongoing urban change in today's rapidly urbanizing world economy.

Two recommended survey texts in urban geography are *Urbanization: An Introduction to Urban Geography,* by Paul L. Knox (Englewood Cliffs, NJ: Prentice-Hall, 1994) and *Interpreting the City: An Urban Geography,* by Truman A. Hartshorn and John W. Alexander (Englewood Cliffs, NJ: Prentice-Hall, 1988). A systematic and structural interpretation of cities in North America is Maurice Yeates and Barry Garner, *The North American City* (New York: Harper & Row, 1980).

The form of the North American city of the 1990s is presented in *Edge Cities: Life on the New Frontier,* by journalist Joel Garreau (New York: Doubleday, 1991).

The role of suburbs in the development of urban America is dealt with in *Borderland: Origins of the American Suburb, 1820–1939,* by John R. Stilgoe (New Haven: Yale University Press, 1988); Peter O. Muller, *Contemporary Suburban America* (Englewood Cliffs, NJ: Prentice-Hall, 1981); and Muller's *The Outer City: Geographical Consequences of the Urbanization of the Suburbs* (Washington, DC: AAG, 1976).

In our opinion, one of the best North American case studies of how towns were founded, how they competed, and how they were influenced by resources and transportation access is John C. Hudson, *Plains Country Towns* (Minneapolis: University of Minnesota Press, 1985).

Beyond the economics and demography of cities are the all-important social and behavioral aspects of urban life. An excellent treatment of these subjects is David F. Ley, A *Social Geography of the City* (New York: Harper & Row, 1983). Also useful is a collection of essays edited by John A. Agnew, John Mercer, and David E. Sopher in *The City in Cultural Context* (Boston: Allen & Unwin, 1984). A journalist's interdisciplinary view of contemporary cultures is Frances Fitzgerald, *Cities on a Hill: A Journey Through Contemporary American Cultures* (New York: Simon & Schuster, 1986).

Districts and neighborhoods in the city at a human scale are investigated by several works. Some of the best include Larry S. Bourne, ed., *Internal Structure of the City: Readings on Urban Form, Growth, and Policy* (New York: Oxford University Press, 1982), and Michael E. Eliot Hurst, *I Came to the City: Essays and Comments on the Urban Scene* (Boston: Houghton Mifflin, 1975). Classic sociological views of neighborhood change through time include Herbert J. Gans, *The Urban Villagers: Group and Class in the Life of U.S. English-Americans* (London: Collier Macmillan, 1982), and Walter Irving Firey, *Land Use in Central Boston* (New York: Greenwood Press, 1968).

Land use patterns within the city have been studied for several decades by urban geographers and land use planners. The pioneer works of Burgess, Hoyt, and Harris and Ullman are summarized in many urban geography textbooks (see the work by Yeates and Garner). More recent analyses of land use patterns in the city are R. J. Johnston, *Urban Residential Patterns: An Introductory Review* (New York: Praeger, 1972), and Brian J. L. Berry, *Commercial Structure and Commercial Blight, Retail Patterns and Processes in the City of Chicago* (Chicago: University of Chicago Department of Geography, 1963).

The urban place in a regional context was introduced by Brian J. L. Berry in *Growth Centers in the American Urban System, 1960–1970* (Cambridge, MA: Ballinger, 1973). The best-known case study of the urban region is Jean Gottmann, *Megalopolis: The Urbanized Northeastern Seaboard of the United States* (New York: Twentieth Century Fund, 1961).

Field trips in cities will be enhanced by reading Grady Clay, *Close-up; How to Read the American City* (Chicago: University of Chicago Press, 1980), and his newest book, published by the American Planning Association, *Right Before Your Eyes: Penetrating the Urban Environment* (Chicago: Planner's Press, American Planning Association, 1987). The latter contains an essay that is particularly helpful, "The Street as Teacher."

Urban design ideas are presented in Grady Clay's fascinating book as well as in *Building Cities That Work* by Edmund P. Fowler (Montreal: McGill-Queens University Press, 1992).

ENDNOTES

1 For more extensive coverage of the history of cities, see L. Mumford, *The City in History: Its Origins, Its Transformations, and Its Prospects* (New York: Harcourt Brace Jovanovich, 1961).

2 V. Gordon Childe, as discussed in Mumford, *The City in History*.

3 V. Gordon Childe, "The Urban Revolution," *Town Planning Review* 21 (April 1950): 3–17; also in Robert Gutman and David Popenoe, eds., *Neighborhood, City, and Metropolis* (New York: Random House, 1970): 116–118.

4 Kingsley Davis, ed., *Cities: Their Origin, Growth, and Human Impact* (San Francisco: W. H. Freeman, 1973): 13–15.

5 Mumford, *The City in History:* 454.

6 J. R. Borchert, "American Metropolitan Evolution," *Geographical Review* 57 (1967): 301–323.

7 R. A. Murdie, *Factorial Ecology of Metropolitan Toronto, 1951–1961* (Chicago: University of Chicago Press, 1969).

8 Kevin Lynch, *The Image of the City* (Cambridge, MA: MIT and Harvard University Press, 1960).

9 Herbert J. Gans, *The Urban Villagers: Group and Class in the Life of Italian-Americans* (London: Collier Macmillan, 1982.)

10 Walter Irving Firey, "Sentiment and Symbolism as Ecological Variables," *American Sociological Review* 10 (1945): 140–148.

11 Brian J. L. Berry, *Commercial Structure and Commercial Blight: Retail Patterns and Processes in the City of Chicago* (Chicago: University of Chicago, Department of Geography, 1963).

12 R. E. Park, E. W. Burgess, and R. D. McKenzie, *The City* (Chicago: University of Chicago Press, 1925).

13 Homer Hoyt, *The Structure and Growth of Residential Neighborhoods in American Cities* (Washington, DC: Federal Housing Administration, 1939).

14 Chauncy Harris and E. L. Ullman, "The Nature of Cities," *Annals of the American Academy of Political Science* 242 (1945): 7–17.

15 Walter Christaller, C. W. Baskin, trans., *Central Places in Southern Germany* (Englewood Cliffs, NJ: Prentice-Hall, 1966).

16 M. Eliot Hurst, *I Came to the City: Essays and Comments on the Urban Scene* (New York: Houghton Mifflin, 1975): 315.

17 M. Eliot Hurst, *I Came to the City:* 312.

18 Buckminster Fuller, *Utopia or Oblivion? The Prospects for the Future* (New York: Bantam Books, 1969): chapters 7 and 8.

10 THE THEME OF HUMAN-ENVIRONMENT INTERACTION

Clearing the tropical rain forest in Ecuadorian Oriente. (Photo courtesy of Stephen F. Cunha.)

Geography was defined in an earlier chapter as the *study of Earth as the home of humankind*. Sadly, geography might not be a permanent science: it will come to an end if Earth ceases to be a home for humankind. This chapter is about how we are in serious danger of losing our global habitat through mismanagement. Abuse of our Earth is no less important a problem than the survival of the human race itself.

Three of the *National Geography Standards* frame the content of this chapter. They are:

Standard 8: *The geographically informed person knows and understands the characteristics and spatial distribution of ecosystems on Earth's surface.*

Standard 14: *The geographically informed person knows*

and understands how human actions modify the physical environment.

Standard 15: *The geographically informed person knows and understands how physical systems affect human systems.*

There are three steps to solving the problem of environmental degradation on Earth. First, we must understand it. Second, we must understand our alternatives for correcting it, and then we must plan detailed solutions with agreement among all participants. The final step is to implement the plan.

Your authors have no quick or magic solutions to Earth's problem of survival under increasing environmental stress. However, we will contribute to the first necessary step—understanding the nature of the problem. As we do so, please keep in mind that formal study of nature-society relationships is recent, even in modern scientific times.

EARLY VIEWS ON HUMAN-ENVIRONMENT INTERACTION

It is important to remember that humankind has been here only 2 million years out of Earth's 4500 million-year history, or a vanishingly small four one-hundredths of a percent of the planet's existence. We are truly late-comers to the stage. Of that 2 million years, only the last 2000 or so show evidence that people have made a significant change in the natural environment.

At the risk of oversimplifying several centuries of Western Civilization, we suggest that, prior to the mid-1800s, human interaction with the environment consisted of adaptation to environmental limits for settlement. Some have suggested that the Western religious teachings of human dominance over nature in favor of a more rewarding hereafter have led to a difficult dualism. The view that nature must be conquered seems to have been part of Western thinking since the time of the Old Testament.[1] This dualism also accounts for a common mechanistic view of the world, even prior to the Renaissance and the Scientific Revolution.

In contrast, pre-nineteenth century views in non-Western cultures, including Native American societies, saw humankind as part of nature. With such a view, it was much more difficult for individuals or groups to wreak havoc on environmental systems, even if the technology to do so had been available (see the box on page 246). Significantly, this view is not merely historical, as witnessed by resource development issues in many underdeveloped nations today.

The Nineteenth Century Brings Some Warnings

Among the first cries for a sensitivity to human impact upon natural systems were those of George Perkins Marsh (1801–1882). As a lawyer, scholar, and diplomat, his lectures were carefully reasoned and accurate, but his was a small audience in an America thought to have inexhaustible resources and untested resilience. Little attention was given to Marsh's documentation of the relationships between the rise and fall of past civilizations and their use and misuse of resources. His warnings in *Man and Nature, or Physical Geography as Modified by Human Action* (1864)

were left to be rediscovered in the latter half of the twentieth century.[2]

Marsh's warnings were echoed with conviction by Major John Wesley Powell (1834–1902). As a geologist, ethnographer, and director of the U.S. Geological Survey, Powell was in a position to recommend policies toward disposition of the public domain in the far West.[3] He spoke against the waste of resources, especially the clear-cutting of forests, overgrazing, and misuse of irrigation. Unfortunately, Powell's warnings and recommendations were unheeded by decision makers in Congress.

Another John, and another prophet of environmental crises to come, was John Muir (1838–1914). Muir's influence on President Theodore Roosevelt and the conservation movement of the early twentieth century should not be underestimated. Creation of resource-management agencies such as the U.S. Department of Agriculture's Forest Service and the building of the National Parks System were new expressions of a conservation ethic that would struggle in competition with many other political and economic values through the rest of the century. Muir is well remembered for founding the Sierra Club and for influencing the creation of Yosemite National Park (1890), but his most important contribution to a common understanding of Earth as the home of humankind was pointing out the value of wilderness as a resource in itself, rather than as a storehouse of other resources to be extracted. Like Thoreau before him, Muir suggested a compatible relationship with Earth that involved neither influences of the environment over humans nor of humans over the environment.

A geographer also remembered for describing relationships between humans and the environment is Ellen Churchill Semple (1863–1932). Along with others at the turn of the century, such as Albert Perry Brigham (1855–1932), Semple stressed her belief that environmental conditions exerted enough influence on human activity that many patterns of migration, settlement, and economic activities were preset or "determined." This view, usually reinforced by selecting examples of societies that had adapted to environmental extremes, held that physical geographic conditions were a major force in shaping civilizations.

As the role of technology came to be seen as important in understanding societal changes, the concept of **possibilism** was endorsed. Possibilists followed the logic that environment laid out the possible alternatives for use of and adaptation to local conditions, but other factors such as culture and use of technology often were at least as important.

The environmental influence upon human activity was a popular topic in classrooms through most of the twenti-

ENVIRONMENTAL PERCEPTION AMONG NATIVE AMERICANS

North American Native American groups generally have an extremely close relationship with the environment. Take, for example, the Konkow Maidu who reside in the Sierra Nevada foothills of north central California. Although the Konkow are acculturated for the most part, they still gather their traditional food sources and worship nature in much the same way as their ancestors.

The Konkow feel that food was given to them by Worldmaker, whom they call Wonomi. In the beginning of time, Wonomi made the Earth and all its inhabitants. For humans he provided an endless supply of food sources, including fish and acorns, the traditional dietary staples. These foods are still gathered by the Konkow. As food was given to the people by Wonomi, it is considered sacred, and therefore the act of gathering is a religious observation.

Before gathering traditional foods, one must be pure in heart and leave an offering of acorn meal, scraps of material, or pennies as a sign of thanks to the Worldmaker for providing for his people. When gathering food sources, a respect toward the gathering area is always observed. Therefore, the Konkow gather only what actually will be utilized, and the area itself is never decimated with trash or any other signs of human intervention. The gathering places are considered sacred, as well as the act of gathering itself. It is important that these areas be left in the natural state. However, the traditional gathering areas are becoming scarce through land development, and thus the Konkow feel it is important to protect the remaining areas from further encroachment.

The Konkow could be considered nature worshipers. Their religion is based on an interrelationship between themselves and their homeland. They are *animistic,* which means they believe the world is filled with spirits, so there are places in nature that they consider to be sacred. These places include prominent landform features, mountain tops, groves of trees, waterfalls, and mountain lakes. Many of these places are mentioned in their mythology, which is considered to be truth and not mere stories. Other places are considered sacred because they are the homes of the ku kini, or guardian spirits who help the Konkow in life. The ku kini are the manifestation of power, which is the "cosmic breath of the world"; it is what makes the world alive, and is considered part force and part object.

These sacred places are scattered throughout their territory. Some places are considered important for showing respect to the gods. They are used in the vision quests or ritual acts, such as the spring and fall feasts held there. Other places are discussed in mythology. For example, there is a place in the Konkow territory where Bat, one of the first people, was said to live. Another mythological area is a lake which is considered poisonous due to the presence of a large snake that is believed to reside there. It is a place to stay away from in the Konkow perception.

The perception of these traditional sacred sites influences the psychological makeup and lives of the Foothill Konkow individuals. It is an aspect of their environmental perception to note these sacred areas and to incorporate these values in their daily lives. The Konkow feel an adherence to their homeland, stabilized by this perception of sacred sites. This sense of sacredness in regard to the landscape leads to a powerful sense of communion between the society and the environment. To the Konkow Maidu nature is not just natural, it is given religious value because it is created from the hands of Wonomi.

eth century, even in the face of increasing evidence that human influences on the environment posed far more critical issues and problems.

Carl Sauer provoked his students and the readers of his many books and articles into considering the role of culture as the foremost influence on human activity. His works demonstrated that early humans had more effect on the environment than previously had been acknowledged, and that diffusion of cultural traits and technology had as much to do with environmental adaptation as they had to do with home-grown solutions to environmental limitations. Sauer was a major participant in a symposium of scholars entitled *Man's Role in Changing the Face of the Earth,* assembled by William L. Thomas in 1954. The papers presented were published in book form with the same title.[4] It was the first comprehensive look in the twentieth century at how human activity has changed almost all Earth systems in almost all parts of the world.

The Conservation Movement

From a philosophical and ethical point of view, the efforts of Aldo Leopold (1886–1948), along with those of John Muir, have probably had one of the greatest mass effects on popular understanding of human-environment relationships. Leopold advocated the role of humans as partners and protectors of nature, rather than as her conqueror. He founded the Wilderness Society and was a director of the Audubon Society in the 1930s. Leopold proposed that, just as societies accepted common standards for social behavior, we should have a common and sincere land use ethic.

Although nurtured by the sensitive writings of Leopold in *A Sand County Almanac* (1949), environmental awareness would not appear in full bloom until the 1960s and 1970s, commencing with Rachel Carson's book *Silent Spring,* which described the effects of pesticides on wildlife and natural systems.[5] Another book of the early 1960s having great impact on the American populace was former Secretary of the Interior Stewart Udall's *The Quiet Crisis,* which echoed Marsh's earlier warnings about resource misuse.[6]

Passage of new environmental laws in the 1960s and 1970s such as the Wilderness Act, the Endangered Species Act, the Clean Air Act, the National Environmental Policy Act, and many others, reflected changed public attitudes about expectations for the quality of life in the physical environment. The Environmental Protection Agency and the Council for Environmental Quality were created, and many state environmental protection laws were passed. The decade of the 1970s and the first half of the 1980s saw large increases in memberships in conservation and activist environmental organizations, increased lawsuits on behalf of the public for protection of resources, and significant growth in environmentally related programs in colleges and universities, including many geography programs. As a result of these laws and the activism of people and organizations, the 1990s has seen significant reductions in air and water pollution nationally, wiser environmental planning for new projects, and increased public support for protection and wise use of land, water, energy, mineral, and other resources. Most school curricula now include major units on environmental education.

Modern Environmentalism

Environmentalism now appears in public forums, courtrooms, places of employment, in church activities, and on political agendas. It has entered the political sphere with the influence of the Sierra Club, Greenpeace, the Wilderness Society, the Green Party, Earth First!, and other groups. This popular attention to environmental concerns is also reflected in changed lifestyles and personal values. Just as households and communities have oriented their lives toward economic and social goals in the past, an increasing number are dedicating their lives to minimizing environmental disruption and to promoting a sustained Earth society. The name sometimes given this approach to life is **bioregionalism.**

The bioregionalist believes that, regardless of how our economic and social agendas are framed, the ultimate decision has to do with reversing the global ecological deterioration that we are now witnessing. Bioregionalists have directed their attention to the development of a culture that is rooted in the natural world, rather than on technology dependence and human mastery over natural systems. Bioregionalists also oppose the idea of nation-states and political boundaries as the means by which people identify themselves. Their logic is that biotic or geomorphic regions are more logical and efficient ways to organize groups, rather than the dictates of ideology.

Bioregionalists believe that the predominant attitudes in most modern societies toward exploitation of resources and disruption of environmental systems are the result of our religious, ethical, and day-to-day working separation from the natural world, compared with traditional cultures such as the Hopi or Navajo. Figure 10-1 shows a bioregional view of North America. It may be compared with other regional maps in Chapter 12. In comparing the maps, note similarities in the locations of the regions. Where is the overlap? What common factors would account for these similarities?

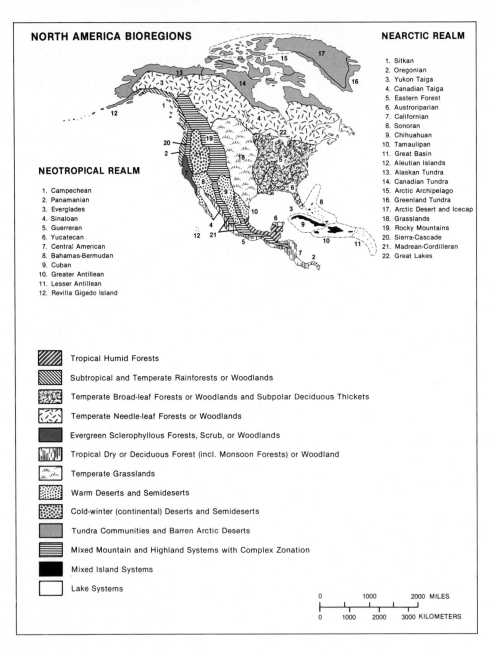

NORTH AMERICA BIOREGIONS

NEARCTIC REALM

1. Sitkan
2. Oregonian
3. Yukon Taiga
4. Canadian Taiga
5. Eastern Forest
6. Austroriparian
7. Californian
8. Sonoran
9. Chihuahuan
10. Tamaulipan
11. Great Basin
12. Aleutian Islands
13. Alaskan Tundra
14. Canadian Tundra
15. Arctic Archipelago
16. Greenland Tundra
17. Arctic Desert and Icecap
18. Grasslands
19. Rocky Mountains
20. Sierra-Cascade
21. Madrean-Cordilleran
22. Great Lakes

NEOTROPICAL REALM

1. Campechean
2. Panamanian
3. Everglades
4. Sinaloan
5. Guerreran
6. Yucatecan
7. Central American
8. Bahamas-Bermudan
9. Cuban
10. Greater Antillean
11. Lesser Antillean
12. Revilla Gigedo Island

Tropical Humid Forests

Subtropical and Temperate Rainforests or Woodlands

Temperate Broad-leaf Forests or Woodlands and Subpolar Deciduous Thickets

Temperate Needle-leaf Forests or Woodlands

Evergreen Sclerophyllous Forests, Scrub, or Woodlands

Tropical Dry or Deciduous Forest (incl. Monsoon Forests) or Woodland

Temperate Grasslands

Warm Deserts and Semideserts

Cold-winter (continental) Deserts and Semideserts

Tundra Communities and Barren Arctic Deserts

Mixed Mountain and Highland Systems with Complex Zonation

Mixed Island Systems

Lake Systems

FIGURE 10-1
North American biogeographical realms. (From Miklos D. F. Udvardy, Whole Earth Review, *1978.
Reprinted with permission of Whole Earth Review, 27 Gates Road, Sausalito, CA 94965.)*

An even more fundamental human-Earth relationship is practiced today as *deep ecology,* the mystical-religious-feminist wing of environmentalism. Followers of this philosophy believe that the root cause of our environmental ills is our belief that human beings stand apart from nature and above it, whether as master or steward. A question on the subject Theodore Roszak asks us to ponder is this: When human beings unilaterally declare their superiority to all other species, who do they think is paying attention?[7]

Progress in environmental consciousness continues to happen at unequal rates throughout the world. While significant strides in conservation practices have been made in North America and Western Europe, only modest progress thus far has been made in much of Eastern Europe and Asia, and improvements have been painfully slow in much of Latin America and Africa.

Present issues and views of human-environment interaction are mixed, depending upon the location and economic circumstances of various components of Earth's population. The next few sections of this chapter present a sampling of events and actions that illustrate a variety of relationships between humans and environmental systems.

CURRENT VIEWS ON HUMAN-ENVIRONMENT INTERACTION

The first Earth Day on April 22, 1970, was not only a news event, but a beginning of environmental awareness by the world's citizens at large. Nearly 20 million people participated in this celebration. With the launching of manned spacecraft, Earth was seen as a unified whole. In 1969, photos of Earth from space had an impact comparable to Columbus's discovery of the New World in the fifteenth century. The term *spaceship Earth* took on a more personal meaning as astronauts described our home from thousands of miles away. In the United Nations, environmental issues were discussed more frequently and research increased on global environmental problems.

Gaia

One expression of this new global awareness is seen in the **Gaia hypothesis,** first proposed by James Lovelock in 1974. Gaia was the Greek goddess of Earth. She is also known in myth as *Anima Mundi,* the great mothering soul of the world. The Gaia hypothesis "suggests that the entire range of living matter on Earth, from whales to viruses, can be regarded as a single entity, capable of manipulating its environment to suit its needs."[8] In simple terms, the entire Earth acts like a living organism. For example, it has been demonstrated that much of the atmosphere is maintained and regulated by the biosphere, and that the biosphere in turn is influenced by the makeup and forces of the lithosphere and hydrosphere.

This holistic view of Earth as a living organism capable of regulating itself was debated by scientists at the 1988 American Geophysical Union and the hypothesis was found to have no scientific validity. While not "provable" to scientists of the twentieth century as its ancient Greek name implies, it was and is a view held by many non-Western and ancient cultures. Today it serves as a useful

metaphor for use in education and public awareness of environmental issues.

The effects of increasing environmental awareness, at least in the developed world, have led to the hope that a sustainable Earth society may someday be realized. The term *sustainable Earth* refers to the concept that a system (as systems are described in Chapters 5 and 6) needs its inputs and outputs of matter and energy to be in balance if its overall equilibrium is not to be upset or destroyed. There is much to learn and do before we can begin to feel confident that our descendants will inhabit a sustainable, balanced Earth system.

"State of the Earth" Report

Like "state of the Union" reports given by our president, some ecologists, geographers, and journalists have presented "state of the Earth" reports. These usually contain good news and bad news. A collective summary of several such reports, most notably Worldwatch Institute's annual report, *The State of the World,* is offered here as a checklist of human-environment issues worthy of our attention and action:[9]

- Earth's population has doubled from 2.5 billion to 5.3 billion people between 1950 and 1990. For the era of 1990 to 2030, it is projected to increase by 90 million per year. During this same time, about 100,000 people starve to death every day.

- Grain production per person has been in decline since 1984, a reversal of nearly four decades of rapidly rising cropland productivity. Environmental degradation such as topsoil erosion continues to weaken the natural systems that support agriculture. About one-third of the world's cropland is losing topsoil to erosion.

- The doubling of population since 1950 has been accompanied by a fourfold increase in the burning of fossil fuels. A secure food supply depends upon stabilizing the climate and lessening crop damage from air pollution, which will require the phasing out of fossil fuels.

- Burning of fossil fuels has increased the amount of carbon dioxide in the atmosphere by 30 percent, which is threatening the stability of Earth's climate.

- Burning of fossil fuels has also contributed acids into the atmosphere that precipitate and fall to Earth downwind, acidifying water bodies and killing forests. Increasingly, strict pollution laws require most new power plants in the 1990s to use equipment that removes the sulfer dioxide emissions that cause acid rain.

- The ozone layer in the upper atmosphere, which shields the surface of Earth against ultraviolet radiation from the sun, is being destroyed through the use of substances in refrigeration fluids and plastic foam products.

- Due to climatic change, accompanied by overgrazing, much of the world's grassland is turning into desert. This limits potential food production from those grasslands.

- Worldwide clearing of forests also is changing climatic patterns by releasing heat. Trapped carbon dioxide is released into the atmosphere causing greater variation in local temperatures. Forest clearing is also causing massive soil erosion and the disappearance of thousands of species of plants and animals.

Most of these changes in our global environment involve human activities that have altered atmospheric or other Earth systems to our long-term detriment for the sake of short-term gain. In the examples of human-environment interaction that follow, however, we can also find cases of benign change or even rather positive changes toward restoring natural environments. Still, the detrimental upsets both outweigh and outnumber the restorations.

CASE STUDIES ON HUMAN-ENVIRONMENT INTERACTION

Agriculture

Possibly no single human discovery or innovation has had as much effect in changing the face of Earth as agriculture. Its beginnings in prehistory made possible a division of labor in early societies so that specialization in crafts, arts, building, and government were possible. With the ability of fewer farmers to support a greater number of soldiers and urban dwellers, the beginnings of centralized authority and political states appeared.

Early dependence on agriculture and the expansion of agricultural systems resulted in massive changes to the landscape. Native vegetation was replaced with domesticated plants, soils were chemically and physically changed through plowing and plant selection, the natural hydrology of places was modified for irrigation, and the secondary effects of settlements such as towns and roads appeared.

Where agricultural surpluses in some societies existed throughout recorded history, developments in science and technology stimulated economic and population growth. The Industrial Revolution, urbanization, and subsequent social-political change would not have been possible without a firm agricultural base. Today, the interrelationships between agricultural systems and Earth are no less important. Obviously, nations with food surpluses possess power, while nations lacking food surpluses do not. Beyond the simple fact of unequal distribution of food resources in the world, agriculture is the key element in the successes and failures of addressing many problem issues in the world today.

In the 1990s, the total world food production is barely enough to feed everyone. Yet each year, more than 12 million people die from starvation, malnutrition, and related diseases. The nature of the problem and potential solutions lie in the nature of foods and the other materials produced, their environmental effects, the distribution of products, and the need for supporting or stabilizing future population levels. An estimated 80,000 species of edible plants are known to exist in the world, yet most of Earth's population depends upon about 30 types of crops. Of these, four make up the vast majority of all diets—wheat, rice, maize (corn), and potatoes.

Through a program that began in the 1970s called the Green Revolution, more productive strains of rice and wheat have been developed, and researchers are now working to improve other crops through genetic engineering. Thus far the effects of these programs have been somewhat successful, but many societies in the less-developed countries have not yet received the newer plant materials. Also, many groups—especially in Africa—subsist on other crops, such as sorghum, millet, cassava, yams, and cowpeas, which have yet to be improved.

Another limitation to solving these programs is the dependence upon energy-consuming farming techniques, including irrigation, fertilizers, and mechanical tilling. If expensive fertilizers, water, and pesticides are not introduced, yields of the new crop varieties decline due to soil depletion; this depletion is caused by the discontinuance of crop rotation and natural soil enrichment from legumes (peas, beans, and clover) when fertilizers are used.

Today, more land is under cultivation than ever before, but most of these new lands are marginal in quality and are profitable only if the prices for their products remain high. Environmental costs for farming these lands often include soil erosion, depletion of groundwater supplies, destruction of natural ecosystems, and disruption of traditional nonagricultural societies. At the same time, total worldwide agricultural land is increasing, but the world's population is increasing at a faster rate, causing the amount of cultivated land per person to decrease every year. By the year 2020, with a global population increasing from 5 bil-

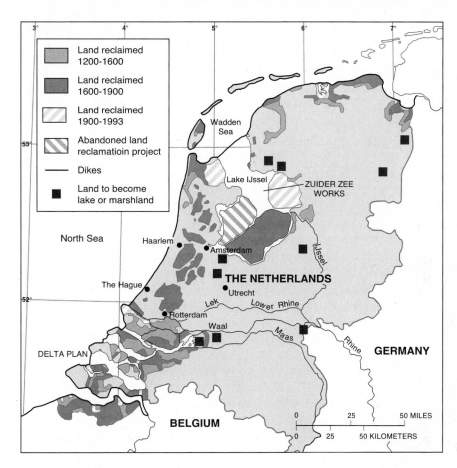

FIGURE 10-2
Reclamation projects in the Netherlands. (From James M. Rubenstein, The Cultural Landscape: An Introduction to Human Geography, *4th ed., New York: Macmillan, 1994, Figure 1-11, p. 27. Reprinted with permission of Macmillan Publishing Co.)*

ion to 7.8 billion, food production will need to be increased by 56 percent over present levels of production!

Most world leaders recognize that food production is tied to many other economic variables and political motivations, but long-term solutions will depend on minimizing damage to the resources and environmental processes that sustain agriculture. This approach to sustainable agriculture on Earth lends itself to more traditional means of farming such as crop rotation, sowing of complementary crops, use of organic fertilizers, water conservation in irrigation, and diversification of preferred diets. These techniques, along with the potential for continued success in biotechnology (including genetic engineering) and improved transport and storage, hold promise for feeding future generations until global population levels stabilize.

Creating New Land in the Netherlands

The people of the Netherlands, numbering about 10 million, live on a land area of about 12,500 square miles (32,375 km²). With a population density of 850 per square mile, the Dutch enjoy a very high standard of living based on agriculture, trade, and some industries. About 60 percent of the land and the population of the Netherlands are below sea level. This is partly true because the land slopes toward a central depression located at Lake IJssel, and additionally true because former sea bottom has been made useful by creating "polders," which are surrounded and protected by dikes. Figure 10-2 locates the Netherlands reclamation projects that have been accomplished to date.

Historical evidence dating back to Roman times enabled researchers to estimate that Dutch land has been sinking at a slow but steady rate of about 0.4 inch (1 cm) per year due to geological downwarping (folding) and from the weight of new sediments brought in from upstream river erosion. By A.D. 1000, defensive structures had been built and broken several times by large storms. By the seventeenth century, windmills were used on a large scale to drain lowlands, islands were built up, and the dikes were reinforced and connected to create a series of interlocking polders.

At the beginning of the twentieth century, draining and diking had reclaimed about 2000 square miles (5200 km²) of land, but geomorphic sinking had lost about the same amount to the North Sea. At great cost, engineers were staying about even in their battle with the sea. From 1918 to 1932 an enclosing dike was built to cut off 1500 square miles (3900 km²) of turbulent, stormy inland sea and to create 850 square miles (2200 km²) of new land from it. The master dike across the Zuider Zee meant that polder dikes could be lower yet safer. In addition, the creation of freshwater Lake IJssel provided a long-needed dependable source of water for irrigation, urban use, and groundwater recharge.

In 1953, the seaward dikes gave way in an intense storm that flooded much of Zeeland (the southwestern corner area labeled "Delta Plan") and drowned more than 1800 people. Damage to property was beyond calculation. It was decided that a repetition of the disaster could best be prevented by enclosing the estuaries of the Rhine River at their mouths, thereby shortening the coastline. Engineering studies considered a bewildering number of variables, including wind speed and direction, tidal range, coastal morphology, and wave characteristics. The Rhine is now channeled through a system of sluices that flush out salt and polluted freshwater while the dam gates remain closed most of the year to keep out the seawater. During times of exceptionally high water in the Rhine, the sluice gates are opened to prevent the river from flooding.

Rather than close the Eastern Scheldt estuary, as had been done with the other three estuaries, it was decided at the urging of conservationists to keep it open, but closable, in times of danger to preserve a very valuable marsh and wildlife area. This is accomplished by a giant storm surge barrier that allows tidal flow into and out of the estuary, except when there is a threat of floods or pollution, such as an oil spill. Benefits of the Zeeland project have included, in addition to flood control, the infusion of capital into the local economy and the shortening of transportation routes due to highway construction on top of the new dikes and dams.

A remaining reclamation project currently debated among economists, conservationists, and government officials is a planned 125,000-acre polder now under Lake IJssel northeast of Amsterdam. This Markerwaard Project would create new agricultural land, which some say is not needed, at the expense of recreational waters and waterfowl habitats. Economic arguments are also being weighed. A planning document published in 1988 shows the Markerwaard as Marker's Lake.

Modification of the Atmosphere: Acid Rain

One of the most serious global environmental problems, especially in the northern hemisphere, is the deposition of acid into forests and lakes. This form of air pollution, often termed *acid rain,* is the result of emissions of sulfur dioxide and nitrogen oxide from the burning of fossil fuels in power plants, industries, and automobiles. When suspended in the atmosphere, the emissions are chemically changed into sulfuric and nitric acids that fall to Earth downwind from the source, sometimes hundreds of miles from their place of origin. The use of tall smokestacks in North America and Western Europe has improved local air quality but has extended the range of downwind pollutant deposition. It has also allowed formation of secondary pollutants such as solid particles of sulfate and nitrate salts. These substances then damage plants directly or indirectly by changing the pH value (acidity) of soils or water.

To better understand acid rain, let us quickly review the *pH scale.* The pH scale is logarithmic and ranges from 0 to 14. The center of the scale, 7, is neutral; progressing from 7 down to 0 is increasingly acidic; and progressing from 7 up to 14 is increasingly more alkaline or basic.

On the pH scale of 0 to 14, normal rain is approximately 5.6 to 6.6, or slightly acidic. Any precipitation or surface runoff of 5.0 or less is considered harmful because the acid can accumulate and easily upset natural chemical balances. Values of 3.0 and 4.0, roughly the pH of vinegar, are now routinely recorded in the New England states.

New England, eastern Canada, and Scandinavia are particularly vulnerable to acid rain effects due to their thin rocky topsoil and shallow streambeds. Fish kills and other effects have been reported since the 1950s, with conditions worsening each year since. Crop damage and other impacts now are reported in the mid-Atlantic states, Colorado, California, and the Great Lakes area. Zones of damage also are increasing in Western and Central Europe.

Major sources of acid rain in North America are the six states of New York, Pennsylvania, Ohio, Michigan, Indiana, and Illinois; they account for about 45 percent of sulfur dioxides emitted into the atmosphere in the United States. Overall, the northeastern sector of the United States accounts for about half of the total emissions. Figure 10-3 locates the major areas of acidic emissions and their downwind areas of deposition and damage.

The geographic nature of the problem is exemplified by the estimate that three to four times as much sulfur moves across the border from the United States to Canada as moves in the opposite direction (on an average annual basis). Likewise, the United Kingdom and West Germany export acid through the atmosphere to Sweden.

The typical sequence of effects as acid increases in a water body starts with increases of acidophilic (acid-loving

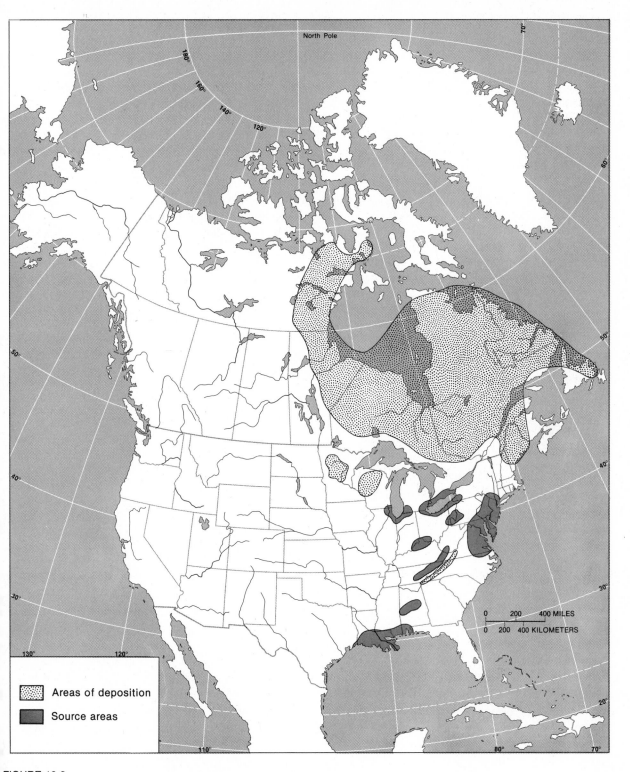

FIGURE 10-3

Acid rain sources in North America and areas of downwind deposition.

mosses, fungi, and algae, which choke out the native aquatic vegetation. Fish die from direct injury or the inability to hatch eggs. Sphagnum moss chokes the water while starving the remaining fauna. The algal and fungal growths form a tight mat on the lake bottom as anaerobic bacteria begin to produce carbon dioxide, methane, and hydrogen sulfide. The lake is dead.

Natural stone, concrete, and mortar also are subject to damage from acid in the environment. Sulfur dioxide attacks are producing sulfate crusts on ancient buildings such as the Acropolis in Athens. The continuing chemical reaction beneath these crusts has resulted in decomposition of the stone and obliteration of sculptural details.

Due to the global nature of the problem, it is obvious that potential solutions must be international in scope. While the United States and Canada have local air pollution standards and controls, they are neither effective nor appropriate in coping with the acid rain problem. There are no international agreements for controls in Europe. As with other atmospheric upsets such as increased carbon dioxide, ozone, and heat, new policies and firm international agreements among affected nations must be the first steps toward stabilization.

Adding lime to acidified lakes has been attempted as a way of rejuvenating some aquatic environments. Thus far it has proven costly but effective on a limited scale. No other means of rejuvenating damaged waters or terrestrial ecosystems have been found, although some research is underway.

The nature of this and similar environmental problems illustrate how our legal systems are not evolving fast enough to keep pace with our technological ability to cause environmental problems. Laws usually are passed to react to problems, not to anticipate them.

Atmospheric Warming and the Greenhouse Effect

We saw in Chapters 5 and 6 that the dynamics of the atmospheric system of Earth depends upon a balance of energy inputs and outputs. Carbon dioxide (CO_2) accounts for only 0.03 percent of the total gases in the atmosphere. However, carbon dioxide, along with other gases such as ozone, methane, and nitrous oxide, plays an important role in temperature regulation. CO_2 admits virtually all incoming shortwave solar energy, but traps and retains much of the reflected longwave infrared energy (heat) that is normally radiated into space. CO_2 thus behaves much like glass in a greenhouse. The resulting heat buildup, known as the **greenhouse effect,** then has the potential of affecting all the global climatic systems, and secondarily, all plant and animal life.

The concentration of carbon dioxide in Earth's atmosphere has been increasing since about 1850 when coal began to fuel the Industrial Revolution. There is a complex debate in the 1990s among scientists about whether or not this relatively short term (in geologic time) situation is part of an ongoing natural Earth process. However, many climatic models indicate that, if fossil-fuel burning continues at today's rates (or greater), and the burning of tropical forests continues at the same rate, carbon dioxide levels will increase to 600 parts per million (ppm) by the middle of the twenty-first century. This increase—from 275 ppm in 1860, to 350 ppm today, to 600 ppm around 2050—may raise the average global temperature as much as 7°F (4°C), with temperatures near the poles rising two or three times this amount. If current emission rates continue, scientists predict that each succeeding decade will see a global warming of 0.5°F and a sea level rise of 2 inches (5 cm) due to polar ice cap melting.

The impacts of such a climatic change will be the redistribution of worldwide rainfall patterns, with some areas having increasing and others decreasing rainfall amounts; the melting of the polar ice caps, resulting in worldwide flooding of low-lying coasts; and the complete relocation of all agricultural systems, resulting in as yet unpredictable political and economic consequences. Figure 10-4 shows areas below 300 feet (less than 100 m) in elevation that could be affected by rising sea levels.

Further impacts will include the semiarid regions of Africa becoming hotter and drier; the humid, tropical areas of Asia becoming increasingly at risk for floods; and the wetlands along the coasts of all continents becoming radically decreased. Wetlands are critical to both marine and terrestrial wildlife, so this change will cause enormous disturbances in fauna.

Possible responses on our part to this global warming include (a) adjusting to its effects or (b) slowing it by discontinuing the burning of fossil fuels and forests. However, to be effective, one-third of all fossil fuel burning per year would have to be discontinued; and stopping the burning of forests would slow development in Third World nations. Thus, either the discontinuance of fossil fuel and forest burning or the long-range planning for climatic change will be expensive and politically difficult. However, these solutions are probably no more expensive and politically difficult than the relocation of people, the loss of flooded urban and agricultural land, and the expense of agricultural adjustments. Reforestation is another long-term solution that could be practiced at almost any scale, from farm plots to plantations.

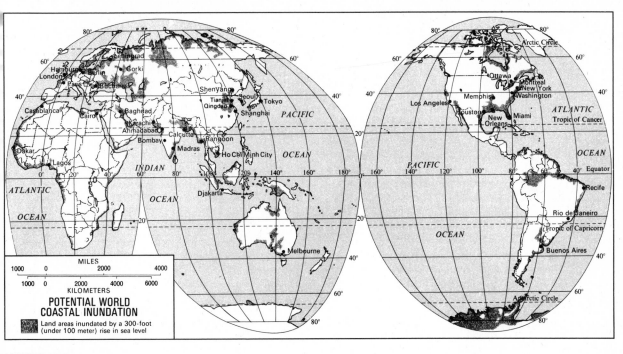

FIGURE 10-4
Areas subject to flooding if sea levels rise. (From Harm J. de Blij, The Earth: A Physical and
Human Geography, *New York: John Wiley & Sons, 1987, p. 125. Reprinted with permission.)*

Desertification

Lands with low or variable rainfall are particularly sensitive
to abuse. When abuse reduces natural vegetation and when
farm or rangeland productivity is lost, such areas become
desertlike. This process of land degradation, most often
called **desertification,** has played a part in the decline of societies throughout history. Areas in Iraq, Pakistan, North
Africa, and Syria were much more productive centuries ago
than today.

In recent years, the process of desertification has increased at an even greater rate. About 2 billion acres (810
million hectares) have become desertified in the last 100
years. (This is equivalent to about the area of the continental United States.) The process can be observed in the
southwestern United States, Central and South America,
Australia, Central Asia, and especially in Africa. About 40
million acres (16.2 million hectares) are added yearly to the
world inventory of desertified lands. Figure 10-5 locates areas in the world with the most serious desertification problems.

These dry or semiarid lands are not natural deserts.
They may have as much as 30 inches (76 cm) of rainfall in
a given year, as distinct from deserts, which are usually defined as averaging less that 10 inches (25 cm) per year. The

rainfall on semiarid lands, however, is usually seasonal, and
varies from year to year. These dry lands are the home of
one-sixth of the world's population. While the production
per unit area may be low in these areas, their total production of meat, cereals, fibers, and hides is very high. Some relationships among rainfall, natural vegetation, population,
and products produced may be seen by comparing maps in
an atlas.

Researcher Alan Grainger identifies four major causes
of desertification:[10] overcultivation, overgrazing, deforestation, and irrigation practices. He further identifies three
contributing factors: population changes, climatic changes,
and changing social and economic conditions.

In traditional societies, adjustments to environmental
changes were made continually. The agricultural system
was really a self-regulating socio-economic system. Non-
irrigated, rainfall-watered crops occurred where rain was
most likely to fall in sufficient quantity. Livestock grazing
occurred in the dryer zones, and seasonal nomadic herding
was practiced where pasture was most variable. Drought-resistant crops were grown. Pastoralists and crop farmers co-
operated by using barter.

A change to cash crop cultivation and increased pressure from greater populations attempting to use the same
area has set the stage for the upset of the balance among wa-

FIGURE 10-5
Areas undergoing desertification in the world. (From James M. Rubenstein, The Cultural Landscape: An Introduction to Human Geography, *2nd ed., Columbus, OH: Merrill Publishing Co., 1989, Fig. 13-8, pp. 468–469. Reprinted with permission of Merrill Publishing Co.)*

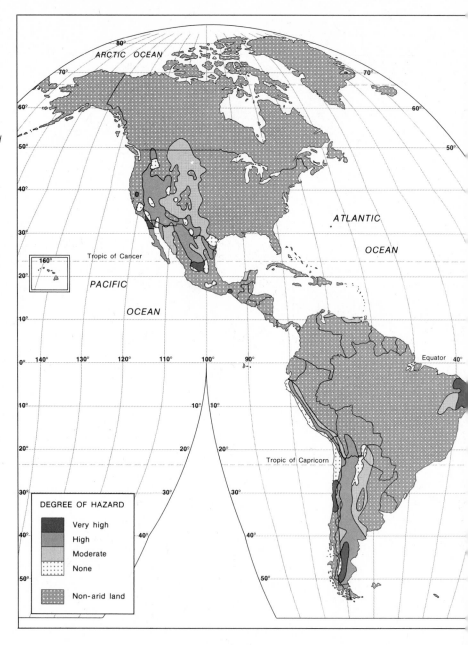

ter, biomass, and energy in these areas. The sequence goes like this:

1. Excessive grazing or plowing reduces the already thin vegetation, exposing more ground surface.

2. The exposed soil humus (topsoil) is broken down, and a thin crust forms.

3. The crust prevents additional water from percolating into the soil, so it runs off the surface, carrying valuable minerals with it.

4. Devoid of replenishment, groundwater levels fall.

5. The denuded soil becomes infertile and dry.

In the case of irrigation, the principle mechanisms for desertification are salinization and alkalinization of soils due to inadequate drainage. Upstream deforestation contributes to the destruction by accelerating soil erosion and deposition on downstream farmland or rangeland.

Population growth in affected areas is 2.5 percent per year or more, and even in the most difficult drought years

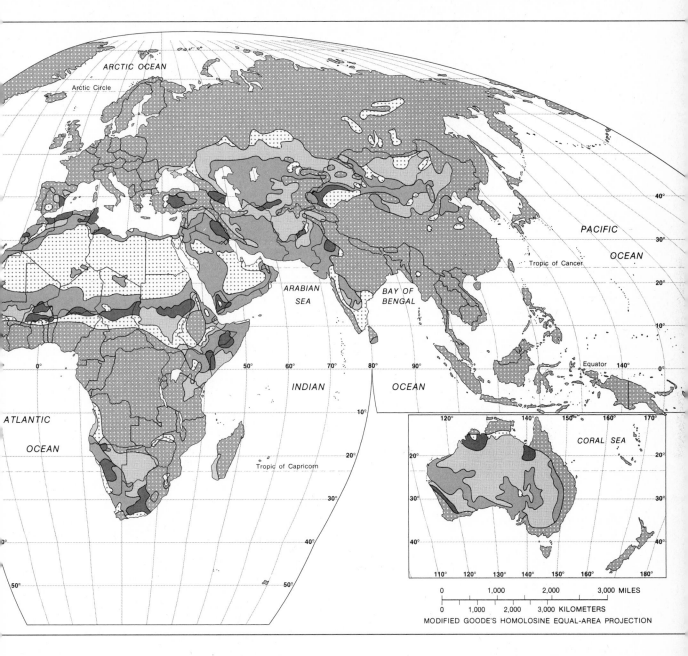

MODIFIED GOODE'S HOMOLOSINE EQUAL-AREA PROJECTION

the birth rate is often ten times the death rate. The birthrate among nomadic groups is also very high at the same time that their traditional trading economy has been destroyed. If the process of desertification continues at the same rate for the next decade or two, over 600 million people may be affected.

Arid conditions alone do not cause desertification. It usually occurs when drought is combined with inappropriate or inadequate human adjustments. For example, areas previously left fallow during drought may be intensively used, not allowing the natural semiarid ecological system to adjust.

Because dry or semiarid lands are those most subject to desertification, and because these lands are often adjacent to natural deserts, it is sometimes assumed that the desert "invades" the adjacent land. This is not true, but it appears to be so because the human-caused desert eventually looks exactly like a natural desert.

Solutions to this land use problem are apparent, but difficult to implement. The death of livestock during drought years is usually due to failure of pastures rather than to failure of the water supply. Fortunately, there is evidence that damage to rangeland from overgrazing is not necessarily permanent. Animal dung and even a little ground cover will initiate recovery of the range in a fallow period. Pastoral economic and social systems must be adapted to allow seasonal recuperation of vegetation and to preclude any use during severe drought years.

Likewise, proper forest and water resource management can reclaim up to half the land that has become desertified worldwide. The limiting factor in the land reclamation process thus far has been a lack of funding, even though it has been demonstrated that the process would begin paying for itself in about five years. Thereafter, the reclamation process would be self-supporting.

Deforestation

Another serious example of human-caused change in natural vegetation that is creating a major ecological upset is the ongoing process of global **deforestation**. Reference to a world map of natural vegetation in an atlas shows that forests cover vast areas in the middle and northern latitudes, as well as great portions of the tropics. However, comparison of the natural vegetation map with a world generalized land use map shows that much of the forest has been modified or totally changed to support human activity. This is particularly true in the tropical forests, which cover about the same area as the continental United States but possess over one-half of all plant and animal species on Earth. The world's most diverse and most sensitive biome is being destroyed more rapidly than any other. One-third of the original tropical forest has already been cleared.

Surveys sponsored by the National Academy of Sciences and the United Nations Food and Agricultural Organization showed that 45,000–56,000 square miles (73,000–91,000 km^2) per year of forest are being destroyed, and an additional 62,100 square miles (100,000 km^2) are being disrupted. Almost all of this deforestation is taking place in the less developed countries, particularly Brazil, Indonesia, Colombia, the Philippines, and Mexico. (This survey information is based upon remote-sensing data.)

In *Tropical Forests: Patterns of Depletion,* Norman Myers reports that about 19 percent of tropical deforestation is from commercial logging, and that fuelwood gathering accounts for about 10 percent. Cattle raising (mostly in Latin America) causes clearing of about 8 percent, or 7700 square miles (20,000 km^2) per year.[11] Small-scale farming, such as that practiced by shifting cultivators and landless squatters, accounts for the remaining 63 percent.

Deforestation by logging is primarily taking place in Southeast Asia and West Africa. Conversion of forests to cattle ranching is most prevalent in Latin America. The greatest loss to cultivation is in Southeast Asia. The areas least affected, at least to date, are in central Africa and western Amazonia.

There are many effects of the deforestation process, and many are globally significant. For example, clearing and burning tropical rain forests contribute large amounts of carbon monoxide, methane, and nitrous oxide into Earth's upper atmosphere, where these compounds become catalysts in reactions that tend to thin the life-shielding ozone layer. Injection of carbon dioxide into the atmosphere also blocks radiation of the sun's heat away from Earth, thereby adding to global warming (the greenhouse effect).

Soil erosion by removal of covering vegetation causes flooding and accelerates downstream erosion. For example, in West Africa, runoff was found to be 20 times greater for cultivated and bare soil than for the original forest.

Deforestation of large areas also decreases the amount of water that enters the atmosphere through evapotranspiration. In the upper Amazon basin, water for as much as 80 percent of the rainfall is supplied by evapotranspiration from the forest. With less water in the air, local cloud cover is diminished, which permits increased solar radiation and drying of the soils, further interrupting the hydrologic cycle. Rain forests have a built-in insulation capacity that maintains constant temperature and humidity conditions. (Trees are often called "nature's air conditioners.") Changes in humidity through forest disturbance further stresses the surviving trees.

El Niño

El Niño is an example of the never-to-be-forgotten complex relationships between dynamic Earth systems and human activity. El Niño is a climatological and oceanic event that occurs irregularly off the west coast of South America around Christmastide (the name *El Niño* means "Christ Child"). Its causes are unknown and its effects widespread. The name refers to a general warming of the equatorial Pacific Ocean and a slowing or discontinuance of deep-water upwelling along the coast (Figure 10-6). The loss of nutrients from the upwelling causes fish to die or move away, thereby reducing commercial fish catches.

El Niño activity is now known to be related to the occurrence of droughts in North Africa, northeastern Brazil, and even Central Europe. It also has been correlated with severe winter weather conditions in North America and Europe. The effects of these unpredictable changes includ-

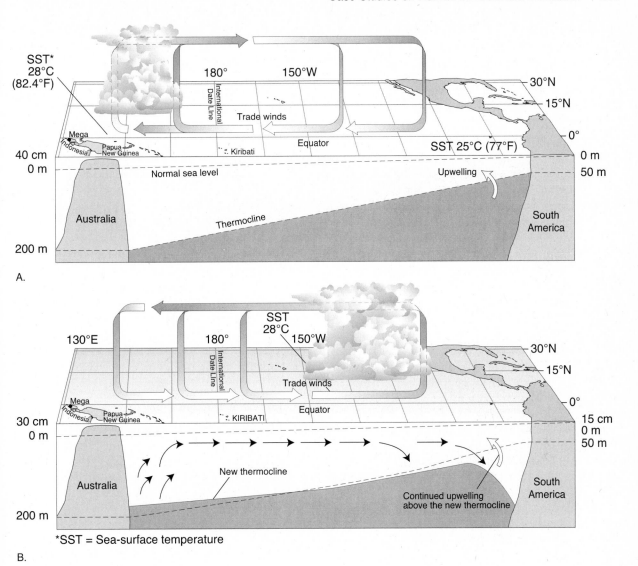

FIGURE 10-6
The El Niño effect has been credited with the creation of dramatic changes in climatic patterns on Earth.   A. Normal oceanic pattern of the gradual upwelling of cold ocean water from the depths of the sea towards its surface.   B. The El Niño effect brought on by more dramatic and continuous upwelling and resultant reversals in surface winds and currents. (From C. S. Ramage, "El Niño," Scientific American, 1985. Copyright © 1985 by Scientific American, Inc. All rights reserved.)

ed loss of fish catches, crop reductions, and drought-induced famines. In other areas, this climatic anomaly has created increased rainfall in central Chile and California, causing flooding. Heavy snowfall in central North America and flooding in Europe have also been related to these changing patterns.

Photographs of coastal devastation, damaged buildings, eroded irrigation systems, and starving livestock were shown in news media during El Niño events in 1972,

1976, 1983, and 1993. Earlier events in this century were recorded in 1911, 1925, 1940, 1957, and 1965. As with other unpredictable climatic and geologic events, we are reminded that much is yet to be learned about Earth's geosystems, and that wise political and economic decisions are required to accommodate them.

As much of our society is attuned to crisis reaction, *recognition* of a problem is the first step in solving it. In years past a crisis might develop over more than one lifetime, making it difficult to detect. At today's accelerated pace of events, our technology can create problems faster, and we become aware of them more quickly. But solutions normally require a much longer perspective.

RESTORING THE EARTH MOVEMENT

It is logical, appropriate, and somewhat overdue that environmental awareness and action to halt degradation should be followed by repair of the damage. The restoring the Earth movement was inspired by John Berger's book *Restoring the Earth,* which contains case studies of successful restoration projects undertaken at the local level.[12] Restoring the integrity of damaged environments is being done at all scales, but it is mostly occurring at localized sites where personal commitment is strongest. A site is usually viewed and renewed biologically, structurally, and functionally. Examples of such citizen-activist projects are restoration of streams, lakes, marshes, forests, and rangelands. Wildlife habitats in cities and suburbs seem to be popular projects for volunteer labor, and visible results are in good supply. An often-seen bumper sticker reads, "Think Globally, Act Locally."

Those who have participated in grass-roots restoration projects usually remark that a major benefit is reversal of the gloom-and-doom fatalism often engendered by concern over global environmental issues. Other results have been more books on the subject; creation of "mitigation banks," where habitats are restored to compensate for losses to development; restoration work on government land such as the national forests; regional conferences where ideas are shared; and media attention, informing the public and promoting commitment. Many public-private partnerships (for example, The Nature Conservancy and U.S. Fish and Wildlife Service) take on projects that maximize cost-effectiveness and public support.

Today's restorers go beyond replacing topsoil. They are replanting entire tropical forests, rebuilding streams, and re-creating wetlands in hundreds of projects from Costa Rica to Maine.[13]

A SMALL-SCALE ENVIRONMENTAL IMPACT STUDY

Since passage of the National Environmental Policy Act (NEPA) in 1969, several U.S. states and other countries have passed legislation to require environmental analysis of proposed projects that may affect valuable resources, endanger certain plant and animal life, or pollute the air or water. Most of these studies are applicable to rather large projects, but what if they were required for *all* projects that *may* have an adverse effect on environmental systems, public and private? Perhaps some planning mistakes would not be repeated and unnecessary environmental upsets would be avoided.

For purposes of understanding the interrelationships among decision makers and their social and physical environments, let us assume that a new shopping center is proposed at the edge of your community. While local regulations vary, normal procedure would be for the developer to file application forms with the appropriate jurisdiction, such as the county planning department. The planners, upon review of the application, might then require an environmental impact study based on their judgment of potential negative environmental effects resulting from the construction and operation of the shopping center.

A list of possible negative effects, in no special order, might be:

1. Construction on poorly drained soils, which could cause foundation weakening

2. Location in a flood plain or area subject to other natural hazards such as earthquake or landslide

3. Removal or modification of a natural habitat for rare or endangered species of plants or animals

4. Increased downstream storm-water runoff due to new impermeable surfaces such as roofs and parking lots

5. Increased traffic congestion along with associated noise and air pollution from automobile exhausts

6. Possible water pollution from parking lot dirt, oil, and tire fragments

7. Removal of productive farm land from agricultural use

8. Increased demand on municipal services and facilities such as water supply, sewage treatment, police and fire protection, and electricity production

9. Inducement of further growth in the area, thereby multiplying all the above impacts

In most such cases a team of planners, engineers, and scientists would be assembled to study the site and the construction plan in preparation for predicting the magnitude of these impacts.

Using essentially the same format as impact studies for major dams and highways, our local analysis would most likely take the form of an *informational* document rather than an *advocacy* document. It would be used as a database by decision makers in conjunction with testimony given at a public hearing of project advocates and opponents.

Perhaps through the public awareness brought about by the decision-making process, many of the potential impacts could be mitigated by the planners. For example, auto traffic could be reduced by making public transit available to the site. Lost open space and wildlife habitat could be compensated for with purchases of suitable land nearby, to be maintained by a public agency. Storm-water runoff and water pollutants could be trapped on-site for treatment. Flood-control improvements and needed public services could be paid for by the developers through exactions levied by the appropriate agencies. If done wisely, the proposed project could be both an economic asset and environmentally compatible.

While this case study may seem idealistic, areas of the United States that are subject to environmental impact review report a marked increase in the quality of public and private development projects. In many cases, the preserved environmental amenities and expressed sensitivity to preservation of community resources have led to public-relations values far in excess of what could be done with advertising.

The key to any environmental review of projects is citizen information and participation in the planning process. This takes commitment. *That commitment can come from attitudes learned in school through geographic and environmental education.*

With a strong geographic and environmental education program in the schools, and cooperative enforcement of environmental regulations in the community, a major reversal of a two-century-long process can occur. The next step will be restoring lost or damaged environmental systems that were the victims of our exploitive past (see the box on page 262).

OTHER MAJOR CONCEPTS IN ENVIRONMENTAL GEOGRAPHY

In addition to the concepts presented in Chapters 5 and 6 dealing with physical geographic processes, the following concepts should be understood. As they involve human-environmental relationships, they each have an orientation toward both the natural characteristics of Earth and human activity upon it. We hope that use of these concepts will eliminate dualistic notions about conflict and incompatibility of humans and nature.

Biosphere

As introduced in Chapter 6, the concept of **biosphere** is a means of separating the living portions of Earth's systems from the three other portions—land (lithosphere), water (hydrosphere), and air (atmosphere). Obviously, this is not really possible because both soils and oceans are partly composed of living things. It is really a matter of emphasis that defines the limits of the biosphere for geographic study.

Within the context of scale, it is useful to visualize the Earth's biosphere as divided into large regional biomes, and each of the major biomes made up of many plant communities depending on their local circumstances or ecosystems. Biogeographers study locations of plant communities as they relate to other plant communities, climate, soil, or other environmental variables within the greater Earth biosphere. The results of these studies are often displayed on maps. Plants are studied more often than animals because they are easier to locate and often determine the nature of animal populations in a biotic community.

Not to be confused with the above concept, visitors may visit the Biosphere II project in Arizona. Biosphere II is a human-built ecosystem maintained inside a four-acre glass shell somewhat like a giant terrarium. Sponsored by a private foundation, it holds promise for ecological research and education.

Ecosystems

Ecosystems are both real phenomena and mental constructs for understanding environmental interrelationships among organisms. A complex of things affects a plant or an animal. Thus, a wild blackberry, or any one of us, exists in an environment with inputs of energy, temperature, various life-sustaining substances such as air, food, and water, and other influences. Figure 10-7 is a diagram of an ecosystem for a baby playing on the front lawn of its home. Like snowflakes, no two ecosystems are exactly the same, but understanding similarities and differences helps us understand underlying processes of energy and matter exchanges in biota.

Resources

A **resource**, quite simply, is anything that humans use. Natural resources are those things used directly from Earth such as water, minerals, and wood. Use of a resource depends on its accessibility and the cultural-economic environment in which it is considered for use. Bison are no longer considered resources to Great Plains residents but were the mainstay of the economic and social system of

GEOGRAPHER AT WORK
Monitoring the Environment

Douglas Richardson, President of GeoResearch, Inc., Billings, Montana.

Doug Richardson received his B.A. degree from the University of Michigan, where he developed an interest in energy resources and agriculture. He learned that there was a great deal to do in the area of natural resources regulation. After earning a Ph.D. from Michigan State University in geography, Doug immediately went to work in Wyoming on the Northern Cheyenne Research Project. This project was organized by the local American Indian tribe to provide assistance and advice in planning their energy resources. Doug worked with tribal leaders, examining their energy options. Through lease cancellations, he made it possible to return over half of the area's coal land to the local tribes. Having established a reputation from this, Doug worked with the Council of Energy Resource Tribes, where he made recommendations for reclamation of Indian lands.

Doug formed his own consulting firm in 1980 with offices in Washington, DC, and Billings, Montana. GeoResearch now employs about 25 researchers and several academic geographers on a consulting basis. Many of their projects continue to be located in the Rocky Mountains and Great Plains in North America, but Doug has also done work in Europe, particularly studying acid rain in Italy and energy needs in the Netherlands.

As president of GeoResearch, Doug is involved in proposal writing, conceptualizing the design of projects, and of course, overseeing the actual research and writing of project reports. Doug says, "Our company is unique in the way geographers are unique. We go for an integrated holistic approach."

When asked where geography is going now, Doug replied, "With geographic information systems, geographers are going to play a broader role in the natural resource fields. These techniques, combined with a well-rounded education, are vital."

As an example of the importance of geographic thinking, Doug cited the major American super-collider project as a site-selection challenge that is "just made for geographers." The research necessary to make a wise site selection decision in just one state filled eight volumes in the first year of study!

Another geography-oriented project handled by GeoResearch is a two-year air pollution emissions survey in Italy. The collected atmospheric and topographic data were used to create a model of pollution dispersion and its effect on soils, vegetation, agriculture, building materials, and monuments in the region. It was a dynamic regional study fully employing the interdisciplinary nature of geography.

Doug intends to continue working at the scientific edge of knowledge in order to take on other interesting and socially useful projects.

Native Americans who lived there over a century ago. Some resources are used directly, such as lumber, fish, and water; others require secondary treatment, such as in the production of paper, metal products, or gasoline.

Resources may be classified as nonrenewable or renewable. The *nonrenewable resource* of most concern is raw petroleum, which, when refined and then burned for fuel, cannot be replaced. Many metallic minerals such as iron, copper, and uranium may be considered nonrenewable because cost-effective methods of recycling them are not yet developed. *Renewable resources* are those that can be successfully recycled or that can be used without degradation and therefore reused. One of the main objectives of resource research and development is to create more of a reliance on renewable resources and less dependence on nonrenewable resources.

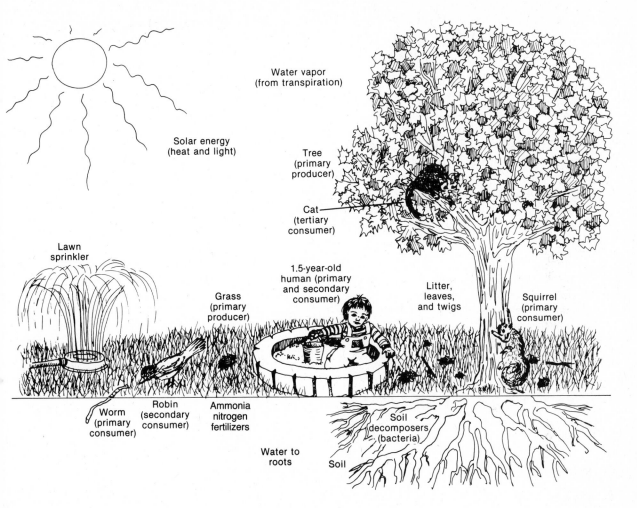

FIGURE 10-7
A baby playing on the lawn is an example of an ecosystem in action.

The present world distribution of resources and resource consumption is very unequal. Many resources are extracted from developing countries and used in developed parts of the world. The major developed countries, with 25 percent of the world's population, use approximately 80 percent of the world's energy and mineral resources. A continuing challenge to geographers is to learn how to equalize resource consumption in an equitable manner without depleting the total resource base for generations that follow.

Conservation

Conservation is a matter of attitude and education. The term originally referred to early twentieth-century attempts to prevent soil erosion and forest destruction caused by overuse of resources. Today, conservation is best thought of

as a perception and frame of reference for decision making. A conservationist attitude matched with wise political and economic choices is insurance for future generations. The most critical resources for sustaining today's economy are those that produce energy. Evaluation of alternative energy sources is an ongoing process of converting our nonrenewable resource base into a conservation-minded renewable resource base, with less dependence upon fossil fuels and more dependence upon hydroelectric, solar, wind, geothermal, or other sources yet to be developed.

Effective conservation policies are difficult to implement in less-developed nations because a "gap of rising expectations" exists between their own situation and what they see in the more developed nations. To catch up, many such nations permit damaging resource extraction and the loss of long-term resource value in order to receive cash-in-hand. Examples of this are the cutting of rain forests, over-fishing, wildlife poaching, soil abuse, and water waste through inefficient irrigation (see the box on page 265).

Hazard and Risk Analysis

Almost daily, news media bring information about drought, floods, earthquakes, hurricanes, or volcanic eruptions somewhere in the world. These sometime violent interactions between the natural processes of the Earth system and people are the subject of hazards researchers and planners. The study of **natural hazards** includes the assessment of economic loss, human suffering, and social disruption caused by these events. In *Perspectives on Environment,* published by the Association of American Geographers, James K. Mitchell summarizes natural hazards research:[14]

> Few people are aware that the Bezmianny eruption of March 30, 1956, was the world's most powerful volcanic event of recent years. Since it took place in an almost uninhabited part of Kamchatka and caused no known casualties, the eruption was largely a scientific curiosity. In contrast, the extrusion of a small volume of lava from a secondary cone on the slopes of Tristan da Cunha became a focus of global interest during October of 1961. By smothering a lobster processing plant and menacing the island's only settlement, the la-

va flow disrupted the economic system and compelled total evacuation of the three hundred residents to England.

To those in positions of responsibility, the *location* and *effects* of events are perhaps more important than knowing their *cause* or their *magnitude*. Mitchell offers the example of American cities that average under 20 inches (50 cm) of snow per year experiencing more severe snow-caused disruptions than other urban places that have considerably greater annual accumulations, presumably because the snowier cities plan for it. Likewise, China has suffered about 40 percent of all recent earthquake-related deaths worldwide, although the total number of quakes is distributed throughout the world. A tropical storm in 1961 in the United States caused 248 deaths, while a similar storm in Bangladesh in 1970 killed over 250,000 people. Figure 10-8 shows historic areas of major hazard events around the globe.

Studies of environmental hazards may emphasize physical causes in an attempt to predict future events. It is not surprising to learn that some floods are at least amplified, if not caused, by human modification of watersheds through vegetation removal, or that some earthquakes are related to groundwater manipulation by humans.

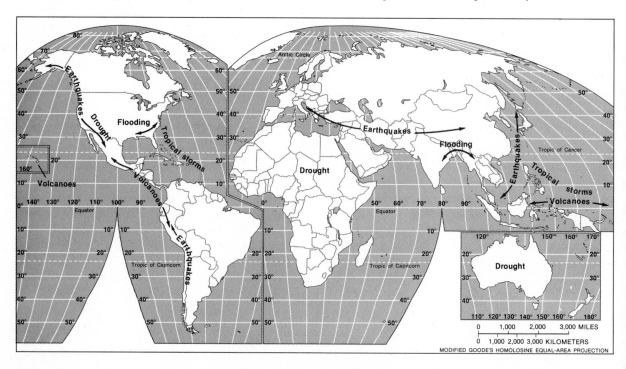

FIGURE 10-8
Historic areas of major hazard events on the globe.

GEOGRAPHER AT WORK
Energy Research in Action

Thomas J. Wilbanks, Corporate Fellow and Senior Planner (seated), with Administrative Assistant Sherry Wright and Geographer Ed Hillsman at Oak Ridge National Laboratory in Tennessee.

Thomas J. Wilbanks is a Corporate Fellow at the Oak Ridge National Laboratory (ORNL) and Senior Planner in ORNL's Energy Division. As an undergraduate student, he became interested in how technology relates to the world's problems. Finding other fields too restrictive, he chose to study geography in graduate school. "I grew up with the *National Geographic* magazine and was always interested in maps and travel. But it wasn't until my undergraduate advisor (a historian) suggested that geography was the field I was looking for that I considered it as a specialty. As soon as I found it, though, I knew it made sense for me."

After completing his Ph.D. at Syracuse University, Tom taught at the university level in the areas of economic geography and environmental policy. After moving to the University of Oklahoma in 1973 as its Geography Department chairman, he found himself in a state where energy was important, just at the time when the oil embargo made energy a high-priority national issue. Besides the subject's timeliness, it turned out to be an issue with strong connections to technology, environment, economic and social development, and the perspectives of geography.

In his new specialty, Tom found his background to be ideal, regardless of labels. As he remarked, "The question is not what aspects are "geographical"; it is what we, trained as geographers, can contribute to the larger issues. If we worry too much about labels, we limit what we can do."

As the energy policy efforts of the United States expanded during the 1970s and a Department of Energy was formed in 1977, the country's national laboratories—traditionally centers of excellence in energy technology research and development—turned their attention to broader programs of energy analysis. Tom was offered an opportunity to help lead this process at the Oak Ridge National Laboratory, a diversified facility of 4500 people with a tradition of independent problem solving. There, he focused upon the environmental and social impacts of energy options—for instance, participating in the development of National Energy Plan II and testifying before the U.S. Congress on a number of occasions. In addition, he helped build a group of more than a dozen geography Ph.D.'s involved in energy-related research and policy advice.

In the 1980s, he focused mainly upon the energy and environmental needs of developing countries in Asia, Africa, Latin America, and the Middle East. Issues being addressed at Oak Ridge include:

- Sustainable energy systems for developing countries
- Environmental management without inhibiting development
- Energy services for rural areas
- Technology choices and how they relate to development of institutions and human resources

According to Tom, "energy is going to be the single most important issue related to environmental policy in the Third World, and decisions made in the Third World will affect the cultures of us all. We need to find energy options that, at the same time, will support development and contribute to a healthy, stable environment. It's one of the biggest challenges of our generation."

WHAT WE LEARNED FROM THE NORTHRIDGE EARTHQUAKE

The quake struck at 4:31 A.M. on Monday, January 17, 1994, with no warning. It lasted approximately 40 seconds. It occurred on an unknown fault twelve miles below the community of Reseda, an urbanized suburb of Los Angeles. Fifty-five deaths (in a population of one-half million) were attributed to the quake, either as a direct result or as an indirect result, such as heart attacks brought on by the stress of the disaster. An early damage estimate was $30 billion. Three hundred schools, thousands of small businesses, and 11 major streets and freeways were damaged to the point of being unusable. The Santa Susana Mountains to the north grew a foot or more.

Lessons learned from the experience, according to the January 31, 1994, edition of *Newsweek,* are the following:

- Search equipment has been acquired (air bags with a lifting capacity of 72 tons, fiber-optic search cameras, and trained search dogs), but doubts remain as to whether enough trained personnel exist to deal with "massive casualties."
- Most reinforced bridges and overpasses in the area survived but more than 600 are awaiting retrofitting. The reinforcement program began after the 1971 Sylmar quake.
- Changes in building codes have been enacted regarding unreinforced masonry buildings and wood-frame homes built over ground floor parking, but large numbers of older ones are in use and have not been fitted with such features as straps to secure water heaters or bolts which secure foundations to wall framing. [Authors' note: The major surprise of this quake was the poor performance of multistory steel frame buildings.]
- Seismic design regulations for buildings and infrastructure have not been completely implemented, in part due to cost.

Understanding the adjustment decisions made by people involved with potential hazards or real hazard events is important. Their considerations may include the extent of human occupance of the hazard zone, the perceptions or estimations of the nature of the hazard, and the behavior of affected people after the event. For example, the most common behavior after a catastrophic storm or flood is for residents of the area to return to the site of the occurrence. Social and economic planners are then faced with comparing the effects of relocation (if possible) with the likelihood that the event will recur soon. Perception studies often have shown that the real nature of a threat is often not consistent with a population's view or understanding of it. Their responses may then be labeled "irresponsible." Common responses to the uncertainty about natural hazards include denying its existence ("it can't happen here"), denying its recurrence ("it's a freak of nature"), making it knowable ("floods come every five years"), or transferring uncertainty to a higher power ("it's in the hands of God" or "the government is taking care of it").[15]

Earthquake hazards in California exemplify the influence of the above factors upon policies and procedures for emergency preparedness. At the state and county levels, the various offices of emergency services prepare contingency plans for the inevitable "big one" that is predicted to occur within our lifetimes somewhere in the state. Citizens are informed about what to do and what materials to have on hand. Strictly enforced building codes include special provisions for seismic shaking. New development within mapped earthquake fault zones is very limited and strictly regulated.

Emergency responses to the 1994 Northridge earthquake near Los Angeles demonstrated that comprehensive disaster planning saves lives and reduces losses. Yet, with

This collapsed parking structure at California State University, Northridge, was one of many buildings damaged beyond repair during the recent earthquake near Los Angeles. (Photo by Christine Rodrigue.)

all of these preparations, it is statistically known that there will be very high casualties if a "maximum" quake were to occur in an urban area during the work day. (See the box, "What We Learned From the Northridge Earthquake" on pages 266–267.) Despite this, California's population continues to increase by several thousand persons per day.

A complete list of topics for the study of human responses to natural events could be quite long. Some nominations in addition to those already mentioned are tornadoes, typhoons, landslides, soil erosion, infestations of plant and animal pests, human and crop diseases, and environmental pollution.

With continued population increases and resultant stresses on Earth's environmental systems, it seems likely that major losses due to natural events will continue until research findings are translated into workable policy in affected areas. Four all-too-common natural hazards regular-

ly affecting people on Earth are shown in the photographs in Figure 10-9.

Pollution

A **pollutant** is the result of change in the physical or biological realms that is undesirable to humans. Pollutants may be directly harmful, such as toxic waste, or may cause a secondary problem, such as global warming due to increased pollution of carbon dioxide in the atmosphere. Some pollutants are mainly an annoyance, such as noise. Others are deadly to humans and other animals, such as DDT, lead, and asbestos. Some pollutants such as sewage and livestock wastes are classified as biodegradable, whereas others, such as heavy metals, are nonbiodegradable. Dealing with pollution should be viewed in the same context as resource utilization; that is, pollutants should be either eliminated by

A.

B.

C.

D.

FIGURE 10-9
The environment as hazard.   A. Tornado.   B. Earthquake.   C. Wildfire.   D. Hurricane.

not creating them, or converted to resources through recycling or reusing.

Work in the environmental sciences invites the collaboration of chemists, geologists, biologists, geographers, and others toward a common understanding of how Earth systems work and how we can best exist within those systems. The world of the geographer is primarily seen as understanding the spatial distribution of physical features, resources, and humans on Earth's surface.

Most important, the geographer's role is to understand the relationships between these and the many other environmental components that make up the planet. It is imperative that both regional and global perspectives be kept within the daily decision-making processes at all levels of human life.

The next chapter continues our view of Earth as a dynamic system by emphasizing movement and change. Views of the relationship between nature and society as they have been expressed over the centuries are presented in the box on page 269 as a conclusion to this chapter's important geographic content on the theme of human-environment interaction.

THE MANY FACES OF MOTHER EARTH

Theodore Roszak has collected three texts widely separated in time and place. Taken together they reflect the long career of the *Anima Mundai* as she has been celebrated in folklore, philosophy, and science.

> Behold! Our Mother Earth is lying here.
> Behold! She gives of her fruitfulness:
> Truly her power she gives us.
> Give thanks to Mother Earth who is here.
> Behold on Mother Earth the growing fields!
> Behold the promise of her fruitfulness.
> Truly her power she gives us.
> Give thanks to Mother Earth who is here.
>
> (Pawnee tribal hymn)

> When the whole fabric of the World Soul had been furnished to its maker's mind, he next began to fashion within the soul all that is bodily and brought the two together, fitting them center to center. And the soul being everywhere in woven from the center to the outermost heaven and enveloping the heavens all around on the outside, revolving within its own limit, made a divine beginning of ceaseless and intelligent life for all time.
>
> (Plato, *The Times,* fourth century B.C.)

> The Earth holds together, its tissues cohere, and it has the look of a structure that really would make comprehensible sense if we only knew enough about it. From a little way off, photographed from, say, a satellite, it seems to be a kind of organism. Moreover, looked at over geologic time, it is plainly in the process of developing like an immense embryo. It is, for all its stupendous size and the numberless units and infinite variety of its life forms, coherent. Every tissue is dependent for its existence on other tissues. It is a creature, or if you want a more conventional, but less interesting term, it is a system.
>
> (Lewis Thomas, in *The New England Journal of Medicine,* 1978)

Used with permission from Theodore Roszak, The Voice of the Earth *(New York: Simon & Schuster, 1992).*

LEARNING OUTCOMES

The authors of *Guidelines for Geographic Education* referred to in previous chapters, define the third of the five themes as "Relationships within Places: Humans and Environments." This theme is explained as follows:

> All places on Earth have advantages and disadvantages for human settlement. High population densities have developed on flood plains, for example, where people could take advantage of fertile soils, water resources, and opportunity for river transportation. By comparison, population densities are usually low in deserts. Yet flood plains are periodically subject to severe damage, and some desert areas, such as Israel, have been modified to support large population concentrations.

Objectives

Students will:

1. Describe several ways in which people inhabit, modify, and adapt culturally to different physical environments.

2. Know that human changes to natural settings are based on cultural appraisals and occupance.

3. Give examples of ways people evaluate and use natural environments to extract needed resources, grow crops, and create settlements.

4. Describe how the human ability to modify physical environments and create cultural landscapes has increased in scope and intensity through the use of technology.

5. Give examples of how human alterations of physical environments have had positive and negative consequences.

6. Be aware that each culture group sees different possibilities and constraints in physical environments.

7. Understand why humans attempt to control the quality of the natural environment and to mitigate the effects of hazardous natural events such as drought, floods, earthquakes, and hurricanes.

KEY CONCEPTS

Acid rain	Desertification	Environmental impact report	Polder
Bioregionalism	Ecosystem	Environmentalism	Pollution
Conservation	El Niño	Gaia hypothesis	Possibilism
Deep ecology	Endangered species	Greenhouse effect	Sustainable Earth
Deforestation	Environmental determinism	Natural hazard	

FURTHER READING

The most popular textbook on the subject of environmental studies is G. Tyler Miller, *Living in the Environment* (Belmont, CA: Wadsworth, 1993). An inspiring history of the North American environmental movement is Philip Shabecoff's *A Fierce Green Fire: The American Environmental Movement* (New York: Hill and Wang, 1993). Historical perspectives on human modification of Earth are presented in the impressive and comprehensive *The Earth as Transformed by Human Action: Global and Regional Changes in the Biosphere Over the Past Three Hundred Years,* by B. L. Turner II et. al., eds. (New York: Cambridge University Press, 1990).

A concise and colorful summary of global environmental problems is contained in several popular atlases of environment issues. They include *Atlas of the United States Environmental Issues,* by Robert J. Mason and Mark T. Mattson (New York: Macmillan, 1990), and Norman Myers, ed., *Gaia: An Atlas of Planet Management* (New York: Anchor Books, 1984). *Saving Our Planet: Challenges and Hopes,* by Mostafa K. Tolba (London: Chapman and Hall, 1992), is another useful source on the world's current environmental crisis.

Another comprehensive book on the topic is Harm J. de Blij, ed., *Earth 88: Changing Geographic Perspectives* (Washington, DC: National Geographic Society, 1988). Others include A. M. Mannion, *Global Environmental Change: A Natural and Cultural Environmental History* (New York: John Wiley, 1991), and *The Human Impact on the Natural Environment,* by A. Goudie (Cambridge, MA: MIT Press, 1993). The annual *State of the World Report* is edited by Lester R. Brown and several others at the Worldwatch Institute and is published by W. W. Norton, New York. A special edition of *National Geographic* magazine (December 1988) devoted to the question "Can Man Save This Fragile Earth?" includes a world map that vividly summarizes environmental concerns.

Numerous sources on the controversial issue of global warming have been published in the past 10 years. For example, see Robert C. Balling, Jr.'s *The Heated Debate: Greenhouse Predictions vs. Climate Reality* (San Francisco: Pacific Research Institute, 1992); *Climate Change and Its Biological Consequences,* by David M. Gates (Sunderland, MA: Sinauer Associates, 1993); and Derek M. Elsom's *Atmospheric Pollution: A Global Problem* (Cambridge, MA: Blackwell, 1992).

The Natural Hazards Research and Applications Information Center at the University of Colorado is a clearinghouse for environmental information, in addition to publishing its own monographs, bibliographies, and working papers. *Natural Hazards: An Integrative Framework for Research and Planning* by Risa I. Palm (Baltimore, MD: The Johns Hopkins University Press, 1990) is a welcome addition to the geographic literature on hazards. Another valuable contribution is a monograph from a 1987 conference on hazard's research by Andrew Kirby, ed., *Nothing to Fear: Risks and Hazards in American Society* (Tucson, AZ: University of Arizona Press, 1990).

Geographic approaches to resource management are offered in Susan Cutter et. al., *Exploitation, Conservation, Preservation: A Geographic Perspective on Natural Resource Use* (New York: John Wiley, 1991). Major information sources on deforestation are Gillean T. Prance, ed., *Tropical Forests and the World Atmosphere* (Boulder, CO: Westview Press, 1986), and D. Poore, ed., *The Vanishing Forest: The Human Consequences of Deforestation—A Report for the Independent Commission on International Humanitarian Issues* (London: Zed Books, 1986). Deforestation also is the subject of Susan Hecht and Alexander Cockburn's *The Fate of the Forest: Developers, Destroyers and Defenders of the Amazon* (London and New York: Verso, 1989).

Among the many special publications about desertification is the October 1987 issue of *Land Use Policy,* edited by Mostafa Kamul Tolba.

A geographic perspective on bioregionalism that compares it with environmental determinism is "Old Theories in New Places? Environmental Determinism and Bioregionalism" by Stephen Frenkel in *The Professional Geographer* 46 (Aug 1994): 289–295.

The information on El Niño was extracted for this chapter from *The Professional Geographer* 36 (Nov. 1984): 428–436. The same issue also contains "Geography and the Lessons from El Niño," by Cesar N. Caviedes.

The promise of building a sustainable economy while improving environmental quality in the near future is eloquently ar-gued by Paul Hawkens in *The Ecology of Commerce* (New York: HarperCollins, 1993).

An undated but recent collection of readings about Ecofeminism (that is, feminist political action for environmental causes) is Judith Plant, ed., *Healing the Wounds: The Promise of Ecofeminism* (New Society Publishers, P.O. Box 582, Santa Cruz, CA 95060).

One of the best-selling books about the environment published in the 1990s is *Earth in the Balance: Ecology and the Human Spirit* by U.S. Vice President Al Gore (New York: Plume, 1993).

ENDNOTES

[1] Lynn White, Jr., "The Historical Roots of Our Ecologic Crisis," *Science* 155 (1967): 1203–1207. This work offers the thesis that the main "cause" of our ecological problems lies in the Judeo-Christian tradition.

[2] George Perkins Marsh, *Man and Nature, or the Earth as Modified by Human Action* (New York: Charles Scribner, 1864); republished in 1965 under the editorship of David Lowenthal (Cambridge, MA: Harvard University Press).

[3] John Wesley Powell, *Report on the Lands of the Arid Region of the United States, with a More Detailed Account of the Lands of Utah* (Washington, DC, 1878).

[4] William L. Thomas, ed., *Man's Role in Changing the Face of the Earth* (Chicago: University of Chicago Press, 1956).

[5] Aldo Leopold, *A Sand County Almanac* (New York: Oxford University Press, 1949), and Rachel Carson, *Silent Spring* (Boston: Houghton Mifflin, 1962).

[6] Stewart Udall, *The Quiet Crisis* (New York: Holt, Rinehart & Winston, 1963).

[7] Theodore Roszak, *The Voice of the Earth* (New York: Simon & Schuster, 1992).

[8] Norman Myers and Gaia Staff, *Gaia: An Atlas of Planet Management* (New York: Anchor Press, 1984). See also James Lovelock, *The Ages of Gaia: A Biography of Our Living Earth* (New York: W. W. Norton, 1988.)

[9] As summarized and discussed by Harold Gilliam, "A State of the World Address," *San Francisco Chronicle* (October 25, 1987): A-1.

[10] Alan Grainger, *Desertification, How People Make Deserts, How People Can Stop, and Why They Don't* (London: Earthscan, 1982).

[11] Norman Myers, "Tropical Forests: Patterns of Depletion," in Gillean T. Prance, ed., *Tropical Forests and the World Atmosphere* (Boulder, CO: Westview Press, 1986): 9–22.

[12] John J. Berger, *Restoring the Earth: How Americans Are Working to Renew Our Damaged Environment* (New York: Alfred A. Knopf, 1985).

[13] Sharon Begley, *Newsweek* (January 18, 1988).

[14] James K. Mitchell, "Natural Hazards Research," in Ian Manners and Marvin Mikesell, eds., *Perspectives on Environment: Essays Requested by the Panel On Environmental Education, Commission On College Geography* (Washington, DC: AAG, 1974): 311–341.

[15] Ian Burton, Robert W. Kates, and Gilbert F. White, *The Human Ecology of Extreme Geophysical Events* (Toronto: University of Toronto, Department of Geography, 1968).

11

THE THEME
OF MOVEMENT

Horse-drawn cart and driver in Udaipur, India. (Photo courtesy of Stephen F. Cunha.)

Geography occurs not only in space, but in time. References to time in today's society often are in units of hours, minutes, and seconds. Birthdays are celebrated annually, as are holidays and festivals in most cultures. The approximate unit of a "lifetime" is understood by all. "Decades" and "generations" are familiar reference points in the chronologies of personal histories. However, units of time that refer beyond these limits become more difficult to perceive. Centuries, eras, and epochs are referred to in history and geology but are little understood by children and by some adults. Their significance at a personal level often is obscured by day-to-day events that take preeminence.

The fourth fundamental theme of geography, *Movement,* refers to Earth's connections, such as trans-portation or communication, over space and time. These two dimensions have changed in relation to one another as technology and societies have changed. Modes of trans-portation and communication also vary among culture groups (Figure 11-1).

The *National Geography Standards* that focus on concepts within the theme of Movement are the following:

Standard 3: *The geographically informed person knows and understands how to analyze the spatial organization of people, places, and environments on Earth's surface.*

Standard 9: *The geographically informed person knows and understands the characteristics, distribution, and migration of human populations on Earth's surface.*

FIGURE 11-1
Village of Jiri in Nepal ... where the road ends and the trek begins to the Solo Khumbu, a Himalayan mountain range. (Photo courtesy Stephen F. Cunha.)

MOVEMENT IN TIME AND SPACE

Understanding movement of time and space on Earth requires a new set of precepts that often are learned somewhere in our formal education. For the archaeologist to explain life in the Neolithic epoch, or for the geomorphologist to discuss the ages of marine terraces, one must understand longer units of measurement and the ongoing evolution of Earth and its life forms.

Chapters 5 and 6 emphasized a dynamic and ever-changing physical Earth. Chapter 7 emphasized changes in human societies and cultures, underlining the similarly dynamic and ever-changing human occupance of Earth. Places change over time, and one of the most dynamic elements of change, both *in* places and *between* them, is *human movement*. This chapter focuses upon forms of movement that take place in economic and cultural activity.

People everywhere interact. They travel from place to place, they communicate, and they depend on others in distant places for products, ideas, and information. Static views of Earth are useful for descriptive purposes, but true understanding of movement and change is necessary for a more complete analysis. The theme of movement must be added to each of the previous patterns, processes, and concepts that we have examined. Four major forms of movement that can be tied to considerations of location, place, and human-environment interaction are:

Migration—movement of people

- Diffusion—movement of ideas
- Trade and transportation—movement of goods
- Communication—movement of information

Each of these forms of movement is examined in the sections that follow.

MIGRATION: THE MOVEMENT OF PEOPLE

Today more than ever, people as individuals and in groups are on the move. We commute from home to work or school; we travel on vacations to distant places; we travel considerable distances to attend special events or to visit special people. In some circumstances, a household's entire life may be mobile, such as the pastoral nomads of Central Asia, the traditional Gypsies in Eastern Europe, and the hunter-gatherers in Central Africa. Urban mobile lifestyles include commuter marriages, traveling salespeople, and homeless street people.

The term **migration** is used to identify the movement of persons or groups who permanently relocate their homes to a new place. Because people carry their cultural customs with them, the process of migration most often has been the mechanism for the spread of cultures and societies. Evidence of this type of movement is visible in the cultural landscape, often telling us where a group of people came from, when they settled the new place, and what kinds of imprints they have made on the new land.

Emigration and Immigration

The migration process may be subdivided into several concepts or generalizations. Two basic conceptual ways of viewing the process of migration are from the perspectives of emigration and immigration. **Emigration** is the departure of people from their homeland, by their own choice, under order of higher authority, or for economic or political reasons. **Immigration** is the entry of people into their new homeland.

"Push and Pull" Factors

Emigration and immigration both depend on an assortment of push-pull factors for their impetus. People leave, or are "pushed," from a place for a variety of reasons, although economic motivation has caused more migration than any other "push" factor. Dissatisfaction with working or living conditions at home is a strong push factor, as is religious or political persecution. For Chinese farmers near Canton in the late 1840s, push factors encouraging them to leave their homeland and travel to California included escaping oppressive Mandarin rulers, famine and starvation, and environmental damage resulting from the Opium Wars. Present-day refugees from Southeast Asia, China, northeastern and central Africa, Middle America, and Eastern Europe all are examples of groups who have left their homes due to political, economic, or religious circumstances.

In our earlier example, however, the California Gold Rush, Chinese migrants moving to northern California were responding to powerful "pull" factors. The discovery of gold and promises of wealth and freedom were an economic attraction that many starving Cantonese farmers could not resist. Similarly, in the present day, both political factors and personal security, along with economic opportunity, are "pull" factors in the migration process. Push and pull factors usually work hand-in-hand, and a successful migration is frequently the result of both working together.

An example of the effects of push and pull factors for workers in the global labor marketplace since 1960 is shown in Figure 11-2. In continental Europe, most of the migrant workers came from the Mediterranean region. Southeast Asians had gravitated to the United Kingdom. The United States primarily draws from Mexico, Central America, and the Caribbean. Saudi Arabia, Argentina, and South Africa are other labor magnet countries.

Chain Migration

Many immigrant groups have been influenced by friends or relatives who already have settled in the new land. The importance of "letters home" cannot be underestimated when considering the massive settlement of Scandinavians near the American Great Lakes in the 1870s or the settlement of numerous other ethnic groups who have migrated since then. The tendency to settle near familiar people of one's own culture has a strong influence on migration patterns.

This "chain migration" effect often results in homogeneous enclaves of similar people residing together in the new homeland, especially first-generation arrivals. "Little Italy" neighborhoods in many North American cities, Dutch farmers in South Africa, Asian Indians in Malaysia, and Chinese people on virtually every continent tend to group together as a result of chain migration. Similarly, Japanese farmers in Brazil formed agricultural colonies for

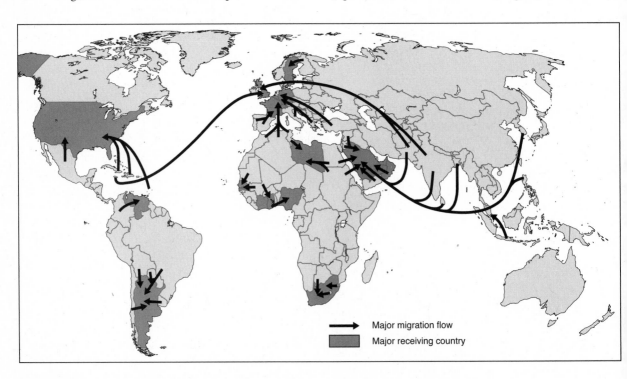

FIGURE 11-2
Major international labor migration flows. (From Population Information Program, 1983, with permission from Guilford Press.)

FIGURE 11-3
A busy shopping street in San Francisco's Chinatown. (Photo by author.)

mutual support and ease in adjusting to their new home. Migrants in many parts of the world have created distinct ethnic cultural landscapes, such as the Chinese in San Francisco as shown in the photograph in Figure 11-3.

Some immigrant groups travel long and sometimes circuitous paths to arrive at their new home. Frequently, many intervening obstacles must be surmounted before the movers reach their destination. Many migrants leave their homeland to settle in temporary places until permanent settlement can be arranged. For example, thousands of "white Russians" (non-Communists) from eastern Siberia escaped their Communist enemies during the Russian Revolution in 1917 by migrating over the Russian-Chinese border and settling in Manchuria. Later, when China itself became a Communist country in 1949, these Russians were again forced to leave.

Traveling by wagon train across China to Canton, the Russian emigrants eventually settled on the Philippine island of Samar to await permission to enter the United States. Early in 1950, these enduring migrants finally arrived in California to begin their new lives in the American West.[1] As a result of this stepping-stone process, the Russian cultural landscape is visible in many California cities today but cannot be found in China or the Philippines (Figure 11-4).

Of course, relocation also occurs among less politically persecuted cultural groups. It is common in today's competitive business world for upwardly mobile young adults ("yuppies") to move from job to job and place to place in search of ever-greater rewards. Others may move up the career ladder with one employer, causing their families to resettle numerous times in new towns or cities. Rituals like the "Welcome Wagon," when newcomers are introduced to a new community, are evidence of how common is the theme of movement in our own culture. The 1990 census revealed that Americans move, on the average, once every four years, although most of those moves are within the same county and state. Jesse O. McKee's book, *Ethnicity in Contemporary America: A Geographical Appraisal,* supplies numerous examples of push and pull factors that illustrate North America's mobile lifestyle.[2]

Acculturation

With the possible exception of some religious groups, it may generally be stated that, as people migrate to new places, their lifestyles and cultural attitudes gradually change in adjustment to the new environment. Most adapt to their new homeland in a variety of ways. Newcomers may need to learn a new language, adopt new political attitudes, incorporate new value systems, learn a new trade or skill for employment, or change their hair and clothing styles. The adaptation of a culture group to a new place by acquiring traits from another culture group is known as **acculturation.** As contact between two groups persists, a new blended culture or lifeway may emerge. In many of today's

FIGURE 11-4
A Russian Orthodox church in downtown Sacramento, California. (Photo by author.)

large cities, a distinguishable "urban culture" is evident in distinctive dialects, clothing, and behavior patterns. Rural migrants to the city may need to learn new vocabulary and new values to become fully acculturated.

Acculturation also may bury many aspects of the original culture and cause it to be dominated by the more powerful group. The culture of Native American peoples in North and South America has been changed drastically by the arrival and subsequent dominance of European peoples. A dramatic example—in fact, one of the most significant culture contacts in human history—was the conquest of Mexico for Spain by Hernando Cortés in 1519.

Likewise, African-American culture in the United States has become more "Americanized" in the last century and has lost much of its uniqueness due to the economic and social pressures to acculturate. When Africans first became forced migrants to the United States from diverse parts of West Africa, they were a heterogeneous group with varying religious beliefs, languages, and lifeways. In less than one generation, much of the slave society was remade into a homogeneous form, based on the dictates of owners and overseers.

In the early part of the twentieth century, new cultural adjustments occurred with African-American migration out of the South. This dramatic story was chronicled recently in a Smithsonian National Museum of American History exhibit, summarized in the box on page 277.

One way to measure acculturation rates among immigrant people is to observe their dietary patterns. Newly arriving migrants usually maintain their homeland cuisine as long as spices or other special ingredients are available. The pungent aroma of Middle Eastern food in a London neighborhood or the smell of Mexican food in downtown El Paso is a reminder of the importance of diet as a cultural indicator. First-generation immigrants are particularly resistant to dietary change. However, as their children attend school, eat in the cafeteria, spend time with friends at local fast-food restaurants, eat in friends' homes, and begin to experience the foods of the dominant culture, taste preferences change. A fully acculturated individual often retains traditional foods for holidays and parties but may eat "mainstream" food on a daily basis.

Assimilation

When all the people in an area give up their original cultural traits and assimilate into the majority culture, we logically label the process **assimilation**. If immigrants are completely assimilated into a local melting pot, which often occurs through intermarriage with members of the majority culture, their original culture is no longer distinctive or even identifiable. Fully assimilated residents of an area eat the same foods, speak with the same accents, and dress in similar styles. The pressure to assimilate is intense for many newly arriving immigrants.

Securing a desirable job often requires giving up one's cultural characteristics. Despite the emotional support given at times to assimilation into the historic American experience, one must question the loss of cultural identities of immigrants in the process. Is the melting pot the ideal model, or is cultural pluralism more to be preferred in today's modern society? Can we have both?

Choosing to acculturate or assimilate—or even to emigrate from one's homeland in the first place—may or may not be an individual decision. Migration by choice, or voluntary migration, usually results in improved economic conditions for migrants. Thousands of Mexican workers and other immigrants from Central America have crossed the North American border in an effort to improve their economic position. This voluntary movement is usually encouraged by strong pull factors into what is perceived as a "land of opportunity." Hispanic culture is becoming more and more visible in the Sunbelt states as thousands cross the border for improved economic opportunities. Likewise, the thousands of European immigrants who settled vast areas of North America, Australia, New Zealand, South America, and Africa, in the mass movement often called the "Europeanization of Earth," moved to these new lands largely by choice.

GONE UP NORTH, GONE OUT WEST, GONE!

An exhibition at the Smithsonian National Museum of American History dramatically reveals the story of the Great Migration, the massive exodus of African-Americans out of the American South between 1915 and 1940. The exhibit begins with this quote:

> The United States has been a haven for millions of immigrants fleeing war, poverty, and discrimination, or seeking freedom. But in some places, at some times, American society has oppressed its own people.

In 1915, more than 75 percent of all African-Americans lived in the South, working mainly as farmers. Few owned land; most were sharecroppers or tenants. Without land, many fell into debt. "Jim Crow" laws segregated African-Americans to their own restaurants, schools, hotels, barbershops, and even drinking fountains. Poll taxes and literacy tests prevented most from voting. Lynchings by anti–African-American groups like the Ku Klux Klan were common.

World War I created labor needs, sending northern factory owners to the South to recruit African-Americans for employment. Letters from relatives already living in the North offered encouragement. Following two natural disasters in the South, the flood of 1915 and the invasion of the boll weevil, thousands of African-American farmers left the land and moved to northern cities.

Life was not much easier in the North. Race riots broke out in Chicago (1919) and Detroit (1925). Wages were low and living conditions miserable. No home gardens were available to supplement a family's food supply, as they had been in the rural South. The nation's first African-American ghettos were formed.

In early 1988, a traveling exhibit of the Great Migration began a tour around the country. Perhaps visitors' reactions were similar to this native Georgian's comments upon visiting the Smithsonian exhibit in Washington DC:

> I just wonder how people could treat one another like animals. What kind of people were they? You're talking about churchgoing people. What was on their minds?

From Jon Cohen, "Gone Up North, Gone Out West, Gone!" Smithsonian (May 1987), pp. 72–82, and a recent visit to the Smithsonian National Museum of American History in Washington, DC. Material from Cohen used with permission.

Amenities

Others may move voluntarily to improve their lifestyles. These migrants may resettle to seek the amenities, or positive features, of a certain place. These amenities may include moving to a mountain home for the view, settling near coastal beaches, or moving to a place with a better climate. The recent rush to the American Sun Belt by thousands of retirees and others interested in escaping snow and severe weather is one example of this type of voluntary migration. Over 12 million people have moved to the Sunbelt states in the past three decades (Figure 11-5).

One of the most demographically surprising of the world's recent voluntary migrations has been the post–World War II population shift into Western Europe. Long considered a region of emigration, not immigration, the nations in Western Europe have found themselves recipients of large numbers of newcomers in the last 30 years. Two changes have caused this region to turn from a people-exporting area to a people-importing area. First, by the mid-1980s, many former colonies of European countries were reversing the old colonial relationship. For instance, migrants from the former colonies were arriving in the United Kingdom to find work. Second, there has been a change in European migration patterns from an intracontinental movement to intercontinental movement. Both

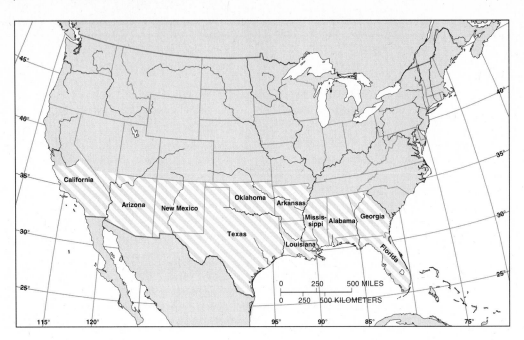

FIGURE 11-5
The U.S. states commonly referred to as the Sun Belt.

changes imply a new European relationship with Third World countries.

An example of such a migration pattern is in Sweden, where this formerly very homogeneous country is now experiencing an immigration trend of unequal proportions. Sweden's high standard of living has attracted so many Middle Eastern immigrants that they now constituite roughly 15 percent of the nation's population. Television in Stockholm often features news stories about Middle Eastern enclaves and other ethnic neighborhoods in the city.

Refugees

Other individuals or groups involved in movement may not leave their homeland voluntarily. Forced migration first brought Africans to the Americas in the seventeenth century to work on the emerging plantations. Political factors forced thousands of South Vietnamese to leave their war-torn homeland after the fall of South Vietnam in 1975. Russian émigrés were forced to leave Russia after the 1917 revolution drastically changed the political system there, creating then Soviet Union. In many countries of the Third World today, refugees are increasing at a rapid rate as forced migration has made them leave their homeland. For example, Afghans resettled in Pakistan by the thousands in the 1980s during the Soviet occupation of Afghanistan (Figure 11-6). Likewise, tens of thousands of **eco-migrants** have been forced to relocate elsewhere by disasters, such as the explosion and meltdown at the Chernobyl nuclear power plant in Ukraine, and the major earthquakes in Armenia.

There appear to be more refugees now than at any time in history. (Figure 11-7 locates refugees and countries of asylum in the world today.) Estimates of the exact number vary, but at least 2000 new refugees have been added to the world *every day* since 1984. Parts of the world that have experienced the most profound social, political, environmental, and economic change in the last two decades are the Middle East, sub-Saharan Africa, and Central America. It is in these regions that most of the world's 10 million refugees are found.

Political threat and persecution are the main reasons for refugee displacement today and are the main reasons for their inability to return home in the near future. Palestinian refugee camps, for example, have become almost permanent parts of the Israeli land problem. Cambodian (Kampuchean) immigrants in the United States are unable to return home to Southeast Asia due to political restrictions there. Ethiopians who have traveled over their border into the Sudan may not be able to return home (or even wish to return) because of their country's severe environmental and economic problems.

FIGURE 11-6
Afghan man in a refugee camp near Peshawar, Pakistan. (Photo courtesy of Stephen F. Cunha.)

Many refugees will remain in their country of asylum for many years or even for the rest of their lives, so it is vital that the global community find some source of support for encouraging economic independence and adjustment to their new homeland. It is possible for refugees to become a positive factor in the future development of their new home. Just as Southeast Asian business owners and professional people contributed to the economy of the United States in the decades of the 1980s and 1990s, so too have former Somalian nomads, displaced by drought in the mid-1970s, resettled and retrained in coastal fishing villages. People can adapt when they must do so for survival. It is important for host societies and the refugees themselves to work together, encouraging successful adaptation to their new lands whenever possible.

DIFFUSION: THE MOVEMENT OF IDEAS

As any victim of a rumor knows, ideas and beliefs may move faster than people. A new idea—whether it be cultural, political, or technical—is termed an **innovation**. Naturally, the locational, spatial, and mobile aspects of innovations are of interest to geographers. We use the term **diffusion** to describe the spread of an innovation outward from its point of origin. This type of movement from place to place has occurred as long as humans have been in contact with each other. The spread of ideas and innovations has been accompanied by an almost constant diffusion of other aspects of ways of life, which are termed *material culture*. Tools, weapons, domesticated plants and animals, or other innovations fall into this category (see the box on page 281).

Unintentional effects of human activities may also be diffused with great consequence. For example, diseases of feral plants and animals have been diffused from continent to continent as cultures have come in contact. The box on pages 282–283 discusses the rapid spread of the disease of AIDS and is one illustration of the high speed of this type of diffusion.

Other examples are the well-intentioned, but now undesirable introduction of starlings, English sparrows, and carp into the eastern United States in the 1880s and their subsequent spread across the entire continent. Other ecologically disastrous introductions and diffusions of undesirable biota are rats, Dutch Elm disease, walking catfish, venereal diseases, killer bees, and water hyacinths. The familiar story of rabbit reproduction in Australia, where rabbits were introduced and flourished in the absence of natural predators, has been repeated worldwide with different species in different host environments.

Geographer Carl Sauer became particularly interested in the spread of plants and animals via the diffusion of agriculture. His work with the culture hearths of domestication pioneered many later studies of both animal and plant domestication and diffusion. Sauer's interest in the Neolithic period and the "agricultural revolution" inspired a whole new approach to geographic inquiry during his tenure at the University of California at Berkeley.

Based on extensive field work in the New World, broad reading, and a curious mind, Sauer postulated certain basic premises about the origin of agriculture. He suggested, for the first time, that the culture hearth of plant domestication was located in areas having a great diversity of plants and animals, which afforded a large supply of genes

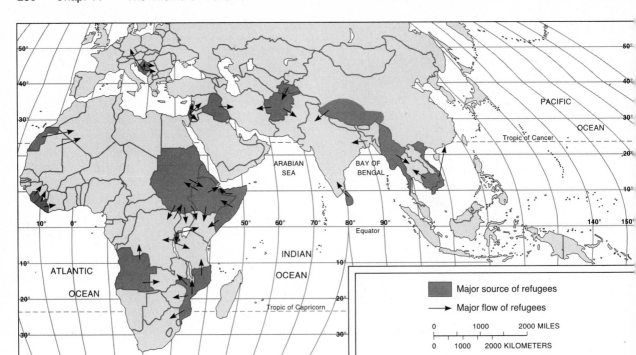

FIGURE 11-7
Refugees and countries of asylum in the world today. (From James M. Rubenstein, The Cultural
Landscape: An Introduction to Human Geography, *4th ed., New York: Macmillan, 1994, Figure 3-2,
p. 90. Used with permission of Macmillan Publishing Co.)*

to work with. These hearths were situated in hilly or moun-
tainous terrain—not in valley bottoms where flooding and
irrigation occurred—and in wooded landscapes where early cultivators could easily dig among the trees for clearing
and planting. Such wooded, hilly places were the most likely sites for innovating early domestication, places where
sedentary people lived who had developed skills with tools
(such as the ax) that would prove useful in cultivation.

Sauer's work on the diffusion of plant and animal domestication has become a cornerstone in cultural geographic thought, and it has provided the impetus and inspiration for numerous theses, dissertations, and
publications on the subject. His work remains vital to geographic interpretations of the theme of diffusion. Figure
11-8 depicts the global diffusion of vegeculture.

Innovation Diffusion

Various elements of culture are in constant flux, such as
language, habits, and customs. As human societies change
through time, some new ideas and inventions are developed
from within; this is known as *independent invention.*
However, most new ideas are brought in from the outside

via *innovation diffusion.* This type of movement may be divided into three major mechanisms:

1. *Contagious expansion diffusion.* This type of movement
   occurs face-to-face from person to person. An idea is
   spread directly from the point of origin outward, like
   ripples from a rock thrown into a pond (Figure 11-9).
   For example, during the campaign to ratify the Equal
   Rights Amendment in the United States in the mid-
   1980s, many women's rights organizations across the
   country held meetings and rallies in support of its passage. The direct contact between speakers and participants at these meetings is one type of contagious expansion diffusion. Another example is the spread of
   some diseases in a wavelike pattern from their source of
   origin.

2. *Relocation diffusion.* People settle in a new area and take
   various elements of their culture with them. These innovations are then spread to the new place; the innovations are relocated. An example of relocation diffusion
   is the spread of a governmental system from one place
   to another (either by choice or by force). When the

DIFFUSION OF PEPPERS: THE WORLD'S FAVORITE ALL-AMERICAN SPICE

Diffusion of peppers worldwide from the Americas has made this condiment the most popular seasoning on Earth. Before Columbus stumbled onto the Americas on his way to the Spice Islands, peppers were entirely unknown outside the western hemisphere. Their subsequent spread to numerous cultures in Africa, India, Indonesia, and China took place in a historical instant and is truly a remarkable story.

When Columbus reached the Caribbean in 1492, he found the natives seasoning their food with a pungent fruit, the taste of which reminded him of the spicy black peppercorns he wished to find. They called the fruit "pepper." Today this term is confusingly used to mean both the red and green fruits we slice into salads and make into hot sauce, as well as the black kernels we grind in our pepper mills.

Peppers are the seed pods of five different domesticated species. They were among the many important plant foods that originated in the western hemisphere and were spread to other parts of the world after European conquest. Maize, amaranth, squash, bottle gourd, tobacco, sweet potato, manioc, tomato, potato, cotton, lima beans, peanuts, rubber, and cacao all are plants that had their beginnings in the New World. Black peppercorns, however, are the dried berries of tropical Old World plants and are no relation to the fruits discovered by Columbus.

Later, the Spanish who came to inland Mexico heard the same fruit being called "chili," an Aztec word. In Mexico today, their term *chili* is a generic name for all types of peppers. By about 5200 B.C. native people had begun cultivating the wild plants. By a process of careful selection and cultivation, the different domesticated pepper species were bred in the upper Amazon Basin, Mexico, the Caribbean, the Andes, and in Bolivia. All modern varieties derive from these original five species.

After Columbus tasted the spicy plants in the Caribbean, they spread like wildfire. The fruits were easily transported in dried form, the seeds remained viable for many years, and young plants found hospitable conditions in almost every country of the world. Peppers are documented to have reached Spain by 1493, Italy by 1526, India by 1542, Germany by 1543, the Balkans by 1569, New Guinea by the mid-1500s, and Melanesia and the Western Pacific before the end of the sixteenth century. No other newly found spice has ever diffused so rapidly.

American pepper production today is big business. California, with more than 19,000 acres in both sweet and hot varieties, produces more peppers than any other state. Florida, Texas, and New Mexico also are important producers. Peppers have come a long way since their domestication in the New World.

From Jean Andrews, "Pepper: The World's Favorite All-American Spice," Pacific Discovery *39 (Jan.–Mar. 1986), pp. 20–33. Used with permission.*

British brought their form of government to Australia, relocation diffusion was the mechanism for its spread from one hemisphere to another.

3. *Hierarchical diffusion.* This type of diffusion is the spread of an innovation from one level of society to another. Typical examples of this type of movement in-

THE DIFFUSION OF AIDS

AIDS has claimed tens of thousands of victims through the work of an unstoppable virus transmitted through blood or through sexual contact. Cases of AIDS have been reported in 123 countries since its discovery in the early 1980s. According to the World Health Organization (WHO), a total of 985,119 AIDS cases have been reported to date. Between June 1993 and June 1994, the WHO reported a 37 percent increase in total cases. The WHO estimates that since the late 1970s, over 16 million adults and 1 million children have been infected, but may have not developed symptoms as yet. As in years past, developing countries, particularly in sub-Saharan Africa and in south and southeast Asia are more significantly affected than are the industrialized countries. The number of AIDS infected adults in industrialized countries is remaining steady—the number of deaths since the middle of 1993 is about even with new reports.

The disease is believed to have started in equatorial Africa in the 1970s. One theory speculates that AIDS was originally a virus that affected wild primates. A mutant form then crossed the species barrier to infect humans. In Central Africa, blood tests reveal that 20 percent or more of the population in cities or along transportation routes have tested positive. Most victims there are heterosexual and are both men and women.

The disease has been transmitted from its African epicenter to other parts of the world via tourists, immigrants, and military personnel, acting as carriers. AIDS appeared in North America in the early 1980s. The initial appearance was concentrated in three areas, New York City, Los Angeles, and San Francisco. By 1985, cases were reported in significant numbers in Florida, Colorado, and Texas. By the early 1990s, it was reported in all parts of the country, but still primarily in urban centers.

There are three general patterns of infection. Pattern one is typical of industrialized countries, with large numbers of reported cases. Most cases occur among homosexual men or intravenous drug users. Pattern-two cases occur most frequently among heterosexuals, and the ratio of afflicted males to females is approximately balanced. Pattern-three areas exhibit very low numbers of infection cases, principally among people who have traveled to the other two pattern areas and have had sexual contact there. The people of many nations in Africa are suffering the most from the epidemic, because all three patterns exist there and medical care is generally less available. The urgency of the situation has caused the creation of a special AIDS program in the World Health Organization.

clude the spread of the popularity among the "masses" of a certain celebrity or a hit song, or the movement of clothing styles from a major metropolitan center outward to smaller towns. A good example of the hierarchical pattern of diffusion is the growth of the Sierra Club. After its initial founding in San Francisco in 1892 by environmentalist John Muir, this activist group spread to Chicago and from this new epicenter to adjoining states (Figure 11-10). By 1960 the organization became national and today is among the powerful environmental lobbying groups in the country.

Often an idea will change as it moves from person to person and group to group. Generally, the farther the idea moves from its point of origin, the more it is changed. This concept is termed **distance decay.** (This term is also applicable in urban economic geography.)

Swedish geographer Torsten Hagerstrand, working at Lund University in southern Sweden, created a theoretical

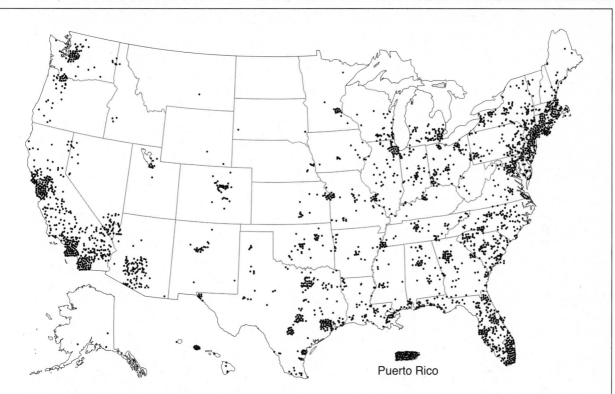

Number of AIDS cases reported per 100,000 people in 1992. (From HV/AIDS Surveillance
Report, *Atlanta, GA: National Center for Infectious Diseases, 1994, p. 3. Used with permission.)*

The map in this box shows the number of reported AIDS cases per 100,000 persons in the United States in 1992. As might be expected, the more populous states exhibit higher numbers of reported cases.

_____

From Jonathan Mann and others, "The International Epidemiology of AIDS," Scientific American
259 (Oct. 1988), pp. 82–98.

approach to the study of these processes of diffusion. Hagerstrand's model suggests thinking about the diffusion process as occurring in four stages:

1. Primary stage: A new idea or innovation is developed and begins to be adopted at its source area.

2. Diffusion stage: The idea is rapidly dispersed outward and is quickly adopted.

3. Condensing stage: The invention spreads to new areas and is adopted there.

4. Saturation stage: The innovation has been diffused to its limit and the spread of the new idea slows and eventually stops.

Modern Diffusion in Action: The Frisbee

To illustrate the mechanism of diffusion waves in the latter half of the twentieth century, it is useful to trace the spread of a common cultural element. We have chosen the Frisbee. This saucer-shaped recreational object is gracefully tossed

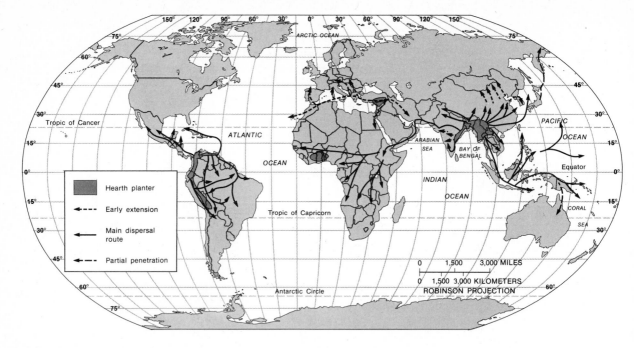

FIGURE 11-8
Global diffusion of vegeculture. (From James M. Rubenstein, The Cultural Landscape: An Introduction to Human Geography, *2nd ed., Columbus, OH: Merrill Publishing Co., 1989, Fig. 9-1, p. 312. Reprinted with permission of Merrill Publishing Co.).*

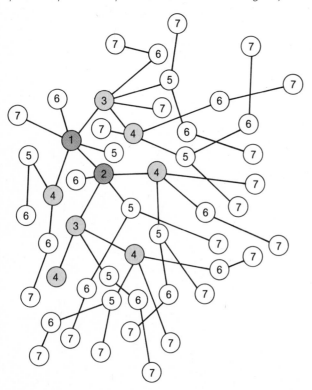

FIGURE 11-9
Contagious expansion diffusion passes from person to person, place to place, like a stone thrown into a pond and the ripples of water it produces.

and caught today in most parts of the world. By the early 1980s, in fact, it had diffused to every continent.

The original culture hearth of the Frisbee appears to have been Yale University in New Haven, Connecticut. As early as the 1890s, students there were fond of throwing Frisbee Pie Company tin pie plates on the campus lawn. To warn their partners of the impending arrival of the pan, throwers would shout "Frisbee!" This popularized the name of the flying object. The popularity of the new Yale game soon spread throughout other New England campuses. In the early 1920s, largely because of a visit to Yale by a southern California moviemaker, the idea of playing Frisbee spread to the beaches of southern California.

Not until the 1960s, however, did the large-scale spread of this collegiate innovation reach other parts of the United States. An interest by the Wham-O toy company (slowed for over a decade by the company's preoccupation with marketing another innovation, the hula hoop) helped diffuse the new game. In the 1960s and 1970s, playing Frisbee subsequently became well known via television reports showing rock concert and beach crowds enjoying the game. By the mid-1970s, the Frisbee had become more of a mainstream innovation in the United States and had spread throughout the world. As evidence of its total acceptance by North American culture, MIT began offering an undergraduate course on the aerodynamics of Frisbee flight. In Japan, a delegation for world friendship was or-

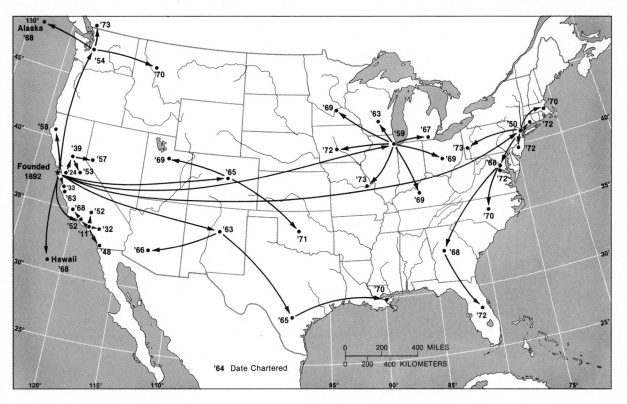

FIGURE 11-10
*Growth of the Sierra Club in the United States, an example of hierarchical expansion diffusion.
(From John F. Kolars and John D. Nystuen,* Geography: The Study of Location, Culture, and Environment, *New York: McGraw-Hill Book Co., 1974, Fig. 7-9, p. 129. Reprinted with permission of McGraw-Hill Book Co.)*

ganized with one goal in mind—to throw a Frisbee off the top of Mount Fuji!

The innovation of the Frisbee in Connecticut and its ultimate spread throughout the world is humorous but it is also a serious lesson in dispersal and diffusion. Certain requirements are necessary for rapid diffusion of modern fads to occur:

- A cultural hearth with noticeable or popular people originating the object

- A receptive group that is interested in adopting the innovation

- An effective system of communication to spread the word quickly to a wide area

Of course, the movement of ideas and inventions occurs much more rapidly in modern times because improvements in communication and transportation allow the rapid transmission of ideas.

Frontier Diffusion

In addition to the forms of movement we have discussed, there are other dynamic spatial interactions that involve movement. For example, settlement forms on the land are often the result of movement into an area. Pioneers are people who move into and settle a **frontier.**

Film and literary descriptions of frontiers suggest that they are areas of recent settlement undergoing rapid physical and social change. To this valid definition we add that a frontier, unlike a boundary, has depth, and it moves as older settlements within it become established and new ones are created at the forward edge. Frontiers have been described in histories of Russia, Australia, Brazil, Canada, parts of Africa, and, of course, the United States.

The significance of the frontier in American history was interpreted in an 1893 essay by Frederick Jackson Turner.[3] He suggested that U.S. culture and the economic development of the nation were best understood by looking toward the West rather than toward Europe. According

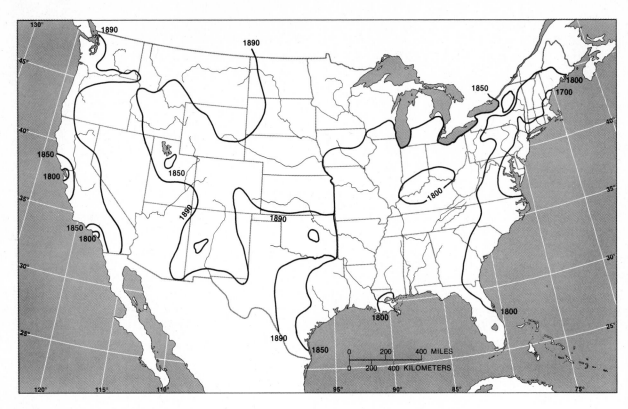

FIGURE 11-11
Westward movement of the frontier in the United States, showing the extent of settled area at six or more persons per square mile. (Data from National Atlas of the United States, 1968, pp. 134–139.)

to Turner, the availability of free or inexpensive land made possible migration out of a rigid social structure to areas of new settlement and opportunity (Figure 11-11).

The Turner thesis has been widely criticized, especially in matters of "political correctness" and detail, but it remains a thought-provoking set of ideas when one looks at the pioneer phase of development in a region.

TRADE AND TRANSPORTATION: THE MOVEMENT OF GOODS

Trade has been a motive for human action since human interaction began. Some of the earliest historical records are accounts of exchange. Some of the first roads discovered by archaeologists were trade routes, and a primary motive for exploration seems to have been the acquisition or exchange of goods. Marco Polo, Vasco da Gama, and Columbus all knew the value of opening new markets.

In the eighteenth and nineteenth centuries, colonial empires were built on the principle of exchanging raw materials in the colonies for finished goods from the mother country. To do this the British built the largest merchant fleet in the world and the largest navy to protect their trade advantage.

Trade also was a motive for settlement of new lands. Many places that were to become major cities were located at crossroads, harbors, river crossings, or other important sites that facilitated trade. Moscow was founded at a crossroads trading post, as was St. Louis. Singapore owed its competitive advantage to location on the Strait of Malacca. Both London and Paris were located on river trade routes. Timbuktu was located along the camel trade route across the Sahara Desert in North Africa.

James E. Vance, Jr., in *The Merchants World: The Geography of Wholesaling*, suggests that much U.S. geographical development was based on American endeavors in trade and profit making.[4] Decision making by speculators, traders, retailers, and consumers saturates the American experience. Trade may be described as the *exchange of value*, usually in the form of material goods. Most often, exchanged materials are transported from one place to another, although there can be trade without transportation, such as in a stock exchange.

When resources are extracted from their original sites, they usually are transported to users. The users in most cases are manufacturers in the secondary sector of the economy, who transform these goods into a large variety of marketable products that in turn are traded elsewhere. When commodities reach the tertiary sector, they most often are distributed by wholesalers, who specialize in the exchange of goods. A commodity may pass through two or more levels of wholesale distribution before it reaches the consumer through a retailer.

The Global Economy: The Development of International Trade

Patterns of today's international trade and commerce rest upon centuries of trial and error and ever-changing economic and political conditions. Most of today's active trading routes were established less than a century ago when it became possible for merchants to reach every part of the globe. The achievement of true world trade became possible with the Industrial Revolution and was facilitated by Western European colonialism. Thoman and Conkling, in their *Geography of International Trade,* suggest that a beginning point was the technical achievement of a steel-hulled, steam-driven ship with a screw-type propeller, which ensured the arrival of large cargoes almost anywhere on the globe.[5]

Western Europe dominated world trade through the nineteenth century and into the early twentieth. After World War I, other nations such as Canada, Japan, and the Soviet Union became important traders in the world economy. Coal was the primary commodity transported, but food, building materials, and minerals also were important.

Major economic changes in the first half of the twentieth century had profound effects upon today's trade patterns. For example, the establishment of Marxist states meant that barriers to exchange would be erected, particularly at the borders of the Soviet Union, Eastern Europe, and China. While these planned economies were growing, colonial systems were declining through national independence movements. After 1945, economic alliances formed in Europe, such as the Common Market, beginning a trend in other parts of the world for the establishment of economic power blocks. In the middle and late twentieth century, the growing economic importance of the Pacific Rim has added another significant new dimension to world trade patterns.

Major events after World War II created other signif-icant economic milestones related to trade. For example, the substitution of petroleum for coal increased the importance of the Middle East in world affairs. Inexpensive labor in the Far East gave manufacturing advantage to Japan, Korea, and Taiwan. Changes in technology, resource availability, and consumption patterns resulted in greater volumes of international trade than ever before as some nations enjoyed an era of prosperity from the 1950s to the 1990s.

Despite many global economic changes, the world pattern of trade routes has not changed significantly over the last century. Figure 11-12 shows the worldwide movement of goods and resources.

Patterns of World Trade Today

World trade routes have traditionally been based upon the economic importance of the nations involved and upon the most efficient or safest route from one point to another. In the northern hemisphere, the most important, and historically newest, route of trade is over the Atlantic Ocean, connecting Europe and North America. International use of routes connecting East Asia to North America and to Europe are increasingly used, as are routes from Middle Eastern countries to East Asia, Europe, and North America. Comparatively speaking, many trade routes in the southern hemisphere remain minor. Although traffic volumes are less, they are important as routes that carry high-value specialty products such as electronics goods, diamonds, copper, and even kiwi fruit.

Our present-day world is opening up to new routes that increase the number of nations participating in the global economy. The most heavily used trade routes tend to focus upon developed nations. The developed countries receive raw materials from the developing world or the economic periphery. One of the most important flows in world trade is the movement of energy sources, particularly petroleum. There are some exceptions to the generalization of raw materials originating only in developing nations. Examples are the U.S. export of lumber to Japan, the U.S. export of food grains to Australia, Canada, and the former Soviet states, and the Swedish export of iron ore to other industrialized nations.

World trade has increased continually since 1945, and this trend is expected to continue with the reduction of economic barriers between socialist economies and capitalistic nations. A current trend is for almost all countries, regardless of political orientation, to expand their global markets. These openings of economic doors frequently have been the beginnings of political accords and agreements. Several ex-

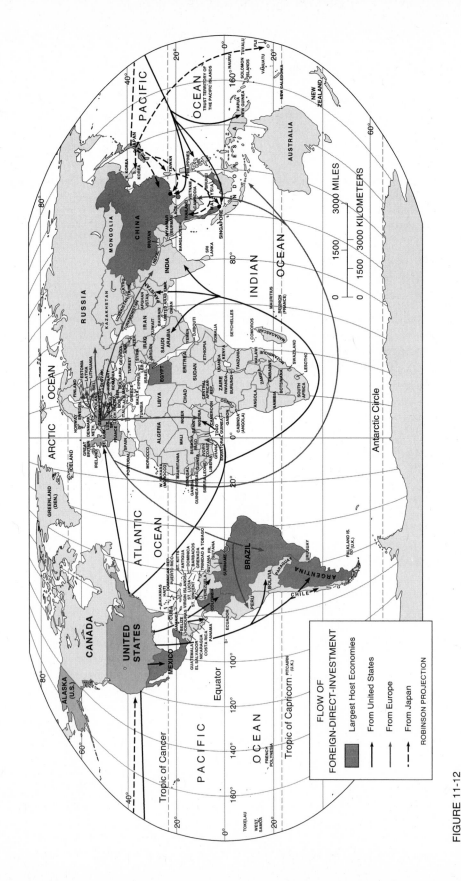

FIGURE 11-12

Flow of foreign direct investment in the world today. (Reprinted from Anthony R. de Souza and Frederick P. Stutz, The World Economy: Resources, Location, Trade, and Development, 2nd ed., New York: Macmillan, 1994, Fig. 12-13, p. 537. With permission of Macmillan Publishing Co.)

amples of global trade regions are offered here for comparison.

Eastern Europe. The participation of new consumers and producers from Eastern Europe in the global marketplace has changed these nations irrevocably. As *Time* magazine reported:[6]

> 1989 will be remembered not as the year that Eastern Europe changed but as the year that Eastern Europe as we have known it for four decades ended.

The demand for Western European and North American products is a current fact of life for almost all Eastern European nations. The desire for cigarettes, denim jeans, computers, and advanced communication systems has caused many companies to "go international."

The 1990s has been a decade of change in patterns of movement in Europe. Policies of international trade, migration, and communication have been reversed. In 1989, unprecedented events unfolded before our eyes as the entire structure of Eastern Europe was transformed by deci-

sions of the last president of the Soviet Union, Mikhail Gorbachev. In 1991, the Soviet Union itself ceased to exist, being replaced by Russia and other nation-states that were part of the former Soviet Union. Today, as European nations are rapidly creating economic unity and as more Eastern nations become part of the European Community and strive to create capitalistic economies, the whole global economy benefits.

Study the map in Figure 11-13 and reflect on how the European nations listed below are trying to change their economic orientation from communism to capitalism.

- Poland freed price controls and ended subsidies at the beginning of 1990. There was a huge redistribution of power and influence. Plans for mass privatization had been made that were based on giving free coupons to millions of people who used them to buy stock in newly privatized industries.

- Russian leaders have decided that mass privatization is the best direction to take. The scale of this task is overwhelming. All Russian citizens received letters inviting

FIGURE 11-13

Since the late 1980s and the opening up of Southern and Eastern Europe, the nations shown on this map have become increasingly involved in world trade and interconnections with the rest of the globe. (From James M. Rubenstein, The Cultural Landscape: An Introduction to Human Geography, *4th ed., New York: Macmillan, 1994, Fig. 7-16 (map only, no language caption), p. 287. With permission of Macmillan Publishing Co.)*

them to visit their local savings banks and to collect coupons enabling them to participate in the privatization of previously state-run industries. While many are distrustful and disillusioned with the idea of a market economy, most Russians are working to solve short-term transition problems.

- In 1993, Czechoslovakia split into two independent nation-states. The separation was caused by a difference of opinion between the Czechs, the majority and the Slovaks about how fast privatization of ownership of land and property should occur.

- Estonia is currently enjoying some economic success with the creation in mid-1992 of a strong currency, and with foreign investment from Scandinavia. Their gross national product continues to increase, industrial production is high, and the private sector is expanding.

With help from the European Bank for Reconstruction and Development many former Communist nations are beginning to move toward market economies. Economic reformers continue to work toward a completely unified Europe with a common currency and a charter guaranteeing the rights and responsibilities of all workers. The General Agreement on Tariffs and Trade (GATT) was formed to mitigate quarrels, ranging from a dispute between the United States and the European Community over government acquisition of industries to complaints about methods of trade in bananas. In 1994, the World Trade Organization ratified GATT and nudged the global economy toward freer trade.

With the unpredictable course of events that occurred in the early 1990s, the map of Europe will never again be the same as trade, travel, and communication continue to erode the old Iron Curtain into recyclable "scrap."

TRANSPORTATION

Transportation is the movement of goods and people through space. *Transportation systems* are the means by which movement occurs. The basic unit of a transportation system is a *trip*, with an origin and a destination. *Trip demand* (that is, *user need*) is assessed regarding the mode of transportation used, the most efficient route to take, the speed of travel, and the reliability of the system. An optimum transportation system becomes a network of interconnected components that provide choices among the above factors. For example, a mode of transit may be walking, riding a bus, shipping goods on trucks, an airline flight, or a railroad trip.

Modes of transportation have changed over the years as technology and need have changed. The route of travel also has been a matter of increasing choice as various modes of travel have been developed and introduced into regional and local settings. Historically, the speed of travel has been an important factor in transportation decision making; it has generally increased over the last century as more choices among modes and routes became available.

The most effective transportation systems coordinate their modes of travel in scheduling, routing, and logistics. The system is based on the least-effort principle, which means that the more choices there are for users, the more likely each user will be to select a suitable type of transportation and route. When all modes and routes are used effectively, it may be said that we are using an efficient transportation network.

The final evaluation of any transportation system comes down to three questions: (1) Does it promptly move the commodity (or person) from A to B? (2) Is the system easy to use and completely reliable? (3) Is it affordable? Try applying these questions to evaluate transportation systems you know, such as the U.S. Postal Service, an airline, a truck freight company, a courier service like Federal Express, and your own travel to school or work.

A transportation **network** is made up of nodes and links. A **node** is a starting or ending point, and a *link* is a travel connection between nodes. Some transportation networks focus at a primary node located at a central place, with radial links and more nodes in decreasing density going out from it. Other networks exhibit a rather uniform coverage of an area to be served, and still others may have unique configurations due to circumstances of topography, political boundaries, or population concentrations.

Development of the Transportation System in the United States

Transportation in the area that is now the United States and Canada was first evidenced in trails used by Native Americans. They did not use vehicles, so the extent of their transportation system was walking and—after the arrival of the Spanish—sometimes riding horses on these trails. Canoes and small boats were used where possible.

Europeans arrived by ship, and the primary form of transport continued to be maritime shipping along the Atlantic Coast between colonial settlements. After a time, these settlements grew and were connected by trails and roads, but it was not until the beginning of the nineteenth century that national roads were planned and built.

Canals became an extension of maritime commerce beyond the navigable portion of rivers and as links between rivers. As traffic increased on the rivers and canals, the energy used evolved from animal power to steam power. By the 1850s, a fleet of steam-powered watercraft used every major waterway in the eastern half of the continent, including the Great Lakes. In fact, the New York State canal system gave the New York City trade advantages over competing Philadelphia.

Travel time by walking, river, or coach in 1800 is mapped in Figure 11-14, based on a journey from New York City outward. This map shows that a trip to Albany took three days and a trip to Syracuse required six in 1800.

Waterway routes were determined by nature and did not always go to desired places, so the river and canal network was supplemented by short rail lines. Steam-powered rail traffic increased as settlement occurred farther from major waterways, especially as the frontier moved westward into arid regions. The steam-rail network reached its peak in North America in the 1940s, both in usage and in track mileage. Today's rail rider is more likely to be a commuter traveling home from work on an urban metro rail system like the one in Washington, DC, mapped in Figure 11-15.

While the rail system was developing in the first half of the twentieth century, automobile use was becoming more common. By mid-century, automobile and truck traffic became greater than rail traffic (by numbers of passengers and amount of goods carried). With development of the interstate highway system, initially funded in 1956 and completed by the early 1990s, virtually all parts of the country became accessible by automobile and truck.

New enabling technologies in the last half of the twentieth century have further reduced travel times, especially by air, so that we are now less than one days journey from any population center in the world. Likewise, developments in electronic communications have made it possible to speak to or to send a fax to anyone in the world with the appropriate receiving device at the speed of light.

Time-Space Convergence

As speed and convenience became more important in moving people and high-value goods, airline traffic increased. Airline routes have been established competitively, primarily for the convenience of travelers. Annual increases in air traffic underline the importance of time in dealing with distance or *space*. This relationship between time and space is conveniently labeled **time-space convergence.**

The importance of this time-space relationship is exemplified by noting that a trip from New York to San Francisco took over 100 days by sea in 1850. A land trip, without any misfortunes or setbacks, to the same destination took over 60 days under optimum conditions. By 1860, the trip could be taken on an overland stagecoach in 30 days, and the Pony Express carried mail communication across the continent in 10 days. By 1900, a rail passenger could make the trip in four days, and by 1940 a commercial airliner could do it in one day with several stops en route. Today, airline schedules promise a successful flight in less than five hours, depending on the jet stream and scheduling hassles.

Linear distance and travel time are not the only important factors in transportation decision making. Convenience is important, and cost usually is the most important factor for most people. Generally, the farther one travels, the more important time becomes. We usually are willing to commit 20 minutes for a 1-mile (1.6 km) walk, whereas a 20-mile (30 km) excursion is relegated to the automobile or to mass transit. A journey over 300 miles (500 km) usually requires pondering the alternatives of ground or air travel. For transportation of bulk products over very long distances, rapid transport by aircraft usually is not practical due to cost; and slower, but less expensive rail or sea modes are used.

It is important to recognize that the availability of a variety of transit modes, routes, speeds, and costs defines a maximum efficiency for the user population. This kind of system is usually developed at considerable capital cost and is more likely to be found in the developed countries of the world. Other countries may be in the formative stages of transportation systems development, perhaps still using animal power or foot traffic.

Evolution of Transportation in Third World Countries

A *Geographical Review* article by economic geographers E. J. Taaffe, R. L. Morrill, and P. R. Gould explored the importance of transportation technology in developing countries.[7] Focusing their attention mainly upon the African nations of Ghana and Nigeria, the authors developed a chronology of transportation processes that is typical of many developing countries.

It is typical for coastal nations to begin their transportation network with a series of scattered ports and small coastal settlements, linked by boat for trade. River mouths are especially favorable locations for these trading villages. The second phase involves an initial penetration line into the interior of the country—a connection often needed to

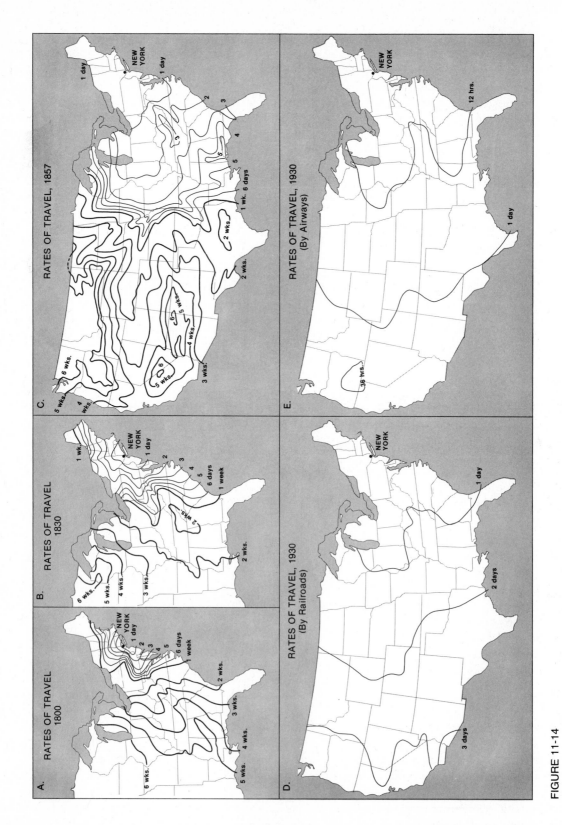

A. RATES OF TRAVEL 1800

1 day
2
3
4
5
6 days
1 week
2 wks.
3 wks.
4 wks.
5 wks.
6 wks.
NEW YORK

B. RATES OF TRAVEL 1830

1 wk.
1 day
2
3
4
5
6 days
1 week
2 wks.
2 wks.
3 wks.
4 wks.
5 wks.
6 wks.
NEW YORK

C. RATES OF TRAVEL, 1857

1 day
1 day
2
3
3
4
5
5
1 wk. 6 days
2 wks.
2 wks.
3 wks.
4 wks.
5 wks.
5 wks.
6
6
5 wks.
4 wks.
5 wks. 6 wks.

D. RATES OF TRAVEL, 1930 (By Railroads)

1 day
2 days
3 days
NEW YORK

E. RATES OF TRAVEL, 1930 (By Airways)

12 hrs.
1 day
36 hrs.
NEW YORK

FIGURE 11-14

Rates of travel from New York, 1800–1930. Contour lines showing travel time are called isochrones. (From Charles O. Paullen, Atlas of the Historical Geography of the U.S., Washington, DC: Carnegie Institution 1932, p. 137. Reprinted with permission of Carnegie Institution of Washington.)

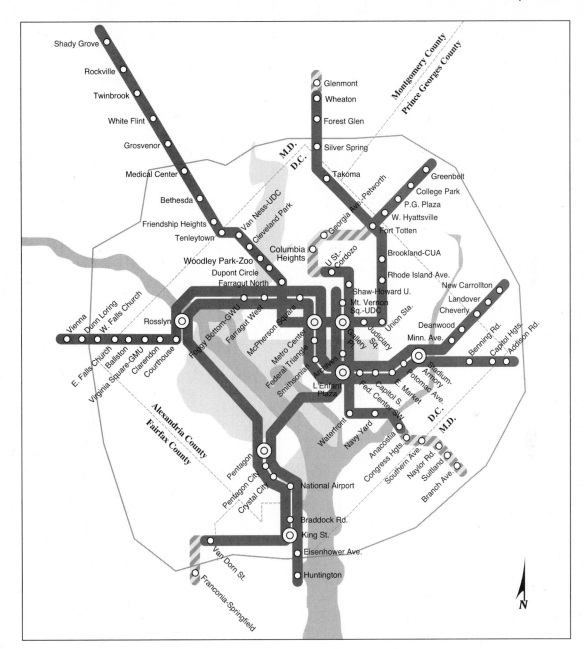

FIGURE 11-15
Metrorail, Washington, DC. (Reprinted from Anthony R. de Souza and Frederick P. Stutz, The
World Economy: Resources, Location, Trade, and Development, *2nd ed., New York: Macmillan,
1994, Fig. 6-15, p. 218. With permission of Macmillan Publishing Co.)*

link the political capital with the interior, or to help in mineral exploration, or to locate places with agricultural potential. The third phase includes development of feeders and
lateral interconnections. The fourth stage establishes high-
priority linkages (Figure 11-16).

The authors found that in Ghana, Nigeria, and other
developing countries, a steady increase in road traffic first
supplemented rail linkages and later became the dominant
form of transportation (Figure 11-17). The final two phases—complete interconnection and the emergence of high-

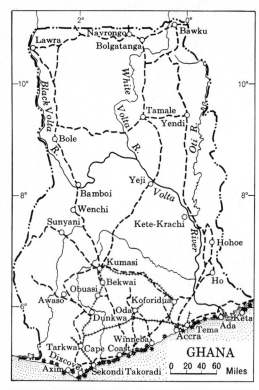

MAJOR TRANSPORT FACILITIES

+++++++++ Railroads

- - - - - - Main roads

* Former trading posts

FIGURE 11-16

Idealized transport development. (From E. J. Taaffe, R. L. Morrill, and P. R. Gould, "Transport Expansion in Underdeveloped Countries: A Comparative Analysis," Geographical Review, *53, 1963, Fig. 1, p. 504. Reprinted with permission of the American Geographical Society.)*

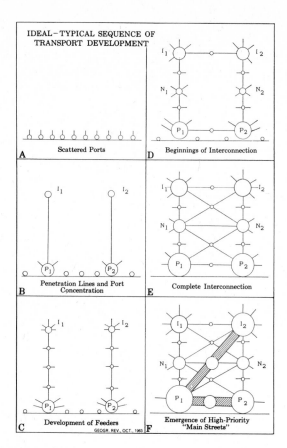

IDEAL-TYPICAL SEQUENCE OF TRANSPORT DEVELOPMENT

A Scattered Ports

B Penetration Lines and Port Concentration

C Development of Feeders

D Beginnings of Interconnection

E Complete Interconnection

F Emergence of High-Priority "Main Streets"

GEOGR. REV., OCT., 1963

FIGURE 11-17

Major transport facilities in Ghana. (From E. J. Taaffe, R. L. Morrill, and P. R. Gould, "Transport Expansion in Underdeveloped Countries: A Comparative Analysis," Geographical Review, *53, 1963, Fig. 3, p. 388. Reprinted with permission of the American Geographical Society.)*

priority "main streets"—was evident in Nigeria, where the seaport of Lagos appears to be developing into an important "main street" between cities in the western part of the country. High-density, short-haul traffic is focused in a high-priority coastal location. (For a description of mass transit in a developing country, see the box on Mexico City's metro system on page 295.)

Relationships between Movement and Land Use

Land use normally is directly related to population density; the more people there are, the more intense is the land use. Also, in industrialized nations, the more people there are, the greater is the need for movement into, out of, and within the region. Therefore, in developed areas, population density has a direct relationship to accessibility. The downtowns of cities usually have both the greatest variety of transportation modes and the greatest density of population to be served. As distance from population nodes increases, the choices and efficiencies of transportation availability become fewer.

Land use problems are intimately related to the transportation system. For example, if a freeway network becomes overcrowded during commuting hours, there may be pressure to develop vacant land on the urban periphery for alternative housing and employment centers. On the other hand, providing extra lanes to relieve a crowded freeway may shorten driving times, encouraging commuters to

---

MOVEMENT IN ACTION: MEXICO'S MAGNIFICENT METRO

Some developing nations have moved through the stages of transportation evolution rapidly and successfully. Mexico, for example, has constructed one of the world's most modern and technologically advanced subway systems in Mexico City. Designed to resist damage from Earth movements, the subway system was almost undamaged by the severe earthquake of September, 1985. By the evening of the same day as the damaging earthquake, its trains were functioning normally.

On a typical workday, Mexico City's Metro carries about 4 million of the city's estimated 17 million inhabitants. Automobile traffic in this crowded city is as bad as anywhere in the world, and gridlock is common during rush hours. The purpose of the Metro, as all subway systems, is to enable residents to go quickly and economically to work and shopping. Metro riders represent a real cross section of the city's population, with workers, students, businesspersons, children, and the elderly sitting together. A high level of order and cleanliness prevails, and this developing nation's subway is one of the safest in the world (along with Tokyo's and Moscow's). According to one rider of the Metro:

> Like the arteries of a giant organism, the Metro lines carry their millions of living corpuscles throughout the world's most populous city to their work and back to their rest. Those of use who ride it feel the vitality of this mighty current, a life force that is found deep in the sediments of prehistoric colonies and that, despite trials and misfortunes, continues undaunted toward the future.

From George R. Ellis, "Mexico's Magnificent Metro," Américas *(Sept.–Oct. 1986), pp. 2–7. Reprinted from* Américas, *the bimonthly magazine published by the General Secretariat of the Organization of American States in English and Spanish.*

---

drive farther, which may cause suburban growth at the edges of the city.

Usually a direct relationship exists between the cost of land (rent) and its accessibility via the transportation system. Accessibility means faster transit, so this saving of time may be important enough to cause land users to pay higher rents. (This is another proof of the saying, "Time is money.") Other relationships between urban accessibility and land use are examined in Chapter 9, "The Theme of Place: Urban Systems."

It may be helpful to return to the section on the development of North American cities in Chapter 9, and to review Borchert's suggested stages in the evolution of movement technology and the transport-network expansion process that shaped the distinctive pattern of urban land use in today's American cities. Borchert's work on the relationship between transportation and the evolution of modern cities underscores the importance of this topic.

COMMUNICATION: THE MOVEMENT OF INFORMATION

Our final topic under the theme of movement is **communication.** Early forms of communication included runners, flags, signal fires, and a healthy shout. With the Industrial Revolution, organized communication was made global with reliable mail delivery. Until electric technology became practical, however, information was only as fast as the fastest mode of transportation. Transmission of messages over long distances required a time investment as well as a physical one until the introduction of the electric telegraph in the mid-nineteenth century.

Geographers John F. Kolars and John D. Nystuen point out that various communications processes constantly occur in today's societies.[8] They classify communications systems by the number of persons involved in the exchange, the time factor (instantaneous versus stored), and the medium of communication (personal or mechanical). For example, one person may communicate with many through the media of radio, television, film, concerts, speeches, editorials, books, and the like. Storage of the message can be achieved with videotape, film, compact discs (CDs), newspapers, books, and such. Examples of group communication, where many communicate with one or many persons, are teleconferences, computer networks, and mail advertising. Internet messaging, mail communication, political lobbying, and telemarketing can be from one to one, one to many, or many to one.

Today's sophisticated communications networks are essentially invisible to most users and travel at the speed of light. With microwave technology, satellite relay stations, optical fiber systems, and a network of worldwide users, the telephone or fax call is the most efficient and least costly method of direct communication. Almost two-thirds of the telephone traffic flowing across the Pacific Ocean is now between fax machines. Optical fiber cables will soon link North America, Western Europe, and East Asia with the ability to carry 100,000 simultaneous messages. Communications costs are also becoming increasingly insensitive to distance. Mass-communication technology likewise has improved to where instantaneous television images are transmitted worldwide. A sporting event or summit conference now requires zero time-space; that is, it is available at the same time to all viewers.

With improved communication among interdependent institutions and agencies, such as financial institutions, government agencies, and education-research centers, their physical proximity becomes less important. **Agglomeration** of insurance companies in Omaha or book publishers in New York is no longer necessary, and the dispersal of "footloose" industries is apparent in many regional landscapes outside of the major metropolitan areas.

The city of the future may rely more upon communication than upon transportation. For example, a greater number of employees than ever before are now in communication with the "office" while doing their work at home on a computer. Perhaps the very definitions of the terms "home" and "office" will require clarification in the near future.

Transportation never takes place without communication. Airline reservations and appointments for meetings are obvious examples of the integration of the two. Airlines, trains, and even automobiles seldom move blindly, and they require preliminary communication to determine routes, the number of people to be served, and their ultimate destination. Cellular car phones and available GPS equipment make it virtually impossible to be lost or "out of touch."

Geographers Thoman and Conkling remind us that communication is important in the geography of economic activity in at least three critical ways:

1. Communication facilitates the knowledge of demand and supply for goods and services. Through communication, we agree on what commodities to ship, when, and where.

2. Communication is increasingly employed as a substitute for transportation. Teleconference calls are now used in remote areas to provide educational and health services where personal contact was previously necessary. Home work-stations are now common where one's efforts are transmitted via computer and modem to office, school, and bank.

3. Electronic communications technology is used to transmit signals from high-altitude and outer-space sensors to receiving stations on Earth, where the data become computer or photographic images of Earth's surface. (The techniques and values of remote sensing were discussed in detail in Chapter 3.)

Quaternary economic activities discussed in Chapter 8 are absolutely dependent on fast, reliable, inexpensive global communication. It is amazing to realize that the geography of communication may cease to have a time-space dimension in the not too distant future.

LEARNING OUTCOMES

Learning outcomes in *Guidelines for Geographic Education* for the theme of Movement are listed here. Students will be able to:

1. Explain why human activities require movement.

2. Describe ways in which people move themselves, their products, and their ideas across Earth.

3. Be aware that movements reflect global patterns of interaction between people in distant and nearby places.

4. Know that few places are self-sufficient and that therefore extensive human networks of transportation and communication are necessary to link places together.

5. Know that movements are concentrated in some areas

and sparse in other areas, thereby creating patterns of centers, pathways, and hinterlands.

6. Describe how changes in transportation and communication technology influence the rates at which people, goods, and ideas move from place to place.

7. Discuss how, over time, movements have increased in speed and frequency.

8. Know that all cultures are interdependent.

9. Discuss how movements can be planned and organized to save energy, reduce travel time, and conserve resources.

10. Explain why improving ease of movement requires careful analysis to determine the best location for each human activity.

KEY CONCEPTS

Acculturation
Agglomeration
Assimilation
Chain migration
Communication
Diffusion

Distance decay
Eco-migrants
Emigration
Frontier
Global economy
Immigration

Innovation
Material culture
Migration
Network
Node
Push-pull factors

Refugee
Time-space convergence
Trade
Transportation

FURTHER READING

Most introductory geography survey textbooks include sections on migration, diffusion, transportation, and trade. One that integrates all these topics is Ronald Abler, John S. Adams, and Peter Gould, *Spatial Organization: The Geographer's View of the World* (Englewood Cliffs, NJ: Prentice-Hall, 1971).

Migration in the context of the North American experience is surveyed in a collection of readings in *Migration and Residential Mobility in the United States* by Larry Long (New York: Russell Sage Foundation, 1988), and by Jesse O. McKee, ed., *Ethnicity in Contemporary America: A Geographical Appraisal* (Dubuque, IA: Kendall/Hunt, 1985); see especially the first chapter, "Humanity on the Move," by McKee. An excellent new book, focusing on the migration and settlement of one migrant group, is Richard L. Nostrand's *The Hispano Homeland* (Norman, OK, and London: University of Oklahoma Press, 1992).

One of the pioneer works to use the theme of movement in the study of cultural geography is Carl O. Sauer, *Agricultural Origins and Dispersals: The Domestication of Animals and Foodstuffs* (Cambridge, MA: MIT Press, 1969), originally published by the American Geographical Society in 1952.

Spatial diffusion is an important component for geographers who specialize in most aspects of human geography. Use of this concept is introduced in Peter Gould's *Spatial Diffusion* (Washington, DC: AAG, 1969). Diffusion as a research focus is analyzed in depth in Everett M. Rogers, *Diffusion of Innovations* (New York: The Free Press, 1983). This work presents a history of diffusion research and analyzes it in the context of decision making and as an agent of change in institutions and organizations. Another survey of the diffusion process and its economic and historic implications is Lawrence A. Brown, *Innovation*

Diffusion: A New Perspective (New York: Methuen, 1981). A case study of how diffusion works and how it becomes a component of major historical events may be read in Gerald F. Pyle, *The Diffusion of Influenza: Patterns and Paradigms* (Totowa, NJ: Rowman & Littlefield, 1986).

Concepts relating to other aspects of the movement theme are surveyed in Richard L. Morrill, *The Spatial Organization of Society* (Belmont, CA: Wadsworth, 1970), and John F. Kolars and John D. Nystuen, *Human Geography: The Study of Location, Culture, and Environment* (New York: McGraw-Hill, 1974). Kolars and Nystuen have written one of the few survey books that devotes special attention to geographic concepts in communication, as well as other forms of movement. The role of computers and communications in the geography of the future is discussed in Yoneji Masuda, *The Information Society as Post-Industrial Society* (Washington, DC: World Future Society, 1981), and Stanley Brunn and Thomas R. Leinbach, eds., *Collapsing Space and Time: Geographic Aspects of Communication and Information* (London: HarperCollins, 1991).

The diffusion of AIDS is discussed in the comprehensive *London International Atlas of AIDS* by Matthew Smallman-Raynor, Andrew Cliff, and Peter Haggett (Cambridge, MA: Blackwell, 1992). Also useful for research on this important topic is a special issue of *Science* 239 (February 5, 1988); and by William B. Wood, "AIDS North and South: Diffusion Patterns of a Global Epidemic and a Research Agenda for Geographers," *The Professional Geographer* 40 (Aug. 1988): 266–279. An article on AIDS by Peter Gould in *The Professional Geographer* 41 (Feb. 1989): 71–77, offers insight into the complicated methodologies of the geographic study of this disease.

A good survey of the field of transportation geography is *Modern Transport Geography* by B. S. Hoyle and R. D. Knowles (London: Belhaven, 1992). Three somewhat dated but still useful books on trade and transportation from Prentice-Hall (Englewood Cliffs, NJ) are James E. Vance, Jr., *The Merchant's World: The Geography of Wholesaling* (1970); Richard S. Thoman with Edgar C. Conkling, *Geography of International Trade* (1967); and Edward J. Taaffe and Howard L. Gauthier, Jr., *Geography of Transportation* (1973). Another useful source on transportation geography is A. F. Williams, "Transport and the Future," in *Modern Transport Geography,* B. S. Hoyle and R. D. Knowles, eds. (London: Belhaven, 1992).

Peter Dicken's *Global Shift: The Internationalization of Economic Activity* (New York and London: Guilford Press, 1992) summarizes recent developments in trade and transportation using numerous case studies of the movement of various products and raw materials, establishing linkages throughout the global economy.

A fascinating book that chronicles the movement of ideas, diseases, and culture into the New World after the voyage of Columbus is Alfred W. Crosby, Jr., *The Columbian Exchange: Biological and Cultural Consequences of 1492* (Westport, CT: Greenwood Press, 1972).

ENDNOTES

[1] Susan Wiley Hardwick, *Russian Refuge: Religion, Migration, and Settlement on the North American Pacific Rim* (Chicago: University of Chicago Press, 1993).

[2] Jesse O. McKee, ed., *Ethnicity in Contemporary America: A Geographical Appraisal* (Dubuque, IA: Kendall/Hunt, 1985).

[3] Frederick Jackson Turner, *The Frontier in American History* (New York: 1920, reprinted by University of Arizona Press, 1986).

[4] James E. Vance, Jr., *The Merchant's World: The Geography of Wholesaling* (Englewood Cliffs, NJ: Prentice-Hall, 1970).

[5] Richard S. Thoman and Edgar C. Conkling, *Geography of International Trade* (Englewood Cliffs, NJ: Prentice-Hall, 1967).

[6] *Time* 134 (November 6, 1989): 40.

[7] E. J. Taaffe, R. L. Morrill, and P. R. Gould, "Transport Expansion in Underdeveloped Countries: A Comparative Analysis," *Geographical Review* 53 (Oct. 1963): 503–529.

[8] John F. Kolars and John D. Nystuen, *Human Geography: Spatial Design in World Society* (New York: McGraw-Hill, 1974).

12 THE THEME OF REGIONS

Plantation house in Alabama, evidence of the special regional setting of the American South. (Photo by author.)

It was the five-year Wiley family reunion, and this year the clan gathered at Uncle Asa's farm in western Pennsylvania. Family members from all over the continent came to the Sunday afternoon picnic, and all were anxious to share their recent experiences and to learn what others had been doing since their last get-together. By the time the apple pie was served, Uncle Dick and Aunt Jane had told everyone who would listen about their recent trip through the former Soviet Union. They provided details and photographs about places experienced in central Asia, especially Kazakhstan and their travels through Siberia, Ukraine, and the dramatic city of St. Petersburg. They exclaimed in amazement about the variety of people, the different languages spoken, the foods tasted, and the climatic differences of each place. Uncle Dick ended his "lecture" with

the final comment that he had previously thought Russia and the other republics of the former Soviet Union to be only forests and snow.

Meanwhile, Grandpa Asa was describing the latest acquisitions to his wine collection, famous vintages from Burgundy, Chablis, Bordeaux, and the Loire Valley. Unfortunately, he offered no samples.

At the same time, Aunt Nancy and Uncle John were relating stories about their recent visit to England and Ireland to visit their son who was working there. Their enthusiasm about the friendliness of the people and the cozy landscapes of western Ireland enlivened the afternoon of family visiting.

Meanwhile, the family's political zealot, up-and-coming young James, worried aloud about the geopolitical aspects of the upcoming election. Realizing that the solid South was no longer unified in its political views, that the Northeast establishment was still in control in many ways, and that the arid West was again underrepresented in popular vote in relation to its resource base and areal extent, James was anxious to predict electoral results.

Cousin Jennie, who transferred recently from an Atlanta to a Reno banking firm, described her new company's connections to the global economy. She noted that most of their major business transactions were now with Pacific Rim countries, and that she might soon find herself traveling at company expense to Singapore, Hong Kong, and Seoul.

For Gloria and John, their first family reunion was somewhat of a culture shock, but they grew quite vociferous when the conversation turned to the effects of urbanization on their remote Appalachian farmstead. They described how air pollution, traffic, and increased crime had appeared in their valley as commuters built new homes on former pastures and corn fields.

The common thread linking all these conversations might have seemed obvious to an unobtrusive listener that summer afternoon. All were discussing places that were larger than local and smaller than global. Without even using the term, the Wileys were discussing **regions.** The concept is used in everyday conversation and a great deal in news reporting. What is a region? How can an understanding of the concept help us in geographic study?

WHAT ARE REGIONS?

The *Guidelines for Geographic Education,* referred to previously, note that the "basic unit of geographic study is the region, an area that displays unity in terms of selected criteria."[1] The *National Geography Standards* reaffirmed the importance of regions in Standard 5 as:

> The geographically informed person knows and understands that people create regions to interpret Earth's complexity.

The criteria chosen to define a particular region determine the usefulness of the concept. Regions may be based on criteria such as physical, cultural, social, political, or urban characteristics. The chosen characteristics set aside this area from others surrounding it.

News reports often refer to global regions such as the Middle East or Western Europe. Other regions are identi-

fied by landform type (the Great Plains, Appalachia). Other criteria are cultural affinity (Latin America, sub-Saharan Africa) or historical, linguistic, or economic similarities (Cotton Belt, Mediterranean Europe). Many regions display more than one identifying characteristic, such as the North American Sun Belt or North Africa.

The regional concept is a way of organizing our knowledge and management of places. Regions are used in decision making by almost all institutions and agencies that have to deal with large areas of land. Most often, they map regions as political units, administrative areas, or service zones.

Some geographers separate regions into formal and functional categories. A **formal region** usually has institutional or political identity and distinct boundaries (New England), whereas a **functional region** is identified by its activities, interconnections, and usefulness (Chicago metropolitan region). Regions sometimes are **homogeneous,** where the criteria are generally uniform (agricultural regions, linguistic regions), or sometimes **nodal,** where a core or vital center is important (urban regions).

A region may have observable boundaries, or it may be set off by more subtle differences. Regions have been called "contiguous communities created by people acting in concert."[2] In essence, dividing the surface of Earth into regions is a geographer's tool for identifying manageable units of study. A brief view of an atlas or the social studies section of your local curriculum library demonstrates the importance of regional thinking. Most geography books are about various regions of the globe. Far fewer books and articles are about the methods of selecting and studying regions. Most school geography textbooks are regional in their design, and geography courses at most universities include a number of regional offerings.

THINKING REGIONALLY: REGIONS OF THE UNITED STATES

Examples of how regional thinking is used to understand a large and complex nation may be found on various maps used in government, business, and universities. For example, Figure 12-1 shows the major physical regions of North America. Landforms are the criteria for delimiting these regions. Other physical geographic criteria used to distinguish regional patterns include climate, vegetation, and soils. (Of course, all criteria are interrelated.) Economic regions in the United States may be based on population, re-

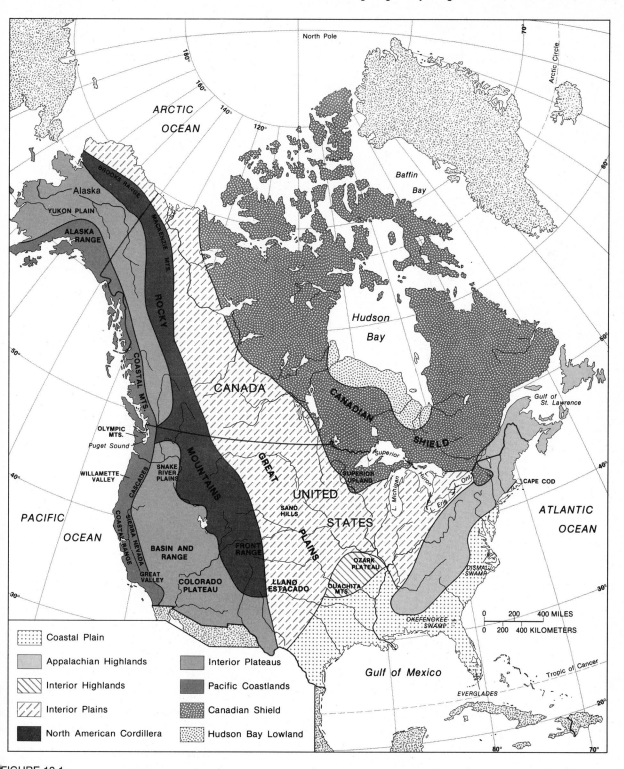

North Pole

ARCTIC
OCEAN

Baffin
Bay

Alaska

YUKON PLAIN

ALASKA
RANGE

Hudson
Bay

CANADA

Gulf of
St. Lawrence

OLYMPIC
MTS.

Puget Sound

CAPE COD

WILLAMETTE
VALLEY

SNAKE
RIVER
PLAINS

SUPERIOR
UPLAND

ATLANTIC
OCEAN

PACIFIC

OCEAN

SAND
HILLS

UNITED

STATES

BASIN AND
RANGE

FRONT
RANGE

DISMAL
SWAMP

OZARK
PLATEAU

GREAT
VALLEY

COLORADO
PLATEAU

LLANO
ESTACADO

OUACHITA
MTS.

0   200   400 MILES

0   200   400 KILOMETERS

OKEFENOKEE
SWAMP

Gulf of Mexico

Tropic of Cancer

EVERGLADES

Coastal Plain

Appalachian Highlands

Interior Plateaus

Interior Highlands

Pacific Coastlands

Interior Plains

Canadian Shield

North American Cordillera

Hudson Bay Lowland

FIGURE 12-1

Landform regions of the United States and Canada. (From James S. Fisher, Geography and
Development: A World Regional Approach, 3rd ed., Columbus, OH: Merrill Publishing Co., 1989,
Fig. 5-1, p. 115. Used with permission of Merrill Publishing Co.)

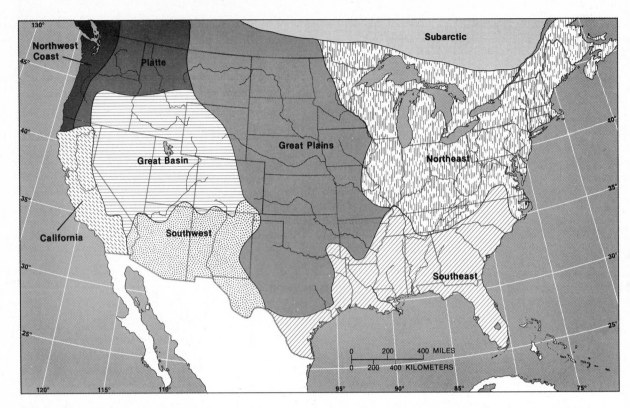

FIGURE 12-2
North American cultural groups before European contact. (Adapted from "Indians of North America," National Geographic, *142, December 1972, p. 739A, map supplement. Used with permission of National Geographic Society.)*

sources, land use, production, or a statistical combination of variables.

Figure 12-2 is a cultural map of Native Americans for the area of the United States before European contact. When compared with the physical regions map (Figure 12-1), there is a fairly good correspondence of boundaries. However, when compared with a present-day cultural regions map that is based upon criteria of language, ethnicity, and religion, there is very little correspondence of boundaries. This is because high-technology North American culture is much less constrained by physical geography than Native American culture was.

Political factors also may be criteria for regions. One of the most creative was drawn by G. E. Pearcy, based upon the interesting if unrealistic proposition that if Congress agreed to abolish historic state boundaries and start over, the political map of the United States might look like Figure 12-3. The new "states" are labeled with regionally identifiable names and are visibly centered on major urban areas.

Continental boundary making even found success in the popular press with publication of journalist Joel Garreau's *Nine Nations of North America* in 1981. Figure 12-4 shows his regional map based upon economic, cultural, and personal/subjective criteria. He suggests about North America:[3]

It is Nine Nations. Each has its capital and its distinctive web of power and influence. A few are allies, but many are adversaries. Several have readily acknowledged national poets, and many have characteristic dialects and mannerisms. Some are close to being raw frontiers; others have four centuries of history. Each has a peculiar economy; each commands a certain emotional allegiance from its citizens. These nations look different, feel different, and sound different from each other, and few of their boundaries match the lines drawn on current maps.

The major message in this map comparison is that regionalization is a technique and a tool for making sense of any large area that is under study for specific purposes. The

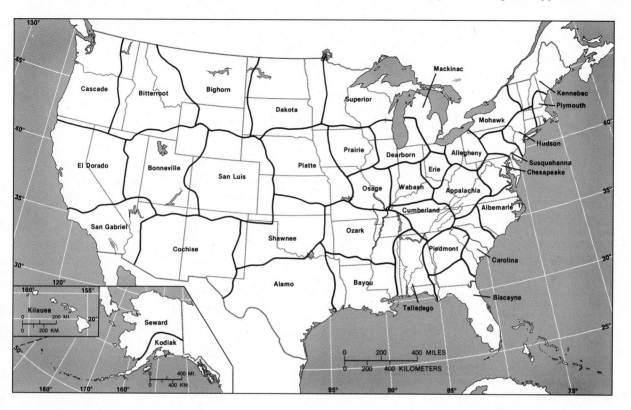

FIGURE 12-3
Pearcy's suggested new states of America. (From G. Etzel Pearcy, A Thirty-Eight State U.S.A., *Redondo Beach, CA: Plycon Press. Used with permission of Plycon Press.)*

techniques of regional analysis include description, mapping, statistical analysis, or landscape interpretation. When done properly, regional analysis is both functional and educational. Regional study is truly a "geographer's tool" because it crosses discipline boundaries and offers a holistic approach to learning about Earth.

SHORTCOMINGS OF THE REGIONAL APPROACH

It must be remembered that the regional approach is just that—an approach. It is not a theory or law. It is a method for categorizing and structuring ideas about places. When conducting a regional study, it is easy to fall into the trap of pure description. Listing only the characteristics of a region, or simply mapping them, sometimes beguiles the researcher into thinking that he or she has accomplished a

complete study and analysis. In fact, analysis of why the region exists, what forces and processes are at work within it, and its likely future need to be investigated and explained. Description is necessary before analysis is possible, but both are appropriate in a good regional study.

A second difficulty in working with regional analysis is choosing the level of detail to present. If the study is too general, it is meaningless. If the study is too detailed, the big picture may be lost. This judgment must be made with the original purposes of the study in mind. One strategy is to create subregions within regions, but this too can confuse logic and organization.

The regional approach has been used by social scientists, particularly during the 1930s and 1940s. It has been borrowed by political thinkers for rationalizing political issues, but there exists the danger of abusing the regional method for the special ends of interest groups and political movements. The status of Basques in northern Spain, the Irish in Ulster, and the French Canadians in Quebec are all inflammatory issues underlain with a strong politi-

FIGURE 12-4

The Nine Nations of North America as defined by Joel Garreau. (From Joel Garreau, Nine Nations of North America, *Boston: Houghton Mifflin. Copyright © 1981 by Joel Garreau. Reprinted by permission of Houghton Mifflin Co.)*

cal agenda. Each group forms a cultural/ethnic region, but recognition of such regions may set them apart for persecution.

A fourth difficulty in working with regions is that they sometimes are very dynamic. Over short periods of time they may change their composition and even their reason for being. Regional thinking often is used in historical study and can carry with it the illusion that a place still exists in a sort of time capsule. A notable study by geographer Donald Meinig of the Great Columbia Plain in eastern Washington State was based on, an important early-twentieth century wheat-producing region.[4] Today, although still a producer of wheat, there is little national or local identification of this place as a separate economic region. In this classic regional study, the historical context of the region was kept before the reader as it evolved in importance, and then faded.

As modernization has taken place in various countries in Africa and in China, regional differences among formerly isolated groups have begun to dissolve. Regional distinctions among peoples in developing nations exist only in the memories of their elders, as cities rapidly become Earth's new regional nodes.

In summary, use of the regional concept must be based upon objective criteria and up-to-date information, and must assume only what can be validated with field observation and intellectual scrutiny. Regional boundaries must not carry the illusion of permanency or precision.

THE REGIONAL CONCEPT

Early Chinese, Greek, Roman, and Muslim travelers were relatively isolated, so their narratives about places on Earth were distinctly based upon their observation of local regions. Early geographers lived in isolation and were unable to draw many comparisons among places or to discuss patterns and processes. Even in the Age of Exploration, Renaissance voyagers had no comparable data of various areas to use in regional analysis. When the Pope drew the line at 50° west longitude to divide the New World into Spanish and Portuguese territories in 1494, he really knew very little about the region's geography. (The legacy of this uninformed decision remains to this day: most Brazilians speak Portuguese, not Spanish.)

It was not until the nineteenth century that the regional approach became the dominant form of geographic thinking. German scientists Alexander von Humboldt and Carl Ritter first mastered the regional method as we know it today.

Von Humboldt, a founder of modern geography, was interested in understanding the interrelationships between Earth features and landscapes. He published vivid, detailed descriptions of areas and compared them with other places. In his monumental work *Kosmos,* von Humboldt's main interest was in physical and biological features. However, he also wrote many books on human geography, commenting for example on the relationship between Asian and Native American cultures. His studies of Cuba, Mexico, and the United States were essentially the first regional geographies.

Carl Ritter was more interested in the human experience on Earth than in physical geography. Although he traveled less than von Humboldt, he was able to use data gathered by others to draw conclusions and comparisons between places. He divided the entire Earth into landform regions and related their features to human occupance. Although Ritter has been criticized by many for letting his spiritual beliefs influence his "scientific" thinking, the development of regional geographic methods was greatly advanced and popularized because of his work.

Using the foundation in regional human geography established by Ritter, French geographer Paul Vidal de la Blache became known as the "father of regional geography." His masterful regional study of France, *Tableau de la Géographie de France,* was a milestone in the study of regional interrelationships and spatial patterns when it was published in 1903. His emphasis was always upon human influences in shaping the landscape.

In the United States, the regional method caught on in the early twentieth century, and its merits and limitations have been debated ever since. In his 1982 presidential address to the Association of American Geographers, John Fraser Hart spoke strongly of the need for well-written regional monographs in the discipline.[5] Hart's interest in regional studies echoed earlier pleas by Preston James for a return to regional geography. In "Toward a Further Understanding of the Regional Concept," James argued that the essence of good geography is identification of interrelationships among apparently disparate phenomena.[6]

The importance of these interrelationships was stressed by another well-known adherent of the regional method in geography, Richard Hartshorne. Hartshorne eloquently defined the region as "simply the device whereby finite minds can comprehend the infinitely variable function of many semiindependent variable factors."[7] These variables are best defined in both spatial and social terms in a complete regional analysis.

GEOGRAPHER AT WORK IN PUBLIC EDUCATION AND REGIONAL SCIENCE

Thomas J. Baerwald, Director of National Science Foundation's Geography and Regional Science Program.

National Science Foundation geographer Thomas Baerwald is a Midwesterner who has been interested in his heritage since childhood. He recalls doing extra-credit projects as early as the second grade that involved collection and evaluation about regions of the United States and the world. Family travel reinforced his wonder about the similarities and differences among places. Tom also began exploring the ways that geography affected historical events. Typical of his student projects in high school was the identification of George Rogers Clark's route to recapture the British post at Vincennes during the Revolutionary War. Tom used topographic maps to identify trails, little realizing that he later would teach others to use maps to understand patterns of movement.

Tom attended Valparaiso University in Indiana, majoring in geography and history. Graduate school at the University of Minnesota followed, where his interests led him to study transportation and urban geography. In the early years of computer modeling, Tom's interest in the versatility of geography helped him to develop models of how cities function and change.

Upon graduation, Tom chose to combine his research skills with his interest in education by joining the staff of the Science Museum of Minnesota. As director of the museum's Geography Department, he helped develop exhibits visited by more than three-quarters of a million people each year. Although public education was paramount to his position, Tom continued to conduct his own research.

After ten years of success at focusing public attention upon geography, Tom was named Director of the Geography and Regional Science Program of the National Science Foundation (NSF) in Washington, DC, where he solicited and screened geographic research proposals. His experience in working with all kinds of geographers and with scientists in many related fields led to his selection as coordinator of the NSF Global Change Research Programs. In this capacity, he has spurred the development of research throughout the U.S. government on the human dimensions of global change—the ways that people alter natural systems and the ways that they respond to changing conditions in the natural environment.

Tom Baerwald's philosophy of geography is that actions speak far louder than words. "Many people talk of what geography can do. I've tried to show its potential through my research and educational activities." Tom has discovered that words can be powerful actors, however. He is co-author of a popular secondary school world regional geography text, and he has contributed to other books that will allow people with varied interests to discover the excitement and importance of geography.

Early studies in regional geography were decidedly nonurban. Studies of rural regions, which emphasized the physical geography of an area, remained unconcerned with land use patterns that were determined predominantly by human decision making. Interest in the complex set of urban variables remained on the outer fringe of the discipline until the 1960s.

After several decades of "regional geography bashing," the study of world regions is once again enjoying popularity among professional geographers and classroom teachers. Today's regional analysis often involves the use of GIS (geographic information system) to interrelate variables over large areas and other computer-based systems of analysis. As the world continues to shrink in the new century, an understanding of the complexity of its many regions will become an even more important focus for geographic study.

For a view of one person's important and interesting career that began with the study of regional geography, see the box on page 306.

CASE STUDIES: TYPES OF REGIONS

A Physical Region: The Amazon Basin

Physical regions may be defined by landform boundaries, climate, soil, water features, or vegetation biomes. One clearly distinguishable and fascinating physical region is the Amazon River Basin in South America. A diligent search through an atlas reveals that various parts of the Amazon Basin can fit within various different criteria. It lies within the boundaries of several countries (Brazil, Venezuela, Colombia, Peru, and others); contains several mineral resources; and exhibits a variety of extractive economies, including plantation farming, grazing, logging, and subsistence farming (Figure 12-5). However, the commonality to all of this area is *physical.*

The drainage of the Amazon River and its tributaries is the largest in the world, encompassing 2.4 million square miles (6,216,000 km²). The basin is set off by high mountains only in the west—the Andes Mountains. Streams in the Andes that headwater less than 100 miles (160 km) from the Pacific Ocean find their way eastward 3900 miles (6275 km) to the Atlantic Ocean! Otherwise, the topography is composed of low hills between rivers and a slightly higher rim of hills around the basin's perimeter.

At its 90-mile (145-km) wide mouth, the Amazon River is by far the largest flow of freshwater on Earth. This great volume is the result of the second commonality within this region: climate. Like other humid tropical climates in Africa, Southeast Asia, and the Pacific Islands, there is very little seasonality. The average annual temperature is about 80°F (29°C) and varies little through the year. In fact, the diurnal (day-night) range of temperature, which can be 10°F to 15°F, is greater than the seasonal range of 10°F! Daytime temperatures are high but not as extreme as those experienced in some desert climates. However, humidity is also high and is closely associated with daily convectional rains. Average annual rainfall ranges from 50 inches (127 cm) in the lowlands to 120 inches (305 cm) in the upper slopes. An accurate summary of this climate is "warm and wet."

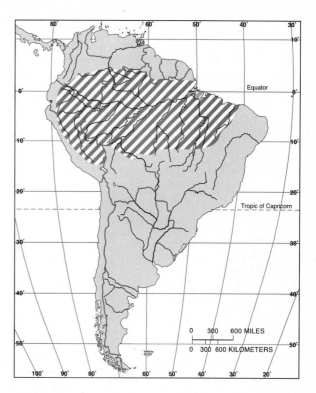

FIGURE 12-5
The Amazon Basin is a watershed, a vegetation zone, and a historically unified cultural realm. It is overlain by several divisive international boundaries and economic activity zones.

Before significant human impact, the natural vegetation of this physiographic region—tropical rain forest, or *selva*—uniformly covered almost all of the Amazon Basin. As noted in Chapter 10, this ecosystem holds more than half of the plant and animal species on Earth and, as such, holds the major potential for the world's future biological diversity, or lack of it, depending upon resource management policies of the countries involved.

Human habitation of the region before European contact consisted of small bands of hunter-gatherer Indians and groups that practiced agriculture by clearing small patches of forest for planting maize, beans, and squash. This ecologically stable system of rotating the forest clearings as the soil became depleted is called slash and burn (see Chapter 8). The abandoned cleared land was in sufficiently small parcels that the forest reentered and regrew to its former composition and density. Also, before European contact, the rotation cycle was long enough to allow the forest to regenerate.

With the exploration and settlement of tropical South America, the coastal zones and mouth of the great river became westernized in population, economy, and culture.

Plantation crops replaced the slash and burn system, with rubber being the major commodity until about 1910. The population became a mix of Portuguese, Indian, and African heritage that today is labeled *Mestizo*.

The major route into the interior during the latter half of the nineteenth century and the first part of the twentieth was the Amazon River system. Since the 1960s, river travel has been augmented by immense road-building projects. Minerals transported by river and the along new roads are iron, bauxite (aluminum ore), manganese, gold, and copper. Few cities were built, but some of the commercial needs of this vast area are met by Manaus (population 1,010,558), Belém (1,246,435), and Santarém (265,105) in Brazil, and by Iquitos (269,500) in Peru. As of yet, no large metropolitan center serves this area, which is as large as the continental United States.

Today the Amazon Basin is being exploited in a frontier atmosphere similar to the settlement of the American West. The native Indians are decreasing in numbers, and their indigenous cultures are being overwhelmed by the invasion of farmers, construction workers, investors, and others from the coastal zones, who are arriving at the rate of 200,000 settlers per year. While providing work for a significant portion of Latin America's population, it is at the cost of large volumes of irreplaceable natural resources. Forest clearing usually results in short-term profit but long-term soil erosion, resulting in abandonment or agricultural poverty. If the present clearing rate of 5000 square miles per year (13,000 km²) continues, it is generally agreed that the virgin forest will be gone by the year 2035. It also has been recently confirmed that massive modification or removal of the selva may affect global climates and even the composition of the atmosphere by reducing this major source of oxygen (Figure 12-6).

New transportation and communication routes are opening the interior from all sides. Oil is now piped from eastern Peru to the more populated western zone, and Venezuela is exporting products into the northern part of the Amazon region. However, most immigration into the basin is from the Brazilian coastal cities of Rio de Janiero and São Paulo.

Although unlikely in the near future, a solution to regional exploitation and long-term ruin would be regional and international cooperation in resource management, planned development, and urbanization of selected areas, while preserving other areas in their natural state. Movement toward a sustained economy that is environmentally sensitive should involve land reform and political stability, as well as support from the more-developed na-

FIGURE 12-6
Destruction of the rain forest in the Amazon Basin affects local economies in the region and may ultimately affect world climates.

tions. At this time, however, this form of comprehensive regional planning does not seem possible.

A Political Region: Poland

The traditional method of treating political geography with a regional perspective is by examining collections of adjacent nation-states. Newspaper references to Eastern Europe, southern Africa, the Middle East, or Latin America imply a discussion of several cultures and economies. This case study of the nation-state of Poland is slightly different because the Polish nation—that is, the group of people held together by heritage, language, and religion—has occupied central Europe for centuries, but the political boundaries affecting these people have changed perhaps more often here than in any other part of the developed world. This antithesis of boundary demarcation and the actual settlement of an ethnic homeland provides a powerful case study with which to illustrate the importance of the cultural imprint of people who are attached to a homeland, and the importance of the overlay of political decision making on maps and on their land.

The flat Eastern European plain that crosses into the vast steppes of central Eurasia has long been a primary route of European and Asian invaders, from the Old Stone Age to the dawn of this region's recorded history in the tenth century. Due to its location, Poland's level land south of the Baltic Sea provided the stage for wave after wave of invading nomads, raiders, and settlers (refer back to Figure 11-13).

Today's nation-state of Poland achieved its autonomous existence only recently. The Poles once ruled a empire that stretched across most of central Europe but, largely because of its defenseless location, foreign powers invaded and conquered Poland until it lost its independent existence altogether. After several centuries of foreign rule, Poland regained its independence as a nation after World War I in 1918.

The Polish empire reached its height in the sixteenth century when it covered large portions of Ukraine and Russia. Poland and Lithuania were united under a common parliamentary government in 1569. This parliament was later to become the cause of internal tension that reduced Poland's internal strength and resistance to attack from outsiders.

In the seventeenth century, Poland lost much of its territory to Austria, Prussia, and Russia. Austria took territory from southern Poland; Russia acquired land from the east; and Prussia took much of its western territory. The political dominance of Poland by its neighbors continued until World War I. At one time, in fact, Russian was made the official language of the nation; later German replaced it. During this long history of partitioning, much of Poland's national identity was based on loyalty to the Roman Catholic church. Polish Catholicism, distinct from Eastern Orthodox and Protestant beliefs common in surrounding countries, has long remained a deep, unifying factor for the region. Likewise, the Polish language served as a bond of unity for its people.

The defeat of Poland's occupying powers during World War I made possible a rebirth of the Polish national state in 1918. Six wars were fought in the aftermath of World War I to shape Poland's borders. In 1921, a compromise was finally reached, and Poland became an independent nation after centuries of foreign rule.

This independent status did not last long. In 1939, at the beginning of World War II, a secret treaty was signed by the Soviet Union and Germany to divide Poland almost in half. Following the invasions of the Soviets and the Nazis, a Polish government was established in exile in Paris and later in London. Despite the lack of a homeland, Polish national identity continued. Fighting alongside British, French, and other Allies, the Poles continued their resistance of German dominance during World War II. At the end of the war, Poland was once again recognized as an independent nation. Its boundaries, however, were moved 150 miles (240 km) west, thereby retaining only 54 percent of the area of prewar Poland.

After centuries of boundary changes, domination by foreign powers, warfare, and destruction, Poland today remains a clearly identifiable region on the map and in the political arena. Communist rule was firmly established in

the 1947 elections and a strong central government ensued, but under the dominance of the Soviet Union. However, in the 1980s, due to Poland's strong national identity and charismatic internal leadership, the nation resisted total control by the Soviet Union. After numerous strikes, riots, and demonstrations, the Polish people today have reestablished their ties with the Roman Catholic church and have begun to improve their economic and political relationships with other countries in Europe and throughout the world. Free at last to decide their own fate, the Polish people where well on their way to establishing a successful free enterprise system by the mid-1990s.

An Ethnically and Culturally Complex Region: Nigeria

Ethnic and cultural subregions often overlap within one political region. A case study of the country of Nigeria provides geographers with an opportunity to observe and study cultural diversity in a regional setting. Nigeria is the most populous country in Africa, with more than 100 million people today. It has a significant national history and a high standard of living when compared with other sub-Saharan countries. It is an important exporter of oil and other natural resources and thus is important in the global economy. Life in Nigeria, however, is not without its difficulties. There has been great stress in the nation due to cultural and economic differences, such as the civil war in Biafra, where over 1 million people were killed and great economic hardship occurred.

Nigeria is a product (or victim) of regional differences. It is made up of elements from 10 physical regions and components of 250 tribal groups that historically have lived in the area. (Over 800 culture groups are differentiated by language in Africa south of the Sahara Desert.) Nigeria thus provides a case study for geographers interested in how regions, like many nations, exhibit centrifugal forces, which tend to be divisive, and centripetal forces, which tend to unify a people.

Three-fourths of the population of Nigeria belongs to one of three large culture groups: the Hausa, the Yoruba, and the Ibo. Each cultural group is centered in a distinctive portion of the country with overlapping populations crossing the national political boundary into neighboring countries. These culture groups may best be identified by language and religion (Figure 12-7).

The Hausa live in the northern arid portions of the country. They are primarily Muslim and pastoral in their economy. Their political allegiances are based on emirates, much like those found in the Arabic states. Decades ago,

FIGURE 12-7
Cultural groups in Nigeria. (From James M. Rubenstein, The Cultural Landscape: An Introduction to Human Geography, *2nd ed., Columbus, OH: Merrill Publishing Co., 1989, Fig. 4-12, p. 144. Reprinted with permission of Merrill Publishing Co.)*

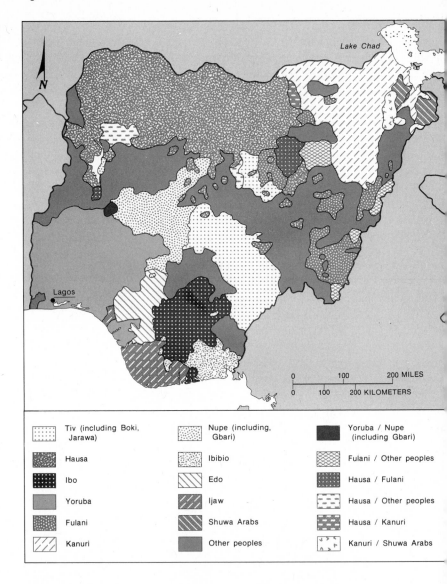

Tiv (including Boki, Jarawa)	Nupe (including, Gbari)	Yoruba / Nupe (including Gbari)
Hausa	Ibibio	Fulani / Other peoples
Ibo	Edo	Hausa / Fulani
Yoruba	Ijaw	Hausa / Other peoples
Fulani	Shuwa Arabs	Hausa / Kanuri
Kanuri	Other peoples	Kanuri / Shuwa Arabs

the Fulani people moved into this area from the north and west and began intermarrying with the Hausa. Today, this blended northern culture group is identified as the Hausa-Fulani.

The Yoruba are found in the southwest portion of Nigeria, mostly in towns and cities. Their language is also unique and their religious beliefs are either Christian or tribal. Much of their identity comes from their belief in a divine king, Oba. The Yoruba are primarily subsistence farmers and are the political plurality. The urban-dwelling Yoruba exercise the most political control in Nigeria.

The Ibo are primarily in the south, central, and eastern parts of Nigeria and are predominantly Christian. They are both farmers and city dwellers. They dominate

white-collar employment, having been clerical and administrative leaders under the colonial British system.

With three divergent ways of life, it is not surprising that there has been considerable political instability in Nigeria. Economic, physical, and tribal regional differences seem to be blurred by the overwhelming dominance of the three cultural groups and their competition with one another.

An Economic Region: The American Corn Belt

Many regions provide a historical perspective about past land use patterns on Earth. Like most other types of regions, economic regions are not static but ever-changing

Regions come and regions go. The North American Corn Belt is a well-known economic region that has changed significantly in recent decades.

Although corn is grown in many parts of the United States, the Corn Belt is traditionally a transition zone of agricultural production between crop-raising land to the east and animal-grazing land to the drier west (Figure 12-8). An area extending from Ohio to the Dakotas, centered in Iowa, this huge region has approximately half its land planted in corn. Much of this high-yielding grain is used to feed hogs and cattle. Pigs usually are raised in local farms and cattle are brought in to be fattened. This system makes available an extremely large amount of meat for consumers and thus responds directly to the high demand for meat protein in the United States.

Economic geographers have been interested in the Corn Belt for many decades. John Fraser Hart, for example, has published numerous well-known articles on the region. Among his many observations, perhaps the most compelling are his comments on the importance of the family farm. Hart has said that the family farm is more than just an economic aspect of this region; it also is an ideological system that is value-laden and "encrusted with emotionalism and partisan politics."[8] Although many people in the United States have never even visited a family farm or thought of it as an economic system, most agree that it is a "good thing" and should be preserved. According to Hart, the following sentiments are connected with the Corn Belt region:[9]

1. Each farmer should own the land he or she farms.
2. Each farm should be large enough to provide a decent living.
3. The farmer and his family should do most of the work.
4. They should receive a fair price for their products.

These sentiments then assume the following values:

1. Hard work is a virtue.
2. Anyone who works hard can make good.
3. Anyone who fails to make good is lazy.
4. A man's or woman's true worth can be determined by his or her income.
5. A self-made man or woman is better than one "to the manor born."
6. A family is responsible for its own economic security.
7. The best government is the least government.

Other considerations of the region known as the Corn Belt include its environmental advantages. These center on its humid climate, level land, fertile soils, and warm summer growing season. Sociocultural and economic advantages include the Corn Belt's relatively dense population, industrious and innovative labor supply, high degree of mechanization, strong purchasing power, well-developed transportation system, and dietary preferences. Due to these positive factors, the region has maintained a strong, stable economy until recent years. Its future is now less certain as myriad unresolved challenges assault this economic region: exorbitant new farm start-up costs, high land costs and high taxes, problems relating to equity versus debt capital, loss of prime agricultural land to urbanization, intergenerational transfer problems, and a change in federal agricultural policies.

An Urban Region: The San Francisco Bay Area

Other regions are distinctly urban. Upon hearing the name San Francisco, many images come to mind. Beyond thoughts of cable cars and bridges and earthquakes comes the understanding that San Francisco is not just a tourist destination, not just a city, but is in fact the core of activity in a large part of central California, with influence reaching to the farthest parts of the Pacific Basin and to other vital centers in North America.

The San Francisco Bay region may be defined by physical criteria as the area west of the Coast Ranges, south of the redwood forests of northwestern California, and north of the agricultural Salinas and Santa Clara valleys (Figure 12-9). The California Air Resources Board officially identifies this area as the Central Coast Air Basin. In economic terms, San Francisco really includes San Jose with its microchip and related electronic industries, the mammoth port of Oakland with its container-ship terminals, the oil refineries at Benicia, Martinez, and Richmond that process petroleum from all parts of the globe, and well-known agricultural valleys, such as Napa and Sonoma. For governmental and political purposes, the San Francisco Bay region is defined as the nine counties bordering on the bay and the several hundred cities and special districts within them.

The urbanized area that includes residential, industrial, and commercial land use is significantly smaller than the regional map shows. It may be seen that this particular urban region has not one node but three, each a city of over half a million people. Oakland, San Jose, and San Francisco are intimately connected by several transportation routes. These distinct nodes are surrounded by their own spheres of influence. Burlingame is most closely asso-

FIGURE 12-8
A historic economic region—the American Corn Belt.

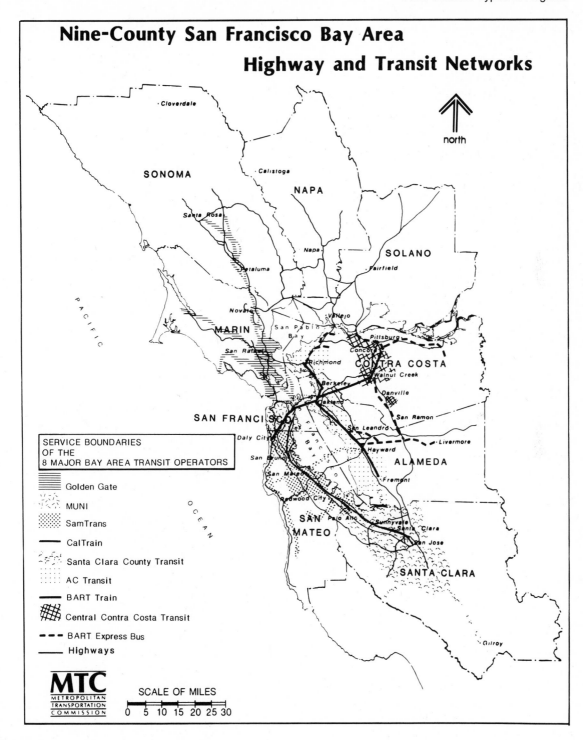

FIGURE 12-9

*An urban region—the San Francisco Bay area. (*Nine-County San Francisco Bay Area Highway and Transit Networks *map reprinted with permission of Metropolitan Transportation Commission, Oakland, CA.)*

ciated with San Francisco and Cupertino is associated with San Jose, but they are within a half-hour drive of one another.

While the communities in the San Francisco Bay region compete for their market share and for housing and employment advantages, they acknowledge that there are many problems and issues that only can be addressed in cooperation. Maintaining air quality, efficient transportation, flood control, crime prevention, and open space require regional planning and cooperation. The Association of Bay Area Governments, along with other specialized agencies, such as the Bay Area Air Quality Management District and the Metropolitan Transportation Commission, represent the formal attempt to address regional problems. In all, there are nearly 1000 local and regional governmental agencies in the San Francisco Bay region that require some degree of coordination in matters of mutual concern. As planning for the next decade is undertaken, the importance of regional thinking is underscored in the news media, at the ballot box, and at the personal level.

THE FIVE THEMES IN ACTION

Learning about Japan through the Five Themes

Japan most often is thought of as a nation-state, but it also may be viewed as a region in itself. In this case, the context of the region is not continental but rather global. All of the common attributes of a region are found in a study of Japan (and thus may be applied to any other nation you may be studying). Each of the following regional attributes is discussed in terms of strengthening or weakening Japan's regional identity.

Location. Japan lies off the east coast of the Asian continent, including four large islands and about 4000 smaller ones along an archipelago that stretches from 24° to 45° North latitude. This span of latitude is roughly equivalent to a span in the United States from the Gulf of Mexico to the Canadian border. Japan's capital city, Tokyo, lies almost at the same latitude as Los Angeles. The islands also form part of the western edge of the Pacific "Ring of Fire," an active volcanic and seismic zone that rims the Pacific Ocean. Administratively, the country is divided into 47 prefectures.

Historically, Japan's relative isolation from the rest of the world, both temporal and spatial, allowed influences from East Asia to be selectively adapted and assimilated into the indigenous culture. However, recent Japanese political and military influence has extended beyond the islands into Korea, Manchuria, several northern islands east of Siberia (currently occupied by Russia), and several islands between Okinawa and the Philippines. Figure 12-10 shows the historical and present extent of Japan's influence.

Obviously, there is no definite boundary surrounding this region. It is more a transition zone of lessening strength as one travels further from the Japanese homeland. Indeed, Japan's economic influence reaches well into North and South America and into much of the rest of the world today.

Japan's location is often referred to as the "Far East" or the "Orient." It is important to realize that these are relative terms used from a European perspective. In fact, Japan lies far *west* of the United States! Only with recent improvements in transportation and communication has Japan's location been less of a barrier to its participation as one of the key players in the economic region known as the Pacific Rim.

Place: Physical Characteristics. The volcanic origin of the Japanese islands is exemplified by one of its best known landmarks, Mount Fuji, one of its 265 volcanic peaks. Frequent volcanic activity along the "Ring of Fire" also places Japan within an earthquake region. A chain of mountains runs down the middle of the archipelago, dividing it into the Pacific side and the Sea of Japan side. This factor has forced the inhabitants to look outward toward the sea, seeing it both as a link and as a barrier. Almost 70 percent of the land is forested. Approximately 15 percent is devoted to agriculture, and only 4 percent is used for housing. Figure 12-11 shows Japan's complex land utilization patterns.

As shown in Figure 12-12, climate in Japan varies with latitude along a north-south axis, ranging from subtropical to subarctic. All of the country is strongly influenced by the westerly winds from Asia, ocean currents, and subtropical storms (typhoons) from the Pacific Ocean. In general, Japan has hot, humid summers but cold winters. The Sea of Japan side of Japan is one of the snowiest regions in the world. Japan also experiences four distinct seasons; this seasonal awareness is an essential element of Japanese culture.

Heavy rainfall and volcanic landforms cause rivers to be short and steep, providing opportunities for hydroelectric development. Natural vegetation as well as farm products and practices are fairly similar throughout the region, with wet rice cultivation as the major crop. Fish is a major

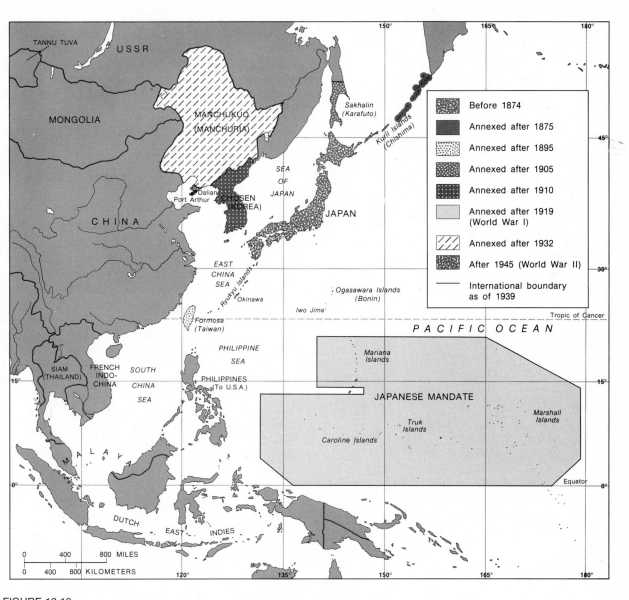

FIGURE 12-10

Japanese historic and present-day boundaries and regions of influence. (From James S. Fisher,
Geography and Development: A World Regional Approach, *Columbus, OH: Merrill Publishing Co.,*
1989, Fig. 14-3, p. 332. Used with permission of Merrill Publishing Co.)

source of protein due to the extensive coastline of the region, and thus fishing and aquaculture are the major resource-based industries.

Place: Cultural Characteristics. Japan is the fourth most
densely populated country in the world, after Bangladesh,
South Korea, and the Netherlands (Figure 12-13). The
original inhabitants of the Japanese islands came from
Siberia, East Asia, and Southeast Asia over 2000 years ago.
They remained isolated from invasion and immigration

until the middle of the nineteenth century, thereby creating a remarkably homogeneous population. However, although Japanese people may seem homogeneous to the
outside world, there is great subregional diversity within
the region itself. For example, urban-rural contrasts are
marked, and dialectical differences are greater than in the
United States, which is larger in population and area.

Also of interest from a demographic perspective is the
rapid aging of Japan's population. It is predicted that peo-

FIGURE 12-11
*Land utilization patterns in Japan.
(From James S. Fisher,* Geography
and Development: A World
Regional Approach, *Columbus, OH:
Merrill Publishing Co., 1989, Fig.
15-1, p. 348. Used with permission
of Merrill Publishing Co.)*

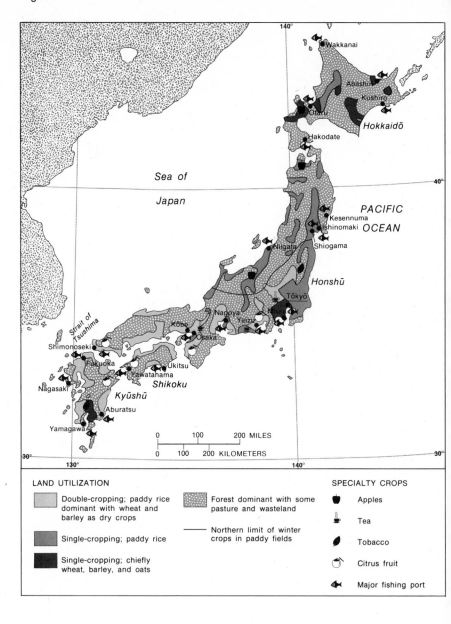

ple aged 65 and over will account for 24.2 percent of Japan's population by 2043.

Human-Environment Interaction. Very little of Japan's original flora and fauna remain today. Almost all areas suitable for settlement have been occupied and modified. Rural forests are harvested on a sustained yield basis, rivers have been dammed and channeled for hydroelectric development and flood control, and all available tillable soil is in cultivation. Likewise, marine resources are extensively used and considerably modified for economic extraction.

Due to rapid industrialization in the first part of the twentieth century, Japan's air and water pollution conditions were among the world's worst. Recent cleanup efforts and a rising environmental consciousness among the population, however, has made conditions very comparable to other industrialized nations. A strong Ministry of the Environment and increasingly strict environmental laws offer promise that more attention will be paid to issues of global resource management such as marine fisheries, forestry, and mining practices.

Japan is not without environmental risk, due to its location on an active chain of volcanic islands. Devastating earthquakes have occurred in the recent past, sometimes

FIGURE 12-12
*Japan's climatic regions. (From
James S. Fisher,* Geography and
Development: A World Regional
Approach, *Columbus, OH: Merrill
Publishing Co., 1989, Fig. 14-4, p.
333. Used with permission of Merrill
Publishing Co.)*

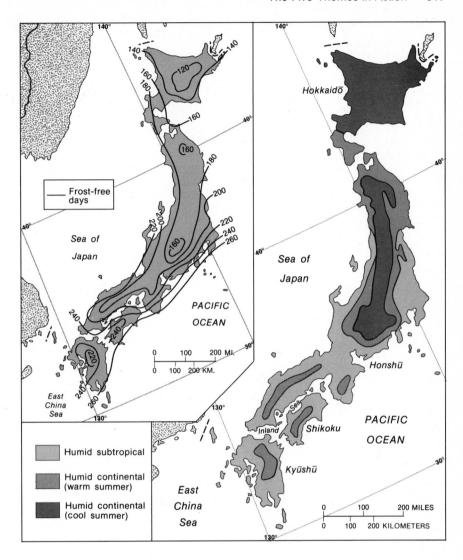

accompanied by *tsunamis* (large seismic sea waves or tidal waves) or fires. Japan's response to the seismic safety problem has focused upon stricter building codes and public safety awareness. In addition, typhoons and flooding are seasonal problems locally.

Movement. Japanese culture had been more constructed than evolved, due to selective borrowing and adaptations of innovations, such as writing, the wet rice-growing agricultural system, and the use of silk. Many other cultural characteristics were indigenous, such as the spoken form of Japanese and the Shinto religion. The net effect of such cultural continuity is a strong national Japanese identity.

Despite Japan's status as a developed nation with a strong economic pull factor, it has had relatively little immigration. The strong Japanese identity is somewhat of a barrier to incoming groups. Japan's largest ethnic minority group is of Korean descent. Likewise, very few Japanese have actually migrated to other parts of the world on a permanent basis. The three largest Japanese-heritage populations abroad live in Hawaii, on the Pacific coast of the United States, and in Brazil.

Place: Economic and Urban Characteristics. To an outsider, the major differences among Japanese landscapes are contrasts between urban and rural areas, rather than specifically locational contrasts. In 1870, only 8 percent of the population lived in a place with 10,000 or more people. Today, 80 percent are classified as urban and only 1 percent actually owe their primary income to farming.

The Tokyo-Kawasaki-Yokohama subregion, a megalopolis of over 23 million people spanning a diameter of 90

FIGURE 12-13
*Population distribution in Japan.
(From James S. Fisher,* Geography
and Development: A World
Regional Approach, *Columbus, OH:
Merrill Publishing Co., 1989, Fig.
14-5, p. 335. Used with permission
of Merrill Publishing Co.)*

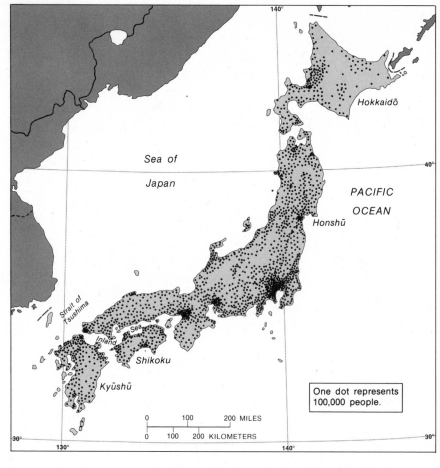

miles (150 km), is the best example of density as the major criterion for a subregional identity, being the largest concentration of humanity in the world. Other densely populated urban subregions include Osaka and Nagoya (Figure 12-14). The entire stretch of coastland between Tokyo and Kitakyushu forms an uninterrupted industrial complex. Some of Japan's major cities, such as Tokyo, Osaka, and Nagoya, developed around towns that once served medieval castles.

Land uses are comparable from one urban area to another, with mixes of residential, industrial, and commercial uses interspersed with transportation links, parks, and government facilities. There is an extreme shortage of residential land and housing in urban areas, however. As one nears an urban center, the landscape of many multistory apartment complexes reflects the population and housing crunch. All urban areas have large subterranean commercial and transportation complexes, similar to U.S. shopping malls. Rural areas exhibit greater subregional differences, based upon distinctive land use patterns, historical tradition, and physical features.

Hokkaido, Japan's northernmost island, is generally regarded as the frontier. The cultural landscape here is noticeably different, with much of the land devoted to dairy farming. The landscape in this region feels much like the North American Midwest because of its rural ambience and agricultural pace of life.

Movement. Due to a strong centralized government, centralized educational system, various transportation modes, and efficient transportation systems, Japan's cultural uniformity is increasingly reinforced. Transportation systems are well developed within Japan, especially within urban areas, where vast networks of train, subway, monorail, bus, taxi, and highway routes crisscross the city. At peak commuting times, trains leave every few minutes and are filled to capacity.

Well-developed external transportation and communication has made possible a worldwide system of exports and imports upon which the national economy is based. The Japanese see their nation as an essential but vulnerable regional node in an international economic system, where

FIGURE 12-14
*Major urban regions of Japan.
(From James S. Fisher,* Geography
and Development: A World
Regional Approach, *Columbus, OH:
Merrill Publishing Co., 1989, Fig.
14-2, p. 331. Used with permission
of Merrill Publishing Co.)*

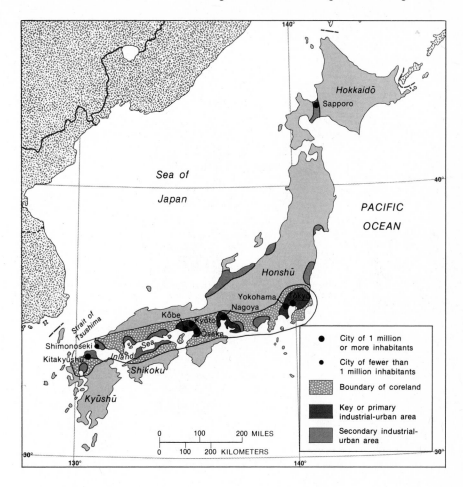

survival depends upon crucial imports of energy and raw
materials. Examples of raw materials imported by the
Japanese include oil, vital mineral ores, lumber, wheat, and
soybeans. The local infusion of labor, technical know-how,
and organizational skill transform these raw materials into
manufactured goods, principally for export.

Region: Japan's Global Setting. Just as Japan built a regional zone of influence in East and Southeast Asia
through military occupation in the first part of the twentieth century, the latter part of the century has been characterized by Japan's economic expansion worldwide through
large foreign investment and trade networks. It is reasonable to expect that the future of Japan as a globally oriented region will be evidenced with more economic
association with China and other parts of the Pacific Rim.
The box on pages 320–321 lists teaching resources and
ideas to help you organize a unit on Japan as a regional
study, using the five fundamental themes of geography. It
also offers a conclusion to this analysis of Japan as a region.

REGIONAL ISSUES
AND REGIONAL PLANNING

Some problems of location, settlement, and development
are regional by nature: flooding occurs in topographic
basins; air pollution concentrates in major cities and downwind from them. Both economic opportunities and difficulties are often based upon resources contained within a
region. Cultural and political patterns may also be understood from a regional perspective, like those offered in the
previous case studies.

Attempts at problem solving on a regional basis have
had mixed results. In general, nation-states with the most
centralized governments have had the most success with regional planning and problem solving. Socialist systems
seem particularly suited to this approach because they are
more able to direct change from a centralized authority
with less resistance from local interests.

TEACHING IDEAS FOR JAPAN UNIT USING THE FIVE THEMES

Location

1. Compare the location of Japan on a typical U.S. classroom map of the world with a world map available from Japanese Consulates in the United States. (Absolute and relative location are the same, but the perspective differs; that is, one is U.S.-centered or Europe-centered as opposed to Japan/Asia-centered.)

2. Using the thematic maps in your classroom atlas, find U.S. cities that correspond in latitude to major Japanese cities (for example, Sapporo, Tokyo, Nagoya, Osaka, Hiroshima, Fukuoka, Nagasaki). Explain similarities and differences in climate.

3. Assign major Japanese cities to groups of students. Have them determine the significance of geographic features.

Place

1. A feature that makes Japan unique is its density of population. Simulate the density in Japan by outlining an area on the floor of the classroom, or outside. Use arithmetic skills to calculate the appropriate ratios for the area created. Have students stand within that area to get a sense of density.

   For example, each student might represent four million people. At a 1 percent growth rate, how many students would need to be added to represent the population in the year 2000? A comparison of population figures and area maps in an atlas should confirm in students' minds the difference between Japan's population density of 5800 persons per square mile and the density of 65 persons per square mile in the United States. However, students should be reminded that in neither country is the population distributed uniformly; for example, populations are much lower in mountain and desert areas.

2. Have students brainstorm a list of things they know about Japan. Categorize the information into physical and human characteristics. As students continue their study of Japan, have them refer to this list and correct, revise, or expand it. Use caution with the items listed until students have verified the accuracy of their statements.

3. Have students research examples of the importance of the seasons in Japanese culture.

Human-Environment Interaction

1. There is a shortage of adequate housing in Japan, yet only 4 percent of the land is used for housing. Debate whether some of the 14.5 percent of land that is used for agriculture should be reallocated for housing.

At the opposite extreme are societies such as the United States that historically are much more participatory at the local level. This fundamental belief in home rule may even be at the expense of regional or national welfare. Regional planning in the United States is usually based on the voluntary joint administration of political entities for a common goal.[10] There are multistate regions, substate regions made up of counties, and regions centered upon metropolitan areas.

Metropolitan regions in the United States often are

2. From available media, have students collect images of Japan that depict how humans have altered the landscape, or how the landscape forces humans to adapt to it. Organize the images into collages or a bulletin board display.

3. Over 65 percent of Japan is forest. Find examples of the importance of wood in Japanese culture.

Movement

1. Have students take a survey of items at home that are manufactured in Japan.

2. Use a classroom student encyclopedia to identify the three most important countries from which Japan imports items and the three most important nations to which they export. (Imports are from the United States, Indonesia, and South Korea. Exports are to the United States, West Germany, and South Korea.)

3. Discuss how the past has influenced and shaped modern transportation routes. For example, the *shinkansen* the high-speed "bullet" train) from Tokyo to Kyoto parallels the *tokaido,* which is the edge of historic settlement along the Pacific coastal strip, rather than following a more direct, shorter route.

4. Fishing nations often are seafaring, which can lead to cultural exchange and migration. Have students speculate about why, in the case of Japan, fishing has not lead to a pattern of seafaring, cultural exchange, and migration.

Region

1. Have students use accessible resources to describe Japan in various regional terms, for example, "Japan is part of the Pacific Rim region," "part of the 'Ring of Fire,'" "part of the region of Asia or East Asia," and so on.

2. Make some generalizations about the regions of Japan depicted on temperature and precipitation charts in your atlas.

3. Using classroom atlases, have students work in small groups to divide Japan into regions. The regions may be of any type, such as topographic, population, economic, or political, but they must be supported by evidence in the atlases.

4. Use atlases to investigate the two distinct regions of Japan: the Sea of Japan side, and the Pacific side.

5. Use information in atlases to explain why Japan utilizes wet rice cultivation.

For more extensive units on different aspects of historical and contemporary Japan, contact The Japan Project/SPICE, Stanford University; the East Asia Resource Center, University of Washington at Seattle; and the United States–Japan Foundation, New York.

called "councils of governments"; they include at least one urbanized area of over 50,000 people, plus surrounding dependent rural territory. They are usually identified in the U.S. census, and representatives from each of the participating agencies meet to work toward the solution of regional problems. Most regional councils issue planning reports for purposes of communication, education, and consensus building. They develop agreements upon projects to undertake in the region, and they decide who will do them and how they will be financed.

Multistate economic regions usually are special-purpose agreements between the federal government and the participating states. One example is the Appalachian Regional Commission (ARC), formed in 1965 to improve economic conditions in an area of dwindling coal mining and inaccessible small communities.

The most successful regional planning effort in the United States has been the Tennessee Valley Authority (TVA), established in 1933. The region was defined as the drainage basin of the Tennessee River, including portions of seven states and at that time one of the poorest parts of the nation. Through the introduction of hydroelectric power and associated manufacturing, TVA became a successful exporter of energy and factory goods. TVA is a semiindependent federal corporation with centralized control over the region's power production, navigation, flood control, recreation, forestry, and agriculture.

Substate regions usually are made up of a group of participating counties having common interests. Many are aided by the outside funding of "economic development corporations," which attempt to promote the labor and natural resources of the region for new business investments.

Special-purpose multistate and substate regions may be organized to protect a unique resource that crosses political boundaries. An example is Lake Tahoe, which straddles the California-Nevada border near Reno. The Tahoe Regional Planning Agency includes membership from counties in both states as well as federal agencies such as the USDA Forest Service. Each brings its own approach to protecting one of the most beautiful and fragile places in North America.

In addition to economic development and environmental protection, regional agencies have been created to improve transportation systems such as mass transit, airport improvement, and highway planning. Some have focused upon health and social services, while others are more general in scope, dealing with housing and general economic planning.

Issues common to almost all regional organizations are delineation of proper boundaries, assignment of proper functions and powers, and organization and funding of mutual efforts. Regional planning offers opportunities for professionals with geography training because it is regional geography being put to work. The potential for planning and problem solving is only beginning to be realized.

LEARNING OUTCOMES

Learning outcomes in *Guidelines for Geographic Education* for the theme of regions are listed here. Students will be able to:

1. Understand that regions are basic units of geographic study.
2. Describe how regions vary in scale from very small to very large.
3. Explain how regions may be defined by cultural or physical features or by a combination of both.
4. Know that the idea of "region" is a conceptual tool, used to help make general statements about complex reality.
5. Describe how the concept of regions relates local places in a system of interactions and connections.
6. Give examples of nodal and uniform regions.

KEY CONCEPTS

Boundary
Formal region

Functional region
Perception

Regional method

Regional planning

FURTHER READING

Something as fundamental to geography as the concept of region carries such importance that almost all geography survey books incorporate it. Particularly worth reference are Roger Minshull, *Regional Geography* (Chicago: Aldine, 1967); Rhoads Murphey, *The Scope of Geography* (New York: Methuen, 1982); Harm de

Blij and Peter O. Muller, *Geography: Regions and Concepts* (New York: John Wiley, 1993); and James S. Fisher, *Geography and Development: A World Regional Approach*, 3rd ed. (Columbus: Merrill, 1989).

A systematic study of what regions are, how they are studied, and the shortcomings of regional thinking is James R. McDonald, *A Geography of Regions* (Dubuque, IA: William C. Brown, 1972). Also useful is *Regionalism and the Regional Question* by Ronan Paddison and Arthur S. Morris (New York: Blackwell, 1988).

There are several regional geographies of North America, including Stephen S. Birdsall and John W. Florin, *Regional Landscapes of the United States and Canada* (New York: John Wiley, 1993); John H. Paterson, *North America: A Geography of Canada and the United States* (New York: Oxford University Press, 1984); John Fraser Hart, ed., *Regions of the United States* (New York: Harper & Row, 1989); and Wilbur Zelinsky, *The Cultural Geography of the United States* (Englewood Cliffs, NJ: Prentice-Hall, 1993). A best-selling regionalization of North America from a journalist's perspective is Joel Garreau, *The Nine Nations of North America* (Boston: Houghton Mifflin, 1981).

The regional concept as a methodology in geography was advocated by several geographers in the twentieth century. For example, see Robert E. Dickinson compiler, *Regional Concept: The Anglo-American Leaders* (London: Routledge and Kegan Paul, 1976); Derwent Whittlesey and others, "The Regional Concept and the Regional Method," in Preston James and Clarence Jones, eds., *American Geography: Inventory and Prospect* (Syracuse, NY: Syracuse University Press, 1954): 19–68; and John Fraser Hart, "The Highest Form of the Geographer's Art," *Annals of the Association of American Geographers* 72 (March 1982): 1–29.

The most useful references to the Amazon region discussed in this chapter are Phillip Fearnside, *Human Carrying Capacity of the Brazilian Rainforest* (New York: Columbia University Press, 1986); Robert J. A. Goodland, *Amazon Jungle: Green Hell to Red Desert?* (New York: Elsevier, 1975); and John Hemming, ed., *Change in the Amazon Basin*, 2 vols. (Manchester: Manchester University Press, 1985).

The discussion on Poland as a political region was assisted by Dean S. Rugg, *Eastern Europe* (New York: Longman, 1986), and George J. Demko, ed., *Regional Development Problems and Policies in Eastern and Western Europe* (New York: St. Martin's Press, 1984).

Useful references to our section on Nigeria as a cultural region include Michael Senior and P. A. Okunrotifa, *A Regional Geography of Africa* (New York: Longman, 1983); Alan B. Mountjoy and David Hilling, *Africa: Geography and Development* (Totowa, NJ: Barnes & Noble Books, 1988), and Alfred T. Grove, *The Changing Geography of Africa* (New York: Oxford University Press, 1989).

A fascinating survey book with excellent photographs about the San Francisco Bay Area is Mel Scott, *The San Francisco Bay Area: A Metropolis In Perspective* (Berkeley: University of California Press, 1985). An earlier, historically oriented book about the region is James E. Vance, Jr., *Geography and Urban Evolution in the San Francisco Bay Area* (Berkeley, CA: Institute of Governmental Studies, 1964).

Background material about Japan as a geographic region is in Donald MacDonald, *A Geography of Modern Japan* (Ashford, Surrey: Paul Norbury, 1985); and in MacDonald's *Japan: A Regional Geography of An Island Nation* (Tokyo: Teikoku-Shoin, 1985). Also see Glenn T. Trewartha, *Japan: A Geography* (Madison: University of Wisconsin Press, 1965); and Edwin O. Reischauer, *The Japanese Today: Change and Continuity* (Cambridge, MA: Belknap/Harvard University Press, 1988).

Regional planning is surveyed in a comprehensive manner by Frank S. So et al., eds., *The Practice of State and Regional Planning* (Chicago: American Planning Association, 1986). An introduction to regional planning in the United States is Gill C. Lim, ed., *Regional Planning: Evolution, Crisis, and Prospects* (Totowa, NJ: Allanheld, Osmun, 1983).

Some of the most thorough historical geographic regional studies, which serve as a model for what can be done elsewhere, are the first two of Donald W. Meinig's series, *The Shaping of America: Atlantic: America, 1492–1800*, Vol. 1 (1986) and *Continental America, 1800–1867*, Vol. 2 (1993) (New Haven, CT: Yale University Press). His classic study *The Great Columbia Plain: A Historical Geography: 1805–1910* (Seattle: University of Washington Press, 1968) is also one of the best regional geographies ever written.

Representative works of three pioneer regional geographers are:

- Vidal de la Blache, *Principes de Géographie Humaine (Principles of Human Geography)*, translated by Millicent Todd Bingham (New York: Henry Holt, 1926).

- Alexander von Humboldt, *Kosmos: Entwurt Einer Physischen Weltbeschreibung* (Stuttgart: Cotta, 1845–1862), and translated by E. C. Otté (London: H. G. Bohn, 1849–1858).

- Carl Ritter, *Die Erdkunde im Verhältniss zur Natur und zur Geschichte des Menschen, oder Allgemeine Vergleichende Geographie als Sichere Grunlage des Studiums und Unterrichts in Physikalischenund, Historischen, Wissenschaften* (Berlin: G. Reimer, 1822–1859), 19 vols.

ENDNOTES

[1] Joint Committee on Geographic Education of the National Council for Geographic Education and the Association of American Geographers, *Guidelines for Geographic Education* (Washington, DC: 1984).

[2] Ann R. Markusen, *The Politics of Regions* (Totowa, NJ: Rowman & Littlefield, 1987).

[3] Joel Garreau, *The Nine Nations of North America* (Boston: Houghton Mifflin, 1981): 1.

4 Donald W. Meinig, *The Great Columbia Plain, A Historical Geography: 1805–1910* (Seattle: University of Washington Press, 1968).

5 John Fraser Hart, "The Highest Form of the Geographer's Art," *Annals of the Association of American Geographers* 72 (March 1982): 1–29.

6 Preston E. James, "Toward a Further Understanding of the Regional Concept," *Annals of the Association of American Geographers* 42 (March 1952): 195–222.

7 Richard Hartshorne, "The Nature of Geography," *Annals of the Association of American Geographers* 29 (Sept. 1939): 440.

8 John Fraser Hart, *Regions of the United States* (New York: Harper & Row, 1972): 271.

9 Hart, *Regions of the United States*: 271–272.

10 Frank S. So et al., eds., *The Practice of State and Regional Planning* (Chicago: American Planning Association, 1986): 133–165.

13 GEOGRAPHY IN ACTION: APPLYING GEOGRAPHY STANDARDS, THEMES, AND CONCEPTS

Vineyard in the Salinas Valley, California. (Photo by author.)

In the previous chapters you have seen how the five fundamental themes in geography provide a structure or skeleton for geographic inquiry. Many of the concepts introduced are appropriate to more than one theme, and in the world outside textbooks, geographic information is more holistic. While each concept and theme may be viewed as a separate topic for geographic study, most places and research topics reflect a combination of them. The following case studies illustrate not only the integrated conceptual nature of geography, but also its close alliance with other fields in the social sciences, humanities, and natural sciences.

We have titled this chapter "Geography in Action" because our realm of study is dynamic and exciting. The four case studies selected here represent only a few of the numerous examples of how geography has affected our lives:

- The case study on the 100th meridian illustrates how a line that differentiates wet and dry climates serves as a focus for comparing historical development and settlement patterns east and west of this line.

- The second case study, on the geography of viticulture, combines elements of agriculture, architecture, culture, and environmental sensitivity into one geographic case study.

- The third case study deals with the causes and effects of migration, including the environmental, economic, and political circumstances influencing human decision making.

■ Finally, we present ideas on the relationships between water as a resource and the development of world civilizations. Control of water was a basis of authority for early culture groups, and it has become a prime resource consideration in modern land use, economic, and political decisions.

Each of these case studies may be viewed as incorporating several of the *National Geography Standards,* themes, and concepts introduced earlier. The last two *National Geography Standards* are particularly important as we consider this last chapter:

> Standard 17: *The geographically informed person knows and understands how to apply geography to interpret the past.*
>
> Standard 18: *The geographically informed person knows and understands how to apply geography to interpret the present and plan for the future.*

CASE STUDY 1: EAST IS EAST AND WEST IS WEST

Regional differences in large nations such as the United States are the basis for discovering geographic diversity. Based on your own experience, where would you draw a dividing line between the regions known as the American West and the American East? This is not a trivia question but an example of basic geographic inquiry.

The 100th Meridian

The map of the United States shown in Figure 13-1 includes a locational grid showing latitude and longitude. Note particularly an imaginary line drawn between the 98th and 100th meridians, from north to south. Closely parallel to this line is the 20-inch **isohyet** or line of equal

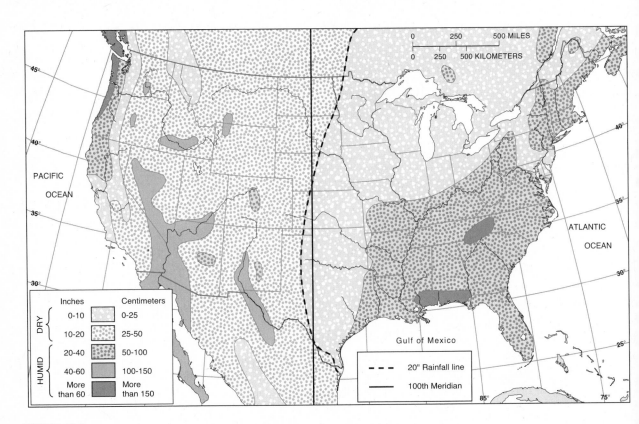

FIGURE 13-1

Dashed line on this climatic map of the United States shows Hollon's approximate limit of desert influence. Solid line shows the extent of the 100th meridian line as it divides the United States into east and west. (Line of desert influence from W. Eugene Hollon, The Great American Desert: Then and Now, *New York: Oxford University Press, p. 2. Copyright © 1966 by Oxford University Press. Reprinted by permission.)*

rainfall amounts. This line marks places where 20 inches (51 cm) of rainfall constitute the average annual precipitation across the continent.

Precipitation Patterns. John Wesley Powell was commissioned in 1881 as director of the U.S. Geological Survey to assess the potential for wise use of land and resources in the American West. This charge was not taken lightly, and after years of research and study based on personal field work, Powell presented to Congress the findings of his monumental work *Report on the Lands of the Arid Region of the United States with a More Detailed Account of Utah.*[1] In this book, he stressed that the simple fact of water availability was the most important consideration in successfully occupying the West.

Agricultural Patterns. To the east of the 20-inch isohyet, pioneer farming could be conducted without irrigation, and the mainstay crop was corn. West of the line, evaporation exceeded precipitation in the warmer months of the growing season. Corn was replaced by wheat and other drought-tolerant grain crops. Farms east of the line could be smaller and support more population per square mile. West of the line, extensive land holdings were necessary for grain farming or livestock grazing.

In addition to the proper choice of crops, three technological breakthroughs had to occur before the Great Plains could be successfully farmed:

1. The steel-tipped plow, introduced to break up the tough prairie sod. Sod is composed of the root network and the organic topsoil created over centuries of thick natural grassland cover.

2. Barbed wire to separate grazing animals from field crops. Wood was not readily available, so wire fences were the necessary alternative.

3. The windmill and its deep pump, which brought life-giving water from the underground aquifer to thirsty gardens and livestock.

These innovations created fundamental differences in land use that remain apparent on current maps a century and a half after settlement.

Settlement Patterns. To pioneer settlers entering the Great Plains, climatic differences were visible on the landscape by a dramatic change in natural vegetation. The first reaction of many East coast American migrants upon seeing the region west of the 100th meridian was one of dismay and disappointment.[2] Called by some the "Great American Desert," the area was seen as a barrier or obstacle to be endured en route to California or Oregon (Figure 13-1). It

was assumed by many that if the land would not support trees, it could hardly support farming or the successful settlement of towns and cities. Homesteads of 40 to 160 acres were not appropriate here, as they were in the East. The minimum land necessary for livestock grazing for one family sometimes amounted to thousands of acres.

Population Distribution. Figure 13-2 shows the distribution of population in the contiguous United States on a dot map. It is readily apparent where a change of density occurs—near the 100th meridian. Figure 13-3 is a generalized land use map which shows differences in products produced. Land ownership among a relatively few people is characterized by large holdings with sparse populations.

Transportation. In the humid East, nineteenth-century America was connected by roads, rivers, and canals. Waterways were particularly effective in transporting heavy goods, and wagon roads connected all local communities. Before and during the Civil War, railroads replaced waterways between the states, and the railroad network, especially in the Northeast, became virtually complete. Figure 13-4 shows the now historic railroad network in the United States in 1970.

By 1870, one transcontinental line crossed the western part of the continent from San Francisco to Omaha.

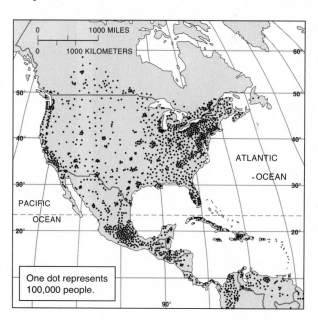

FIGURE 13-2
Population distribution of the contiguous United States. (From James M. Rubenstein, The Cultural Landscape: An Introduction to Human Geography, *4th ed., New York: Macmillan, 1994, Fig. 2-1, pp. 52–53. Reprinted with permission of Macmillan Publishing Co.)*

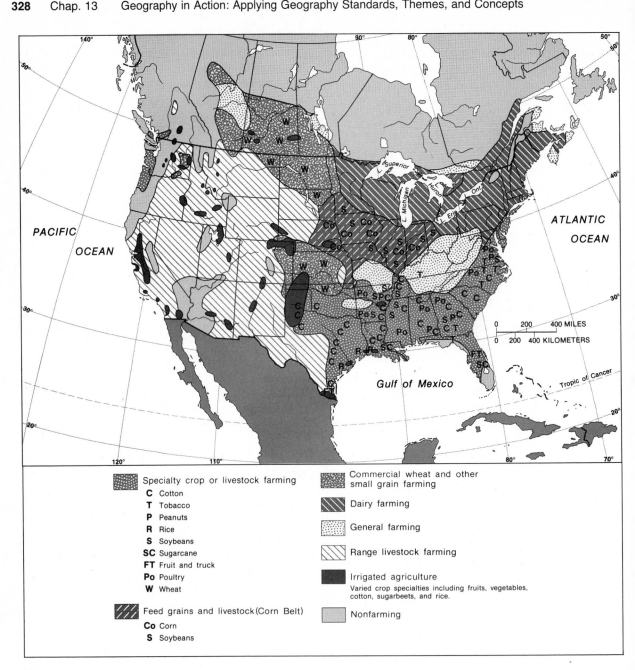

FIGURE 13-3
Land use patterns in the contiguous United States. (From James S. Fisher, Geography and
Development: A World Regional Approach, *Columbus, OH: Merrill Publishing Co., 1989, Fig. 6-1,
p. 147. Used with permission of Merrill Publishing Co.)*

Others were planned but the vast majority of land west of
the 100th meridian was accessible only by walking, pack
animal, or wagon. Those who crossed the western half of
the continent, or who lived there, experienced a land of immense vistas, isolation, and necessarily independent living.

Success in this pioneer environment did not come to everyone. Today's roadway system connects all of the region's
towns and cities, but the journey to a high school basketball game or farm equipment dealer still may take an hour
or two in some areas.

FIGURE 13-4

Railroad network of North America. (From James S. Fisher, Geography and Development: A World Regional Approach, *Columbus, OH: Merrill Publishing Co., 1989, Fig. 5-14, p. 140. Used with permission of Merrill Publishing Co.)*

Urban Patterns. The spread of settlement was not entirely rural. Farmers needed towns or cities for support services and markets. Pioneer craftspeople, artisans, and merchants also occupied the West, based on the number of agriculturists who would support their endeavors. The fewer the farmers, the fewer and smaller the urban places. Large cities were, in effect, oases with immense regions to serve. Salt Lake City, Denver, Santa Fe, and San Francisco carried most of the responsibility in the late nineteenth century for the urban functions of western American society. These cities had the advantage of initial location and settlement, and today they remain important regional centers in the West.

Wallace Stegner's classic biography of John Wesley Powell was entitled *Beyond the Hundredth Meridian* for a very good reason.[3] In this book, he describes some of Powell's recommendations to Congress for determining the future of the West. For example, Powell recommended that irrigated farms be no smaller than 640 acres and that non-irrigated ranches be a minimum of 2500 acres in order to support enough livestock for a family operation. Powell stressed that the management of water, particularly for irrigation, would be the key economic and political issue in the successful development of the West. This has come to pass. Stegner, in the more recent book *The American West as Living Space,* confirms Powell's prediction. Stegner lists a long succession of mistakes made by the U.S. Bureau of Reclamation and private irrigation districts in the management of this all-important resource west of the 100th meridian.[4]

A journey through a thematic atlas of North America will show many more east-west differences, although they are not necessarily separated by the edge of the Great Plains. Some comparisons are found in the amount of government ownership of land (75 percent in the western states and 25 percent in the East), the ethnic diversity of the population (American Indian, Hispanic, and Asian in the West; African-American and European in the East), and in the varieties of economic activities (more primary activities in the West; more secondary activities in the East).[5]

CASE STUDY 2: THE GEOGRAPHY OF VITICULTURE

One of the most romantic relationships between people and Earth is in the enjoyment of what some would say is its ultimate gift, fine wine. Requirements of high-quality wine grapes are very environmentally specific. The cultivation of grapes into select wines is a function of history, and humankind has a cultural affinity with the cultivation of these vines. The study of viticulture presents a virtual geographic textbook in understanding relationships between human decisions and physical Earth systems.

Origin and Diffusion of Viticulture and Winemaking

Fossilized grapevines have been found in North America, Europe, and Asia as far north as Iceland and northern Europe, near the Arctic Circle. There are more than 6000 species of the genus *Vitis* (grapevine) in the world today. More species of wild grapes are found in North America than in other places, and thus it is not surprising that Nordic Vikings named this land "Vinland." However, there are no records or fossil evidence to indicate that people used these vines other than as a source of fresh fruit.

Our best guess for the hearth of grapevine domestication is the Caucasus Mountains in what is now the republic of Georgia. The scientific name of the grapevine of history is *Vitis vinifera,* a species which grows wild in this region. Domestication is the process of selection, care, and reproduction of plants to serve humankind. In this case, the grapevine was most likely planted from cuttings and pruned to produce maximum fruit. Clones were selected and preserved for future use. This particular mountain environment was probably suitable because it contained a variety of ecologic conditions and a temperature regime that was compatible with the needs of the plant, that is, warm and dry summers with cool (but not cold) winters.

Grapevines are easy to reproduce and transport in the form of cuttings, so it probably was not difficult to introduce grape growing into the irrigated Tigris-Euphrates River valleys, already well known for domestication of other Old World plants and animals. This transport was no doubt easier than one might think because the climate was more moist then and the lowland valleys were managed and irrigated by experienced farmers. The hillside terrace gardeners of the Hanging Gardens of Babylon and the agriculturalists of Ur no doubt cultivated grapes by at least 4000 B.C. It must be remembered that trade was extensive even in these early times, and the diffusion of this fruit product was rapid throughout the eastern Mediterranean Basin and even to Egypt by 3000 B.C. (Figure 13-5).

Egypt had early connections with Phoenicia, and as shipping in the Mediterranean became more common, the Greek islands took up the practice of grape growing. By the time of the classical Greek contributions to world civiliza-

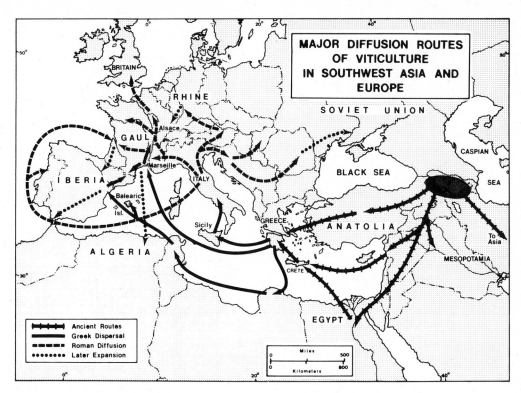

FIGURE 13-5

Major diffusion routes of viticulture in southwest Asia and Europe. (From Harm J. de Blij, Wine: A
Geographic Appreciation, *Totowa, NJ: Rowman & Allanheld, 1983, Fig. 3-1, p. 42. Used with permission of Rowman & Allanheld.)*

tion, the Mediterranean Basin was cultivated with domesticated wheat, olives, grapes, citrus, and other fruits.

It is not known exactly when grapes were first made into wine and stored as such. Earlier fermented beverages
such as palm wine in Egypt and beer in the eastern
Mediterranean indicate that the fermentation of grapes was
not difficult. The storage of wine was another matter, as it
is subject to spoilage. Archaeological evidence from Crete
and other Greek islands indicates that clay vessels (am-
phoras) were used to ship and store wine before 600 B.C.
Ancient wine ships have been discovered sunken in the
Mediterranean with their cargoes still intact. Any reader of
Homer's *Iliad* and *Odyssey* knows of the references to wine
therein.

The Roman contribution to wine and grape production was the establishment of large commercially operated
vineyards, selection of particular grape varieties, and transport of wine in wooden casks. Wherever the Romans went
in what is how Spain, France, Germany, Austria, Italy, and
the Balkans, as well as North Africa, they took their grapes
and enjoyment of wine with them. Just as Roman heritage

is evident today in architecture, law, language, and agriculture, it is no surprise that the world's largest wine-grape
production today comes from Italy and France.

Wine as a fundamental part of Catholicism went with
the French, Portuguese, and Spanish to the western hemisphere and the southern hemisphere. It was introduced
with Cortes into Mexico in 1521, and into South America
and South Africa thereafter. Later still, grape growing and
winemaking would be introduced to suitable climates in
Australia, New Zealand, and eastern and western North
America.

Current World Patterns
of Grape Growing and Winemaking

Figure 13-6 shows the global distribution today of wine
grape growing. A comparison of this map with the world
climate map in Chapter 5 shows that the European wine
grape is primarily produced in Mediterranean climates. It is
no coincidence that optimum growing conditions for *Vitis
vinifera* are cool (but not cold), rainy winters and warm,

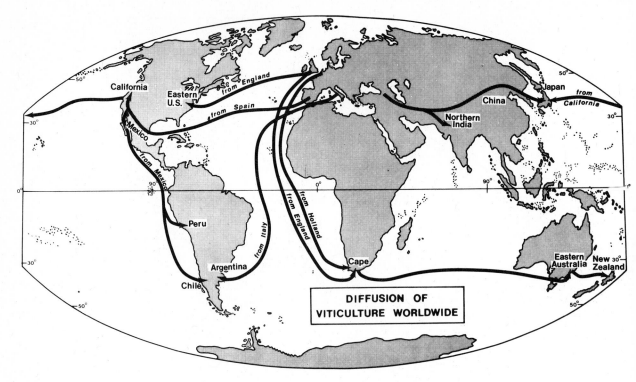

FIGURE 13-6
The diffusion of viticulture worldwide. (From Harm J. de Blij, Wine: A Geographic Appreciation,
Totowa, NJ: Rowman & Allanheld, 1983, Fig. 3-2, p. 58. Used with permission of Rowman & Allan-
held.)

dry summers. This climatic pattern is particularly evident on the west coasts of continents within the westerly wind belts, and in eastern Australia, on the tip of South Africa, in most of California, and throughout Mediterranean Europe. Through hybridization and clonal selection, wine grapes now are grown in more continental climates such as eastern North America, Central Europe, parts of the former Soviet Union, and even Japan.

The world's most important producers of wine today are, in decreasing order, Italy, France, Spain, the former Soviet Union, the United States, and Argentina. Each is recognized for its unique grape varieties and winemaking techniques. Products from these countries are sent to the rest of the world, and this trade has become an important form of global commerce. Table 13-1 compares the major wine countries by area under cultivation, wine production, and per capita wine consumption.

The regional concept is exemplified in wine geography by noting the strict attention given to the grape varieties grown and the regulation of winemaking practices within special districts called "Appellations." These regional regulations are practiced particularly in France, Italy, Germany, and the United States. The famous Bordeaux of France,

Chianti in Italy, and the select varietals of California's Napa Valley are all examples of distinct appellations. Many of the French names of these regions are recognizable on the labels of New World wines because the select grape varieties have been introduced to new locations.

Wine Landscapes

The three wine landscapes shown in Figure 13-7 are remarkably similar because agricultural practices, winemaking techniques, and winery architecture were diffused along with the grapes and winemakers themselves. The landscape of Bacchus is a reminder of the continuity and dispersal of Mediterranean civilization around the globe.

CASE STUDY 3: NEW LIVES AND NEW PLACES

One of the most tragic events in the history of western civilization was the poverty, famine, and emigration that oc-

TABLE 13-1
International comparisons of vineyard area, wine production, and wine consumption (1986)

Country	Vineyard Area 100,000 Acres	1000 Hectares	Percent
1. Spain	3936	1593	16.8
2. USSR	3304	1337	14.1
3. Italy	2711	1097	11.6
4. France	2592	1049	11.1
5. Turkey	1962	794	8.4
6. Portugal	964	390	4.1
7. United States	823	333	3.5
(California)	(701)	(284)	(3.0)
Other countries	7129	2885	30.4
World total	23,421	9478	100.0

Country	Wine Production 100,000 Gallons	1000 Hectoliters	Percent
1. Italy	2,028,850	76,798	23.3
2. France	1,934,352	73,221	22.2
3. Spain	907,458	34,350	10.4
4. USSR	898,872	34,025	10.3
5. United States	500,542	18,947	5.7
(California)	(453,791)	(17,177)	(5.2)
6. Argentina	490,609	18,571	5.6
Other countries	1,960,348	74,200	22.5
World total	8,721,031	330,117	100.0

Country	Wine Consumption per Capita Gallons	Liters
1. France	20.71	78.4
2. Italy	19.34	73.2
3. Portugal	18.70	70.8
4. Argentina	15.63	59.2
5. Luxembourg	14.27	54.0
6. Switzerland	12.44	47.1
7. United States	2.43	9.2
Worldwide (1986)	1.51	5.7

Source: Prepared by Economic Research Department, Wine Institute, from reports of U.S. Bureau of Alcohol, Tobacco, and Firearms and the U.S. Department of Commerce (1987).

A.

B.

C.

FIGURE 13-7
Wine landscapes.   A. Napa Valley in northern California. B. Wine country of Australia.   C. French wine region. (A, photo by author; B, photo by Australian Tourist Commission; C, photo by Ron Sanford.)

curred in Ireland in the 1840s. This event had several causes and effects, each with a geographic connection.

The Irish potato *(Solanum tuberosum)* was neither discovered nor domesticated in Ireland. It was brought from the Andes Mountains in South America to Europe by 1570 and introduced sometime in the 1580s into Ireland. Sir Walter Raleigh is credited with introducing the potato into Europe and was known to have had land estates in Ireland. By 1650, the potato had become the most important food source in that country.

The disaster of famine is recorded in earliest history. Famines were reported in Europe as early as A.D. 700, caused by plant and animal diseases, climatic irregularities, and frozen crops. The potato famine of the 1840s was primarily caused by the blight *Phytophthora infestans,* a fungus that creates brownish spots on plant leaves, eventually killing the plant. The potato blight also originated in South America; it appeared in New England in 1840 and in Europe a year or two later. The fungus is favored by humidity and wind. In Ireland, the blight severely affected the 1845 potato crop and completely devastated the 1846 crop. At the time, one-fourth of arable land in Ireland was planted in potatoes; the effects of this loss were devastating.

The potato famine first struck in the autumn of 1845, reaching a peak in 1846-1847.[6] The effects of the famine were disastrous in Ireland for a variety of cultural and economic reasons. The nutritional value of the potato is very high, which meant that the cost of feeding a large family with potatoes was very low. Europe's population had increased from 125 million to 250 million between 1750 and 1850, a twofold increase; but in Ireland during the same time period, due to early marriages and large families, the population increased from 1.5 million to 8.5 million, a more than fivefold increase. Over four-fifths of these people were rural, and even the "urban" areas were mostly tiny villages (see Figure 13-8).[7] Thus, the highly nutritive potato was abruptly removed from the Irish diet at the very time it was most critically needed to support a burgeoning population.

Compounded with this high population density was the fact that in the nineteenth century most Irish farmland was subdivided into tiny holdings rented by tenants or "cotters". By the mid-nineteenth century a system of rental contracts had evolved in Ireland between absentee landlords and tenants. Contracts were auctioned every year. This meant that each year a renter could be outbid, and his family could be homeless. A secondary effect of this system was that no investment of labor or cash went back into the land for fertilization or other field maintenance. Exhaustion of the soil resulted. Dependence on the potato allowed other crops, particularly grain, to be sold for cash that was saved for the annual rent payment.

Then, it grew worse. When the English Corn Laws were repealed in 1846, particularly affecting wheat sales in continental Europe, the demand for Ireland's grain exports dropped. Irish farm labor was no longer needed because wheat fields were turned into pasture for grazing livestock. Whereas other European nations (for example, Denmark) were able to abandon grain sales to sell dairy products, Ireland's system of land rental—rather than land owner-ship—was not so flexible. It is estimated that as many as 70,000 tenants did not have their contracts renewed in the mid-1840s. After 1846, 3 million people were on the government welfare system, and many thousands more could not be helped.

From 1840 to 1850 the population of Ireland decreased from 8.5 million to 6.5 million. Over 1.2 million migrated to North America, with approximately three-fourths going to the United States. The rest went to Canada. The most conservative estimate of the number who starved and died from disease in Ireland is 300,000. Other estimates suggest that as many as a million people did not survive the starvation and the related epidemics caused by the potato famine.

Out-migration from Ireland has continued ever since the mid-nineteenth century. In the United States today, about 20 million people claim Irish ancestry. The present population of Ireland is 3 million, or 35 percent of what it was in 1840. Ireland's chief export today continues to be its young men and women, despite government-instituted economic development programs to create employment for younger citizens.

The circumstances affecting this major event in human history were a combination of economic, physical, and cultural processes. Comparison of the same circumstances for a different population can be made by viewing the experience of the Moravian Brethren, a Protestant sect from Moravia (central Czechoslovakia) that migrated to Ireland.

Due to religious persecution, some Brethren migrated to the British colonies in America in the late 1600s, settling in Pennsylvania, as did other Protestant groups such as the Mennonites. Of those remaining in Europe, 10,000 were rescued by the Duke of Marlborough and sent to Britain during the War of Spanish Succession. Of these, 5000 settled in southwestern Ireland about 1709.

Travelers to Ireland in the 1850s noted different effects of the potato crop failure on the Moravians as compared with the native Irish. First, the Moravians owned their land, which was divided into small holdings of 10 to 15 acres. Each landowner practiced crop rotation, every four years alternating potatoes, wheat, barley, and oats. Healthy cattle were kept in barns and manure was used on the fields. The Moravians' diet included grain products, poultry, and some meat, along with potatoes and milk. The Moravians suffered very little in the blight due to their different cultural practices, smaller family size, and land-ownership patterns.

The momentum of history is exemplified by noting that potato production is still an important activity in Ireland. Although much land reform has taken place, rural poverty remains a serious concern to the Irish government.

FIGURE 13-8
Rural population distribution in Ireland, 1841. (From Anthony R. Orme, Ireland, *Chicago: Aldine Publishing Co., 1970, Fig. 40, p. 152. Used with permission from Aldine Publishing Co.)*

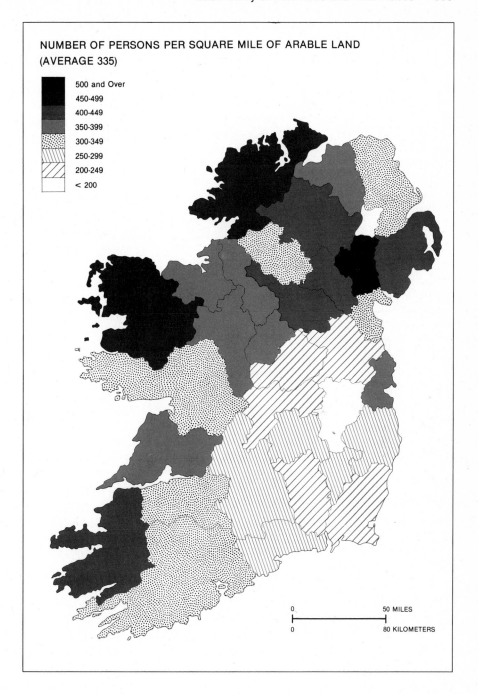

NUMBER OF PERSONS PER SQUARE MILE OF ARABLE LAND
(AVERAGE 335)

- 500 and Over
- 450-499
- 400-449
- 350-399
- 300-349
- 250-299
- 200-249
- < 200

0   50 MILES

0   80 KILOMETERS

The push factor in the massive Irish migration to North America was the need to escape poverty and famine. The pull factor was U.S. industry's demand for employees and the immigrants' hope for ownership of fertile, inexpensive land in the United States and Canada. Most Irish immigrants came by way of New York and settled initially in the Middle Atlantic states, New England, and eastern Canada. Most remained in urban areas where employment opportunities were best. Of Irish-born residents of the United States in 1920, 87 percent were urban dwellers.[8] Canal and railroad building provided a steady demand for labor and helped diffuse the Irish population across the continent.

CASE STUDY 4: WATER: EARTH'S FUNDAMENTAL RESOURCE

The earliest remains of Old World civilization demonstrate the importance of water and its management. The early civilizations of Egypt, Babylonia, Assyria, western Pakistan, and central China were located in river valleys, and their inhabitants understood the importance of living with these arteries of life. Historian Karl A. Wittfogel called these water-dependent settlements the "hydraulic civilizations."[9]

Wittfogel noted that water is one of the few resources that people can agglomerate in bulk and control as a commodity to be rationed. To control its distribution and use, large-scale enterprises were undertaken, such as canals, dams, aqueducts, and flood-control facilities. These works required centralized control of construction labor, thus creating water-oriented political power. The central authority controlled the water and public works, so it also controlled the agricultural process and its products. Thus, the central authority also controlled the populations that depended upon these support systems. Armies could be raised and empires expanded. Such "hydraulic despotism" made possible record keeping, census taking, centralized armies, and, of course, taxation.

By comparison, agrarian societies having sufficient rainfall most of the year tended toward more localized governments and/or feudal systems. Similar patterns observable in the New World are found in the Maya, Inca, and Aztec civilizations. All had aqueducts, irrigation systems—and centralized control. These "hydraulic civilizations" endured over 5000 years in some places, illustrating their effectiveness before the Industrial Revolution and Scientific Revolution were to change the relative importance of basic resources on Earth. Proportionately, hydraulic civilizations covered vastly larger portions of the inhabited globe and larger portions of its population than did all the agrarian civilizations taken together.

Irrigation on Earth

There is some irony in the fact that 71 percent of Earth is covered with water but only 0.5 percent of the total volume of lakes, rivers, and streams is available to us for agriculture and urban areas. Our appropriate participation in the hydrologic cycle becomes more important as water quality in many areas worsens due to pollution and as the per capita supply decreases due to population growth. There are generally four major sources of water for human use—surface water, groundwater, atmospheric water, and oceans. Of these, only two are used to a great extent—surface water and groundwater—due to the cost and technological limitations in collecting water directly from the atmosphere or oceans.

Most of the water used directly by humankind irrigates crops. Traditional irrigation is as old as agriculture; it was at first small scale, based on human or animal labor. As technology improved and populations became organized to perform public works, the extent of irrigated farmlands increased. Today irrigation is used in arid or desert climates, in climatic areas with seasonal precipitation, and in some humid climates where increased production may be obtained by applying a constant source of water. Table 13-2 lists the world's major irrigated areas. Note that most are in zones of moderate rainfall, not in deserts or in the tropics.

Almost all irrigation systems have the same basic elements. The first is a water source, which may be a surface-water drainage area, a groundwater basin, or a diversion from a lake or river. After acquisition, the water must be stored in reservoirs, tanks, or underground aquifers. Then it is distributed, usually through canals, aqueducts, or pipes through a network of smaller diversions, headgates, check dams, and ditches, in a treelike way, until it reaches its ultimate destination. Intermediate storage facilities may be located en route.

Upon application to the land through furrows, ditches, flooding, sprinklers, or piped drip systems, the water is available for absorption by plants. In most cases, however, less than half the water actually goes into the crop biomass. The remainder is lost through drainage, evaporation, or transpiration. After application, a critical but historically oftentimes neglected element of the irrigation process comes into effect. Adequate drainage of surplus water, salts, and excess fertilizer is essential to the long-term working of the system. Drainage may be implemented through pumping, ditches, or subterranean pipes, and must be done primarily to prevent salt buildup.

Conservation of water in irrigation systems may take place at any stage. Water loss by evaporation and seepage can be lessened by covering and lining the facilities. Water quality may be maintained by preventing siltation into the system. Pollution must be guarded against, particularly from agricultural fertilizers and pesticides.

Major irrigation projects may function along with other water-related projects such as electric power generation, flood control, navigation facilities, and recreational uses. Human landscapes created through water management often are dramatic and have far-ranging effects on regional physical, economic, and political systems. For example, the Aswan Dam project on the Nile River changed a depen-

TABLE 13-2
Irrigated countries of the world

Country	1970 Irrigated Area (1000 Ha)	1985 Irrigated Area (1000 Ha)	1985 Percent of Total Irrigated Area Worldwide
1. China	38,123	44,461	20.2
2. India	30,440	40,100	18.2
3. U.S.	16,000	18,102	8.2
4. Pakistan	12,950	15,440	7.0
5. U.S.S.R.	19,951	11,100	5.0
6. Indonesia	4370	7059	3.2
7. Iran	5200	5740	2.6
8. Mexico	3583	4890	2.2
9. Thailand	1960	3600	1.6
10. Spain	2379	3217	1.5
11. Italy	2561	3000	1.4
12. Romania	731	2956	1.3
13. Japan	3415	2931	1.3
14. Afghanistan	2340	2660	1.2
15. Egypt	2843	2486	1.1
16. Brazil	796	2300	1.0
17. Turkey	1800	2150	0.9
18. Bangladesh	1058	2073	0.9
19. Vietnam	980	1770	0.8
20. Iraq	1480	1750	0.8
World	167,399	220,312	

Source: Food and Agricultural Organization (FAO) of the United Nations, Production Yearbook vol. 40 (1986).

dence on flood irrigation into a reliance on pumped-water distribution systems. The effects were seen in land ownership reform, improved techniques in farming, and the increased population densities that could be supported downstream from the dam. Unforeseen consequences of the project were a reduction in siltation and a consequent lowering of fertility in downstream ecosystems, including the Mediterranean Sea.

Water projects historically have caused disputes about resource use, particularly where rivers flow between two or more countries. Some of these situations have been successfully resolved into cooperative agreements, such as between Sudan and Egypt and between Canada and the United States.

Water-Transfer Systems

Throughout human history the best use of water for the greatest number of people has involved relocation of one or the other. People can choose to settle near water resources, or capital and energy may be invested to bring the water to them. As technology and engineering capabilities have increased in the twentieth century, several massive water-transfer systems have been proposed and developed. Some of the largest water-transfer systems on Earth have been developed in the American West, a land of sparse rainfall and seasonal shortages.

Extreme environmental contrasts in this huge region range from rain forests in the Pacific Northwest that receive over 100 inches (about 250 cm) of rainfall a year, to the bottom of Death Valley, which receives less than 2 inches (5 cm). The choice in the region has been to move the water to the people, especially to those involved in agriculture. Some control of water movement was practiced by Native Americans of the Southwest and the Spanish in the mission system along coastal California. Early Anglo-Americans in Utah learned quickly how to irrigate the desert in the Salt Lake Basin.

Several water-transfer systems are planned or under construction, especially in North America. Some move water from one major basin to another, such as the Central Utah

Project. It transfers water into the Great Basin from the Colorado River Basin, most of it going to the Wasatch Front Range, supplying the Utah cities of Ogden, Salt Lake City, and Provo. An even more ambitious plan to move water from the Yukon River to Texas, the North American Water and Power Alliance, has been discussed but not authorized.

Conflicting Demands for Water

Wallace Stegner, who wrote over a lifetime about the American West, reminded us that "we have tried to make the arid west into what it was never meant to be and cannot remain, the 'Garden of the World' and the home of multiple millions…"[10]

As with most resources, when the demand increases and the supply remains constant, increasing value of the resource may lead to increasing conflict. The use of the Colorado River in western North America is an example of international and regional disputes over water use. The Colorado begins as hundreds of tributaries fill its northern basins in the Rocky Mountains of Wyoming, Utah, and Colorado. It flows through seven states before entering Mexico and the Gulf of California. Each jurisdiction historically has claimed rights to the water; in total, these claims far exceed the actual flow of the river in the wettest years.

One of the first major projects on the Colorado was construction in 1901 of the Alamo Canal, which flowed from the California-Arizona border through Mexico, and back to California. After disastrous flooding in 1905, a new "All American" canal was authorized and built.

From 1922 to the present, disputes over use of the Colorado centered on arguments of "supplier" states, where the tributaries begin as snowfall and rain, versus the dry "user" states to the south and the Republic of Mexico (which is supposed to be assured of 1.5 million acre-feet of water per year by international treaty). These conflicting claims culminated after an 11-year court battle with a 1963 U.S. Supreme Court decision that provided increased allocations to Arizona at the expense of California, which was judged to have had a 1 million acre-foot per year oversupply of water from the system.

As the Central Arizona Project put new land into agricultural production after 1964, the Metropolitan Water District in California found itself with a mandatory cutback of 600,000 acre-feet of water per year. An era of water conservation had begun.

On the Great Plains, drought has caused similar concerns for regional agreements on the use of water for agriculture, particularly use of water from the Ogallala Aquifer, which stretches from South Dakota to Texas. "Hydro bat-

tles" throughout the West reminded decision makers that water supplies were finite and that water use must be more rational and less wasteful. Agribusiness consumes 85 percent of all western water, so water savings are most effective in this sector.

Meeting Future Water Needs

Weather modification through cloud seeding has increased snowfall in the high mountain ranges, but it provides only a minimal increase toward the growing demand downstream. Many open aqueducts have been sealed to reduce leakage and evaporation. This measure, however, is still not cost-effective in many areas where government subsidies maintain a very low cost of water to farmers. The problem of water allocation in many areas of the West is based on the principle of "use it or lose it"; that is, this year's allocation is based on last year's use. This promotes waste, with users forced to consume all their allocation whether they need it or not, for fear of receiving a lesser allocation the next year. Many agencies are now advocating a free-market exchange of water, where surpluses can be sold rather than wasted. Development of drought-resistant plants and more efficient irrigation systems (such as drip irrigation) will help reduce water demand in the fields.

In cities, water conservation is promoted by advocating drought-resistant landscaping and municipal recycling of sewage effluent. All these measures are currently keeping pace with increased demands, but no long-term scheme is foreseen for saturating the West with water to everyone's satisfaction. Other nations are looking to the North American experience for ways to meet their own future water needs.

EXPLORING STANDARDS, THEMES, AND CONCEPTS WITHIN THE CASE STUDIES

In each of this chapter's examples, the study of geography is exemplified in the approaches taken to the topics at hand. All focused upon the land and human occupance of it. In each of the case studies, each of the five fundamental themes appeared in varying degrees: *Location, Place, Human-Environment Interaction, Movement, and Region.* Likewise, some of the geographic knowledge and skills contained in each of the 18 *National Geography Standards* appeared in every case study discussed here.

To illustrate our emphasis on the Standards and five fundamental themes of geography, the story of the 100th meridian began with concepts emphasized in the first Standard, that is, with reference to the 100th meridian's absolute location. Its *relative* position became more important as migrants crossed it. A sense of place in the American West was formed in reality by experience, and in the media by a dry landscape. "Western" books and movies, just like the real cowboy experience, took place in the semiarid climate of the Great Plains. In addition to creating an unforgettable sense of place, the arid landscape also reflected human-environment interaction in the adaptations to drought conditions that both animals and humans made.

The section on transportation west of the 100th meridian illustrates the theme of Movement. Identification of regions is possible by noting the importance of urban nodes as islands within vast areas of low population density. The entire region also is identified by the same unifying criteria so important in the other themes; in this case study, drought is a unifying characteristic of the West.

Location in the story of viticulture is exemplified in the location of the original domestication process and its necessary wild grapes. Wine landscapes throughout the world are quite striking and certainly reflect the dominance of this activity in all places where they occur. Human-environment interrelationships are quite sensitive, even delicate, in trying to extract the best grape quality possible from the soils and microclimates within wine landscapes. Movement is illustrated in the historical diffusion of the vine and in today's international trade of wine. Region is exemplified by a very famous wine region on California's north coast, where over 500 wineries await millions of annual visitors to "Wine Country."

The story of the Irish potato famine features the theme of Movement, but the importance of Ireland's location and its characteristics of place also are important. American urban cultural regions in the form of ethnic neighborhoods were formed in the migrants' usual destinations of Boston and New York. Human-environment interrelationships are seen in the circumstance of the fungus infestation and the climatic influences on Irish agriculture.

The concepts and themes of Region, Landscape, and Place are dramatized by comparing places having water with those that lack it. Drainage basins are often the most appropriate criteria for water-resource decision making, and these basins sometimes become political or economic regions. Movement is illustrated by water transfers. Human-environment interaction is very apparent from the comparison of irrigated and nonirrigated landscapes. A view of a golf course in Palm Springs, California, where average annual precipitation is only 2 inches (5 cm) per year, capsulizes all of the themes in one scene.

The Case-Study Approach to Learning Geography

We believe the use of case studies is one of the most effective ways to learn the *National Geography Standards,* and geographic themes and concepts. An even more effective (and exciting) way to study geography is to travel. If you have the opportunity to travel, try orienting your trips to this conceptual and thematic approach to observing geography in action. Photograph examples of alternative land uses, cultural or environmental differences, or distinctive landscapes. Orient your visit to a theme. Your authors have done such case studies on city planning in Scandinavia, aquaculture in Japan, salt production in California, ethnic conflict in Russia and Ukraine, and studies of ethnic retail businesses in western North America. Many other such studies abound in the real world, awaiting your investigation.

Comparative or cross-cultural thematic studies are also interesting and educational, such as comparisons of the methods used by traditional Amish farmers versus those used in agribusiness, high-rise urban living versus rural lifestyles, architectural characteristics of houses in various regions, and even home landscaping choices. One geography educator of our acquaintance makes it a practice to visit scenes of news events related to the land or environment (after the event is safely over!). Videotapes of floods, earthquake damage, and social unrest highlight her collection.

Through travel and inquiry about places visited, organized around the *National Geography Standards,* and the themes and concepts of geography, you can establish an appreciation for the landscape as a superior teaching and learning medium. We wish you the best in your travels and hope this text has provided motivation and guidance as you explore and analyze some of Earth's diverse peoples and places.

=== **FURTHER READING** ===

No understanding of the transition zone on the Great Plains between eastern and western North America can be complete without reference to one of the best regional historical geographies ever written: Walter Prescott Webb, *The Great Plains* (Lexington, MA: Ginn, 1931, reprinted 1971). Other indispensable references on this topic include Wallace Earle Stegner, *Beyond the Hundredth Meridian: John Wesley Powell and the Second Opening of the West* (Boston: Houghton Mifflin, 1954), and W. Eugene Hollon, *The Great American Desert: Then and Now* (New York: Oxford University Press, 1966). An "action geography" about land use and misuse in the West, which the reader can compare with eastern perceptions of the land, is John B. Wright's *Rocky Mountain Divide: Selling and Saving the West* (Austin: University of Texas Press, 1993).

Visual impressions of the region are available in *Atlas of North America: Space-Age Portrait of a Continent* (Washington, DC: National Geographic Society, 1985). Excellent photos and reference maps are in the *Historical Atlas of the United States,* edited by Wilber Garrett and published by the National Geographic Society in 1988. Ralph H. Brown includes an excellent section on the Great Plains and bordering regions up to 1870 in *Historical Geography of the United States* (New York: Harcourt, Brace & World, 1948). Another view is in David J. Wishart's chapter on the Great Plains in a collection of invited essays edited by Robert D. Mitchell and Paul A. Groves, North America: *The Historical Geography of a Changing Continent* (Totowa, NJ: Rowman & Littlefield, 1987).

Two of the best works by geographers about the geography of wine and wine landscapes are Harm J. de Blij, *Wine: A Geographic Appreciation* (Totowa, NJ: Rowman & Allanheld, 1983), and Dan Stanislawski, *Landscapes of Bacchus: The Vine in Portugal* (Austin: University of Texas Press, 1970). The most essential reference on wine regional studies is Hugh Johnson, *The World Atlas of Wine* (New York: Simon & Schuster, 1985). The regional geography of wine in the United States and Canada is the subject of *The Wine Regions of America: Geographical Reflections and Appraisals* (Stroudsburg, PA: Vinifera Wine Growers Journal, 1992). Finally, the relationships between the environments of grapevines and the labels of the wines they produce is explained in "The Wine Appellation as Territory in France and California" by Warren Moran in *Annals of the Association of American Geographers* 83 (Dec. 1993): 696–717.

A geography of contemporary Ireland, Anthony R. Orme's *Ireland* (Chicago: Aldine, 1970), includes an excellent analysis of the circumstances of the potato famine in the mid-nineteenth century. An analysis of the potato famine from the perspective of the

introduction of economically valuable plants into new cultures is in Henry Hobhouse, *Seeds of Change: Five Plants that Transformed Mankind* (New York: Harper & Row, 1986). Other references on the topic include Cecil Woodham-Smith, *The Great Hunger: Ireland,* 1845–1849 (New York: Harper & Row, 1962), and Redcliffe Salaman, *The History and Social Influence of the Potato* (London: Cambridge University Press, 1949).

The process and effects of migration of the Irish to North America is included in David Ward, *Cities and Immigrants: A Geography of Change in Nineteenth Century America* (New York: Oxford University Press, 1971), and is illustrated in James P. Allen and Eugene James Turner, *We the People: An Atlas of America's Ethnic Diversity* (New York: Macmillan, 1988).

The major argument for the importance of "hydraulic civilizations" in ancient history is presented by Karl A. Wittfogel, "The Hydraulic Civilizations," in William L. Thomas, Jr., ed., *Man's Role in Changing the Face of the Earth* (Chicago: University of Chicago Press, 1956): 152–164. Three recommended references for understanding the importance of irrigation are Raymond Furon, *The Problem of Water: A World Study* (New York: American Elsevier, 1967); Hitoshi Fukuda, *Irrigation in the World: Comparative Developments* (Tokyo: University of Tokyo Press, 1976); and Leonard Martin Cantor, *A World Geography of Irrigation* (New York: Praeger, 1970).

Water in the American West is critically analyzed by Donald Worster, *Rivers of Empire: Water, Aridity, and the Growth of the American West* (New York: Pantheon Books, 1986), and Marc P. Reisner, *Cadillac Desert: The American West and Its Disappearing Water* (New York: Viking, 1986).

California's interesting and elaborate water plan is explained in several pamphlets available from the California Department of Water Resources in Sacramento, including *California Water: Looking to the Future* (Sacramento, CA: DWR Bulletin 160-187, Nov. 1987), and in the *California Water Atlas* (Sacramento: State of California, 1978). A survey of water resource management and planning is *Water Resources Planning* by Andrew A. Dzurik (Savage, MD: Rowman & Littlefield, 1990).

ENDNOTES

[1] John Wesley Powell, *Report on the Lands of the Arid Region of the United States with a More Detailed Account of Utah* (Washington, DC: 45th Congress, 1878, Executive Document 73).

[2] W. Eugene Hollon, *The Great American Desert: Then and Now* (New York: Oxford University Press, 1966).

[3] Wallace Earle Stegner, *Beyond the Hundredth Meridian: John Wesley Powell and the Second Opening of the West* (Boston: Houghton Mifflin, 1954).

[4] Wallace Earle Stegner, *The American West as Living Space* (Ann Arbor: University of Michigan Press, 1987).

[5] While these are very generalized patterns, they may be seen along with others by examination of the *Atlas of North America: Space-Age Portrait of a Continent* (Washington, DC:

National Geographic Society, 1985), and the *Oxford Regional Economic Atlas of the United States and Canada* (New York: Oxford University Press, 1975).

[6] Anthony R. Orme, *Ireland* (Chicago: Aldine, 1970): 153.

[7] Thomas Sowell, *Ethnic America: A History* (New York: Basic Books, 1981): 21.

[8] David Ward, *Cities and Immigrants: A Geography of Change in Nineteenth Century America* (New York: Oxford University Press, 1971): 56.

[9] William L. Thomas, Jr. ed., *Man's Role in Changing the Face of the Earth* (Chicago: University of Chicago Press, 1956): 152–164.

[10] Stegner, *American West:* 60.

EPILOGUE

WORD TO TEACHERS OF GEOGRAPHY

By reading this far, you have become aware of the need for geography in the schools, and are knowledgeable about the most important themes and concepts commonly presented within the discipline. Although it was not part of our mission in writing this book, we also hope you have gathered ideas about how conceptual or thematic geography should be taught. In sharing our many combined years of experience in teaching geography, we are convinced that an understanding of the *underlying principles* of a discipline is much more valuable than memorizing unrelated lists of facts about it.

Even more important is the payoff to students who have learned a new way of thinking. Thinking geographically may mean developing a spatial sense about the reasons for the patterns and processes of people and places on Earth. It may appear in personal concerns about land use decision making on a local or global scale. Thinking geographically may be reflected in exercising political choices. Geographical awareness easily becomes global awareness when viewing the news or reading about it.

Recall that in Chapters 1 and 2 we presented a need for thinking geographically. At this point, we suggest that you consider incorporating examples of what we have termed "geography in action" into your teaching. With the skeleton of themes, concepts, and case studies offered here, we hope you will be able to add your own selected substance of geography into the social studies curriculum or wherever else it may be appropriate.

Geography is an approach to the study of the entire planet, and the *National Geography Standards,* five fundamental themes, and related concepts have steered our presentation. Other texts and references use the regional approach. Still others are based on traditional subfields, such as those taught in the university curriculum. All approaches are welcome and they collectively demonstrate the true universality of geography. There is enough candy in the store for all who appreciate the fascination and diversity of our home planet!

The main reason that we have not presumed to present teaching methods in this text is that we believe methods to be very personal, and that they can best be developed from your own travel and classroom experiences. Our advice to teachers and student teachers is to participate in the Geographic Alliance and in other geography workshops and institutes, to take courses in geography at your local community college or university, to travel, to get to know places well, and to bring back what you can to the classroom to share with your students. Continually remind yourself and your students that geography is not just a subject studied out of a textbook. The entire planet is our textbook.

Our own personal rewards in teaching geography have come from a need to convey our concern for Earth, its people, and their future. The major reward is seeing students at any age or grade level reach new conclusions about topics they thought they already understood. We recall the time when urban students from an East Los Angeles high school social studies class were taken for the first time to a redwood forest on a class field trip and felt the awe, wonder, and some fear at being in a new place that they had never known existed. We clearly remember the discovery of some students that their hometown was much the same as a sister city in New Zealand. Likewise, we feel that we have accomplished something when students express a desire to see different but ordinary landscapes in their travels, rather than tourist traps and theme parks.

In an age of technological overspecialization, the generalist approach used in geography, history, and social studies may seem diluted and overrated. We disagree! The attempt at holistic thinking offered in this text is absolutely critical to survival in an ever-changing and dynamic environment that promises even more complexity in the near future. We teachers and teachers-to-be must erase the artificial boundaries between school "subjects" and what we have called "real world" topics.

Many states and local school districts now emphasize thematic teaching and interdisciplinary curricula. Because of this integrated approach, we envision an opportunity for geography as well as other social sciences and humanities to take a more prominent place in the school classroom. Even the classroom itself is rapidly changing, with the introduction of video, laser disc, CD-rom, the Internet, and other computer-based technologies. Integrated teaching lends itself particularly well to technical instruction.

While the fear that computers would replace real teachers in future classrooms was expressed more than a decade ago, it has become evident in recent visits to model technology schools that humans and technology work well together. In many ways, technology is becoming the great equalizer, taking us away from the old classroom seating chart into new formats for learning and teaching. This presents an excellent opportunity for geography, because it is essentially visually based and appropriate to several forms of media presentation.

We hope that this text has made the value of geographic thinking apparent in the decade of the 1990s, where world peace, the state of the global environment, and the significance of local communities have become so important to our collective survival. If, at this point, you wish to continue your study of geography, we recommend three routes:

1. Consider graduate study of geography. It is offered in most public and some private universities. Most such programs are growing and are welcoming educators.

2. Subscribe to one or more of the several journals aimed at the teaching of geography, such as *The Journal of Geography.* Most include good teaching ideas as well as interesting geographic content.

3. Direct your own independent study of geography by using the references at the end of each chapter of this book and other lists of readings.

We wish you well in whatever form your commitment to the importance of geographic education may lead you!

GLOSSARY

absolute location   Exact position on Earth's surface or on a map, usually given in degrees of longitude and latitude, although several other coordinate systems can be used (cf. *relative location*).

accessibility   Ease with which one location may be reached from another. For urban places it is usually true that the greater the accessibility, the greater the value.

acculturation   Process of one culture overwhelming another; also adjustments to differences in conflicting cultures.

adiabatic heating/cooling   Refers to the heating and cooling of air as it descends or ascends in the atmosphere without other sources of heat.

advection   Horizontal movement of air across a land or water surface, usually associated with the formation of advection fog.

agglomeration   Tendency of businesses to locate together for mutual advantage.

agribusiness   Agricultural operations conducted by corporations rather than families.

agriculture   The process of planting, maintaining and harvesting plant crops or husbanding livestock.

albedo   Reflective quality of a surface to solar light and heat energy.

analog   Likeness between places or circumstances. An analog model usually is a simplified construction of a place or process.

aquaculture   The practice of farming aquatic plants or animals.

artifacts   Materials that reflect elements of a culture, such as tools or artwork.

assimilation   Process of absorbing one culture or activity into another.

atmosphere   Layer of gases enveloping Earth, composed mainly of nitrogen and oxygen.

atmospheric pressure   Pressure exerted by gas molecules as measured by a barometer.

basic/nonbasic   Economic activities that bring in new money or value to a place are called basic activities. Those activities that recirculate money in a region are termed nonbasic.

bid rent   Principle that states that rent or land value decreases with increasing distance from an economic center.

biome   Complex of plants and animals common to a designated large physical region.

bioregionalism   Belief that economic and political decision making should be based on environmental criteria.

biosphere   Earth's life zone, containing all of its plants and animals in terrestrial, atmospheric, and aquatic environments.

biotic community   Distribution and interrelationships of organisms within a limited physical area.

boundary   Line separating one political or mapped area from another. Reference is sometimes made to conceptual boundaries between ideas.

break-in-bulk   Transfer points where cargoes are changed from one transportation mode to another; e.g., harbors, where ships are loaded from rail or truck.

carrying capacity   Capacity of region or place to function without loss of quality of life or other selected values.

cartogram   Graphic depiction of space that is not true to scale and therefore not a true map.

cartography   The art of mapmaking.

centrality   Attraction of a central place to its surrounding area, usually measured in terms of economic activity.

central place theory   Explanation of city size, based upon distances of cities from one another. A hierarchy of city sizes and number of functions is based on relationships with neighboring places.

centrifugal forces   Events and circumstances that tend to create disunity within the population of a nation-state.

centripetal forces   Events and circumstances that tend to unite or bind a nation-state or culture together.

chain migration   Migration of people from the same source area to the same new place of residence along certain specific lines of travel.

climate   Average atmospheric conditions over a selected time period.

climate modeling   Simulation of the workings of the atmosphere, usually done with use of computers.

commercial   Land use and activities involved in trade or exchange of goods or money.

communication   Transmission and reception of messages, ideas, or information.

concentric zone theory   One of three explanations of North American city form. The theory observes that greatest land use density and highest land values are in a central zone, while lesser densities and lower values lie in concentric rings around the center.

conduction   The movement of heat through a substance from warmer to cooler areas.

conflict   Disagreement between individuals or among groups, which may have geographic consequences.

conformal projection   Map projection specifically designed to minimize distortions of area shapes.

conservation   Wise use of resources.

consumption   Purchase of goods to be used by the purchaser and not resold to others.

contour   Line connecting all points of equal value on a map. For example, on a topographic map, a line might connect all points having an elevation of 1200 feet above sea level. Such contour lines generally are called "iso" lines, from the Greek prefix meaning "equal." (See in this glossary the terms *isobar, isohyet, isotherm, and isotim.*)

convection   Vertical movement of air often causing precipitation.

core area   Portion of a region in which density and level of activity is highest.

Coriolis effect   Influence of Earth's rotation on freely moving objects on or immediately above Earth's surface.

cultural ecology   The study of the relationships between a cultural group and their physical environment.

cultural landscape   Physical evidence of culturally based decision making on the surface of Earth.

cultural pluralism   Coexistence of several subcultures within a society.

culture   A total way of life held in common by a group of people. It includes, for example, a shared language, religion, value system, customs, livlihood, and common historical experience.

culture hearth   The estimated point of origin of a particular cultural group or a particular set of innovations.

declination   Angular difference between magnetic north and true north.

defensibility   Circumstances of location or design that enable a place to be defended from invaders.

deforestation   Destruction or removal of forests without replanting them.

demographic transition   A change in population growth that occurs when a nation moves through stages from rural agriculture to urban industry in which both birthrates and death rates decline at a progressive rate.

density   Number of items per unit area; e.g., density of population expressed in people per square mile.

desertification   Process of land degradation in which land having low or variable rainfall is abused, reducing natural vegetation with a consequent loss of farm or rangeland productivity, resulting in a new desertlike area.

detrital system   The complex of bacteria and other organisms that break down organic matter into soil components.

diffusion   Spread of an idea, innovation, or activity through space and time.

dispersion   Degree of separation among mapped items.

distance decay   Decrease of activity or importance of a function (usually economic) as distance increases from a center.

distribution   The arrangement of mapped items that give meaning to their relationships with one another.

ecosystem   Complex of evolving and interacting plants and animals living within a particular natural environment, usually mutually dependent upon each other through energy exchange.

edge city   A type of city in North America that is located on the periphery of the older metropolitan area and is characterized by a mix of residential and employment-generating land uses.

emigration   Moving or relocating from a homeland to a new place (cf. *immigration*).

enclave   Territory that is surrounded by land belonging to a different political unit.

environmental determinism   Viewpoint that the physical environment influences human decision making.

environmentalism   Belief that a consideration of protecting the physical environment should influence all human decisions.

equilibrium   Balance in a dynamic system.

equinox   Twice a year, in spring and fall, the vertical rays of the sun fall onto Earth's Equator, creating days and nights of equal length in both hemispheres (cf. *solstice*).

equivalent area projection   Map projection designed to minimize distortion of area shapes.

ethnicity   Identity with a certain group, based on common national background or ancestry.

evapotranspiration   Movement of water from plants, soil, and other sources into the atmosphere.

exclave   Portion of a nation-state separated by geography from another nation-state.

exploration   Process of discovery and recording of places not previously known.

export   Movement of goods out of one place to another place.

flows   Movement of material, people, energy, information, etc. within a system.

food chain   Transfer of food energy from producers (usually plants) to consumers (usually animals).

footloose industry   An economic activity that is not locationally constrained by resources, accessibility, labor supply, or cost.

formal region   Area distinguished by recognized boundaries and uniform major characteristics.

front   Leading edge of an advancing air mass leading to a change in weather conditions.

frontier Edge of settlement, often a zone of economic, demographic, and cultural change.

functional region Area distinguished by the interrelatedness of activities or functions that occur there.

Gaia hypothesis Belief that Earth is a self-regulating system that reacts to environmental change caused by life processes.

geographic information system (GIS) Computer-based mapping and database system that interrelates variables.

geographic literacy Basic operating knowledge of geographic concepts and place locations.

geopolitics Use of geographic information for political decision making.

globe A spherical representation of the planet Earth.

global city City that is economically connected with distant places around the globe more than with local economies or resources.

global economy Worldwide system of trade not constrained by international borders.

gravity model Mathematical formula for determining the relationship of several points to their distributional center; e.g., the market area of a shopping center.

greenhouse effect Effects of atmospheric warming due to carbon dioxide buildup.

grid system Coordinate system of intersecting lines used to identify locations.

Hadley cell Feature of Earth's low-latitude circulation system in the lower troposphere, first suggested by George Hadley in 1735 that suggests the interaction of rising warm air from the warm tropics and colder sinking air from the mid-latitudes.

heartland Zone in eastern Europe and Asia believed to be the center of global power by early twentieth-century geographer Sir Halford Mackinder. Term is now applied to large, inland political regions worldwide.

hinterland Less-developed service area of a central place, connected by exports and imports.

home Place of refuge or an operation center for our personal interactions with surrounding places.

homogeneous region Area distinguished by uniformity in most of its significant characteristics.

Horse latitudes Zone of low or no prevailing winds, found between the trade winds and the westerlies.

humus Organic material in soil usually at the A horizon.

hydrologic cycle Continuous movement of water in a cycle by precipitation and evaporation; the mechanism by which water is continuously transferred among the "reservoirs" of Earth (oceans, atmosphere, and land).

hydrosphere Zone in which Earth's water exists as liquid, vapor, or ice, including the phenomena of evaporation, condensation, precipitation, runoff, and storage.

iconography Symbols and other shared expressions that link a population culturally or politically.

immigration Entry or relocation of persons into a new homeland (cf. *emigration*).

imperialism The policy of nation-states that encourages the acquisition of territory and resources.

import To bring in goods or resources from other places.

industrial location theory Principles of location decision making for industries, based on economic criteria.

information Assembled facts or ideas that are passed on from one party to another and considered in decision making.

information highway Popular term for computer networks used for communications.

information society That segment of a population affecting or being affected by the continual flow of electronic information.

infrastructure Necessary support systems (sewerage, water systems, etc.) for the functioning of a city or other large facility.

innovation Creation of a new idea or the modification of an old one and its dispersal through space and time.

insolation Solar radiation that is intercepted by Earth.

involuntary migration The movement of people against their wishes.

isobar Lines connecting points of equal atmospheric pressure on a map.

isohyet Lines connecting points of equal precipitation on a map.

isolation Influence of separation or distance on a biotic community due to separation by distance or barriers from other influences.

isotherm Lines connecting points of equal temperature on a map.

isotim Lines connecting points of equal cost on a map.

jet stream The high-level, high-velocity movement of westerly winds in both hemispheres.

landlocked Interior country or state that is surrounded by land and has no coastline.

landscape Visible features created by humans on the surface of Earth.

land use Human activity and patterns of settlement on the surface of Earth.

language family Group of languages with shared origins, grammatical structure, or vocabulary that may or may not be spatially linked.

latitude Measurement in degrees north or south of the Equator; parallel lines running east-west around the Earth that measure distances north or south of the Equator (cf. *longitude*).

legend An explanation of the use of symbols on a map.

lifestyle Habits and daily routines of a particular segment of society.

linkage Connections between nodes in a network; e.g., a highway may link several sites in a region.

lithosphere Thin layer at the surface of Earth (the crust) that makes up less than 2 percent of its total volume.

loam A mixture of silt, sand, and clay in soil; usually considered ideal for seasonal crop farming.

longitude Measurement in degrees west or east of the Prime Meridian (which passes through Greenwich, England); lines running pole-to-pole used to measure distances east and west of the Prime Meridian (cf. *latitude*).

Malthusian doctrine General belief that population grows at a geometric rate, while resources are developed at a linear rate (or arithmetical) and thus will not keep up with the needs of the growing population.

map A representation of the surface of Earth usually drawn to scale.

mass wasting Movement of ground materials by gravitational force (e.g., landslides).

Mediterranean agriculture Crops and growing techniques particularly suited to a dry summer and wet winter as well as a mild, temperate climate.

megalopolis Adjacent groups of population centers that form one, large overlapping, interconnected urban area.

mental map Intangible images and memories of a place, based upon one's socioeconomic status, gender, ethnic background, and life experience.

migration Permanent relocation of people from one place to another.

model Simplification of a system or process, expressed in mathematical, graphic, verbal, electronic, or mechanical form.

monsoon Annual shifting winds and rainfall produced by changing atmospheric systems in south and southeast Asia.

multiple nuclei theory North American city pattern in which a city has more than one economic node.

nationalism Feeling of commonality by virtue of culture, language, religion, or politics that tends to bind together a population.

nation-state Political territory occupied by a group having a common identity and a certain degree of unity.

natural hazards Naturally occurring events such as earthquakes, fires, and floods that may cause damage to people or their created landscapes.

neighborhood Local area with which residents identify, often having a center or focal point for identification.

Neolithic The Agricultural Revolution; the time on Earth when plants and animals were first domesticated for human use.

network System of relationships among selected, interconnected "things."

nodal region Area distinguished by a focus of activities on some central point.

node Center point of activity or density in a mapped network or spatial pattern, as can be seen in street intersections, railroads converging upon a city, or sites of cultural significance. A city may be a node in a region.

orographic lifting Uplift of airmass due to physical barrier resulting in cooling and perhaps increased precipitation.

pattern Arrangement of features on Earth's surface as viewed from above.

perception Personal impressions or interpretations of observed phenomena based on past experiences, gender, ethnicity, etc.

personal space Sensitivity to invisible boundaries that encompass the space around a person.

photosynthesis Production of oxygen from carbon dioxide and sunlight by plants.

planning Process of anticipating and guiding future land uses in a selected area.

plantation agriculture Form of agriculture, usually tropical, in which one major crop is grown for export; e.g., coffee or bananas.

plate tectonics Geological theory that explains movement of portions of Earth's crust over geologic time.

pollutant Solid, liquid, or gaseous substance often damaging to ecosystems or to humans; often invisible.

possibilism Viewpoint that human decision making shapes their own destinies and cultures more than the physical environment.

pressure gradient Difference between areas of high and low pressure in the atmosphere. *Steep* pressure gradients refer to areas with major differences in pressure (areas of high wind); *weak* gradients refer to areas with less extreme differences of pressure (areas of light winds or calm).

primate city A city of large size and dominant influence within a country.

primary production Those forms of production dependent upon natural resources, such as farming, mining, fishing, or forestry.

production Process of entering goods into an economic system for sale or consumption.

productivity (biological) Rate of biomass production within an ecosystem.

projection (map) Representation of Earth's spherical surface on a flat map.

push-pull factors Unfavorable and favorable conditions that affect the migration decision-making process.

recycle To reuse discarded material by turning a pollutant or waste product into a resource.

region Area on Earth's surface defined by selected criteria.

relative location A position on a map or on Earth's surface as compared with other positions (cf. *absolute location*).

remote sensing Interpretation of information from high-altitude photographs or orbital imagery.

resolution Level of detail projected, as seen on a map, photograph, remotely sensed image, or computer screen.

resource Anything used by people, including naturally occurring materials (e.g., soil, minerals, trees, and water), labor, or energy sources.

resource extraction Removal and use of natural resources as in mining, fishing, or forestry.

revolution Movement of Earth around the sun in space in an elliptical path (cf. *rotation*).

Robinson projection Map projection developed by Arthur Robinson; generally accepted as a useful global representation that minimizes distortion.

rock cycle A model representing the interrelationships among the three rock-forming processes: igneous, sedimentary, and metamorphic.

rotation Turning of Earth on its own axis, which causes day and night (cf. *revolution*).

scale Relationship between distance on a map and distance on Earth's surface.

seafloor spreading The mechanism driving the movement of continents over geologic time.

seasons Distinct times of the year, characterized by differences in the amount of solar energy received at Earth's surface and the consequent differences in climate and biological activity.

sector theory North American city pattern characterized by pie-shaped wedges of similar land use that follow transportation corridors outward from city centers.

segregation Separation of people or things by selected criteria that may be either mandated or self-imposed.

sense of place Subjective and sometimes emotional connections to our immediate environment.

sequent occupance The successive settlement of several groups of people over time in the same place.

shifting cultivation Movement of small gardens or agricultural plots in a region, such as in a tropical area. Abandonment of fields allows for regeneration of soil.

site Selected place on the surface of Earth.

situation Relationship between the characteristics of a site and surrounding circumstances; e.g., the site of a city may be important in trade relationships among settlements within a region.

soil profile A rendering of layered soil types in a specific place, usually labeled as soil horizons.

solar altitude Elevation of the sun in the sky; the vertical angle between the horizon and the sun.

solar cascade Light and heat from the sun entering and reentering the Earth-atmosphere system.

solstice A point in time around the 20th or 21st of December and June when the sun's *vertical* rays fall on the Tropic of Capricorn (in December) and the Tropic of Cancer (in June) causing seasonal differences in temperature in the northern and southern hemispheres (cf. *equinox*).

spatial behavior Relationships between behavior (personal or group) and environmental influences, such as distance.

spatial interaction The social, economic, and other connections between and among places.

species Unique population of a plant or animal capable of reproducing itself and containing similar genetic characteristics.

spheres of influence Zones surrounding an urban place that are politically and economically dependent upon it.

state Politically organized area administered by a sovereign government.

subsistence Farming and other economic activities undertaken to sustain a single household or small village without economic exchange.

succession Sequence of plant communities that occupy a place over time.

sustainable agriculture An agricultural practice that does not require large imports of water, fertilizer, or insecticides from outside of the local ecosystem.

system Organized collection of component parts that function as an integrated whole.

taiga A northern latitude needle-leaf forest.

technology The ability to accomplish human goals by use of tools, equipment, or harnessed energy.

Terra Incognita Literally, "land unknown"; unrecorded or blank spaces on early maps depicting unexplored places.

terracing Grading of hill slopes for ease of construction or agriculture.

territoriality Sense of privacy, ownership, or occupance of particular areas.

territorial sea Offshore zone claimed by an adjacent nation-state as part of its sovereign territory.

threshold Limit or capacity of a measured characteristic, e.g., a population threshold or carrying capacity.

time-space continuum Relationship between the dimensions of time and space as expressed in travel times between points.

time-space convergence Time taken to travel between known points. Technological change allows greater travel distance in shorter times.

time zone Pole-to-pole slice of Earth's surface, delineated by meridians approximately 15° apart that represent one hour of time passage.

trade Exchange of goods having value.

traditional society Reference to a society not greatly affected by new technology or "Western" social values.

transhumance Seasonal migration of people with their livestock from valleys to mountains, and then returning.

transportation Planned movement of goods or people.

trophic levels Classification of organisms in an ecosystem according to how they are nourished, such as primary producers, herbivores, carnivores, and decomposers.

tundra Vegetation located over frozen ground in high altitudes or high latitudes.

urbanization Conversion of people and places from a rural situation into towns and cities.

vernacular region A region known better by popular acceptance rather than by political or economic criteria.

voluntary migration Migration of individuals or groups without coercion or force.

von Thünen's theory Interpretation of land use patterns considering costs of land (rent), value of products produced, and cost of transportation over distance.

water budget Assessment of water inputs and outputs in a particular area at a certain time.

weather Day-to-day changes in atmospheric conditions in the lower troposphere.

weathering Breaking down of large rocks into smaller materials by natural forces.

INDEX

Tableau de la Géographie de France, (de la Blache), 305
Tahoe Regional Planning Agency, 322
Tan, Amy: *The Joy Luck Club,* 74
Tarbuck, Edward J., 149
Taylor, Peter J., 190
Teaching Geography Project, 22
Tectonic processes, 116–18
Temperate forests, 142
Temperature:
  and climactic controls, 93
  highest and lowest recorded, 105
Tennessee Valley Authority (TVA), 322
Terminal calving, 128
Terracing, 198
Terra Incognita, 31, 65
Terrain model, 52
Territoriality, 71
Tertiary economic activities, 217–18
Theory of the Location of Industries (Weber), 212
Thermosphere, 93
This Remarkable Continent: An Atlas of United States and Canadian Society and Cultures, 168
Thoman, Richard S., 221, 287, 296, 298
Thomas, William L., 247
Thomas, William L., Jr., 340
Thompson, Morris M., 62
Thoreau, Henry David, 74, 245
Time-space continuum, 67
Time-space convergence, 291
Time zones, 39, *40*
Tito, Joseph, 184
Tokyo: street map of, *57, 58, 59*
Tolba, Mostafa K., 270
Toponyms, 168
  of Quebec, Canada, *174*
  of West Virginia, 170–71
Tornadoes, *268*
Toronto: Murdie's "communities" in, *231*
Townsend, Janet, 63
Toxic wastes, 199
Toynbee, Arnold, 4
Traction, 127
Trade:
  development of international, 287
  patterns of world, 287–89
  and transportation, 286–90
Trade winds, 96, 98
Transform faulting, 117
Transhumance, 195, 196
Transpiration, 104–5
Transportation, 272
  dependency, 217
  development in United States, 290–91
  evolution in Third World countries, 291, 293–94
  and idealized transport development, 294
  major transport facilities in Ghana, *294*
  rates of travel from New York, *292*
  and trade, 286–90
Travels with Charley (Steinbeck), 76
Trellis pattern, 127, *129*
Trewartha, Glenn T., 323
Trophic level, 136
Tropical air mass, 99
Tropical forests:
  burning of, 254
  clearing of, 258
  rain, 142
Tropical Forests: Patterns of Depletion (Myers), 258
Tropic of Cancer, 38, 88
Tropic of Capricorn, 38, 88
Troposphere, 93
Tuan, Yi-Fu, 67, 71, 75, 80
Tundra, 141–42
Turner, B. L., II, 270
Turner, Eugene James, 190, 340
Turner, Frederick Jackson, 285
Turtle Island (Snyder), 78
Twain, Mark, 74

Ubiquitous raw materials, 213
Udaipur, India: horse-drawn cart and driver, *272*
Udall, Stewart: *The Quiet Crisis,* 247

Ullman, E. L., 236, 242
United Nations Food and Agricultural Organization, 258
United States:
  development of transportation system in, 290–91
  ecoregions of coterminous, *145*
  geography education in, 12, 17–23
  high school geography project in, 20
  landform regions of, *301*
  land use patterns in contiguous, *328*
  New Yorker's idea of, *70*
  population distribution of contiguous, *327*
  wheat exporting by, 205
United States Bureau of Alcohol, Tobacco, and Firearms, 148
United States Bureau of Reclamation, 330
United States Bureau of the Census, 236
United States Comprehensive Soil Classification System, 134
United States Department of Agriculture, 199
United States Department of Agriculture—Forest Service, 4, 32, 245, 322
United States Department of Commerce, 4
United States Department of Defense, 4, 32
United States Department of Education, 3
United States Fish and Wildlife Service, 260
United States Geological Survey, 4, 32
  topographical maps of, 49
United States State Plane Coordinate system, 50
United States Township and Ranger survey system, 50
Universal Transverse Mercator system, 50
University of California, Davis, 199
University of Miami, 10
Urban accessibility: and land use, 295
"Urban Geographer, The," 187, 189
Urban geographers, 236
Urban growth: planning for future, 240–41
Urbanization:
  five epochs in American, 229–30
  development around Mediterranean Sea, 225–27
  rate of, 222
Urban land use patterns, 233–36
Urban planning: role of location in, 55–56, 71
Urban systems, 222–23
Urban Villagers, The (Gans), 233
USA Today, 215

Vale, Thomas R., 149
Valley glaciers, 129
Valley of the Moon, The (London), 74
Vance, James E., 242, 298, 323
  Merchants World, The: The Geography of Wholesaling, 286
Variable gases, 92
Viles, Heather, 11, 221
Viticulture:
  diffusion of worldwide, *332*
  and major diffusion routes in Asia and Europe, *331*
  origin and diffusion of, 330–31
Volcanoes and volcanism, 117, 120–21
  hazards of, 123
  in Japan, 314
  world distribution of, *121*
von Humboldt, Alexander, 5, 305, 323
  Kosmos, 3
von Thünen, Johan Heinrich, 214, 234
  Isolated State, The, 200
von Thünen's theory, 201, *203*
  isolated state model, *202*

Wagner, Philip L., 189, 190
Wallerstein, Immanuel, 190
Walter, Heinrich, 149
Ward, David, 340
Ward, Peter L., 149
Warm front, 99, *101*
War of Spanish Succession, 334
Water:
  budget-analysis of, 111–12
  conflicting demands for, 338
  in Earth-atmosphere system, 102–10

as Earth's fundamental resource, 336–38
  erosion, 127–28
  evaporation of, 104
  global budget for, 112
  incredible substance of, 103–4
  and land differences, 93–94
  meeting future needs for, 338
  regional systems of, 111–12
  transfer systems of, 337–38
  transpiration, 104
  vapor, 104
Weather: and climate, 93
Weathering and weathering system, 121–26
  chemical, 122–25
  mechanical, 122
Weather maps: United States, *35*
Webb, Walter Prescott, 339
Weber, Alfred, 212
Weber, Max, 175
Weber's Location Triangle, 212–13, *213*
Wegener, Alfred, 118, 119
Weight-losing materials, 213
Weiss, M., 56
Westerlies, 96, 98
West Virginia: toponyms of, 170–71
West Virginia Gazetteer of Physical and Cultural Place Names, 170
Wham-O toy company, 284
Wheat:
  farms, 205
  world-wide distribution of, *206*
Wheeler, James O., 220
White, I. D., 114
White, Rodney, 80
Whittaker, Robert H., 149
Wilbanks, Thomas J., 220, 265
Wilderness Act, 247
Wilderness Society, 247
Wildfires, *268*
Wilford, John Noble, 80
Williams, A. F., 298
Williams, Colin H., 190
Willmott, Cort J., 11
Wilson, Woodrow, 179
Winds, 87
  denudation by, 131
  highest speed recorded, 105
  and jet stream, 100
  local, 100
  and monsoons, 100
  and pressure, 95–96
  Santa Ana, 101
Winemaking, *333*
  current world patterns of grape growing and, 331
  origin and diffusion of, 330
  wine landscapes, 332, *333*
Wine production: geography of world, 207, *207*
Wishart, David J., 340
Wittfogel, Karl A., 336, 340
Wood, William B., 297
Woodham-Smith, Cecil, 340
Woodwell, George M., 114
World Trade Organization, 290
World War II, 4, 19
Worldwatch Institute, 162, 249
Worster, Donald, 340
Wright, Austin Tappan: *Islandia,* 75
Wright, John K., 65
Wright, John B., 340
Wright, Sherry, 265
Writing: four types of, 165

Yeates, Maurice, 242
Yosemite National Park, 245
Young, J. E., 221
Yugoslavia, former nation of, 184
  new ethnic nations of, *186*

Zelinsky, Wilbur, 151, 189, 323
"Zen Macrobiotic Diet," 174
Zero population growth: in Sweden, 161–62
Ziegler, Alfred, 119